Plant Pathology

Series Editor
John M. Walker
School of Life Sciences
University of Hertfordshire
Hatfield, Hertfordshire, AL10 9AB, UK

For other titles published in this series, go to
www.springer.com/series/7651

METHODS IN MOLECULAR BIOLOGY™

Plant Pathology

Techniques and Protocols

Edited by

Robert Burns

Scottish Agricultural Science Agency, DMB, Edinburgh, UK

Humana Press

Editor
Robert Burns
DMB
Scottish Agricultural Science Agency
Edinburgh
EH12 9FJ, UK

Series Editor
John M. Walker
School of Life Sciences
University of Hertfordshire
Hatfield, Hertfordshire
AL10 9AB, UK

ISBN: 978-1-58829-799-0 e-ISBN: 978-1-59745-062-1
ISSN: 1064-3745 e-ISSN: 1940-6029
DOI: 10.1007/978-1-59745-062-1

Library of Congress Control Number: 2008940835

Cover illustration: From Figure 1 in Chapter 1

Printed on acid-free paper

9 8 7 6 5 4 3 2 1

springer.com

Preface

The diagnosis of infections in plants has changed immeasurably over the years. Early references to disease in barley crops date back to the mid-1500s and probably represent the first true records that we have of symptoms being noted. Early records merely described the physical appearance of plants and were usually associated with catastrophic crop failure. By 1870 fungal and bacterial diseases of plants were diagnosed, but it was not until after 1900 that virus diseases were identified. The science of plant pathology in its true sense came into being when it became possible to treat plants to control pathogens or to use husbandry to avoid pathogen problems. Physical symptoms on plants can be diagnostic, but very often they can be caused by several pathogens and a more scientific approach is required. Diagnostic methods mainly developed for the medical and veterinary sciences have now been applied to diseases of plants and we now have a bewildering assortment of methods at our disposal. Plant diseases still account for heavy losses in many parts of the world where total crop failure due to disease can lead to human misery.

In the western world, much emphasis is now placed on effective disease control by the use of clean seed and appropriate chemical intervention, but both rely on good diagnostics to establish disease status prior to action being taken. The development of quick and cheap methods for disease detection ensures that crop plants remain free of pathogens.

Plant pathology techniques fall into three categories: traditional, serological and nucleic acid, although some span more than one discipline. Traditional methods include the use of indicator plants to produce visual symptoms of disease on susceptible hosts and the use of synthetic media to encourage the growth of microorganisms which can then be identified by colony morphology. These methods can be coupled with more advanced techniques where additional information is required for diagnosis.

Serological methods are all based on the unique ability of animal antibodies to bind to small target areas known as epitopes on the proteins that elicited their synthesis. Antibodies to plant pathogens can be raised in the serum of animals quite easily by immunising them with preparations of microorganisms and then collecting the serum by live donation. Antibodies can be recovered from serum by simple chemistry or chromatography and reagents developed by adding markers such as chromogenic chemicals or fluorescent dyes. Monoclonal antibodies can be made in cell lines, using tissue culture obviating the use of animals for serum production. Such methods have long been established in the medical world and now provide rapid robust tests for pathogens in plant material.

Nucleic acid methods are the newest of the technologies but provide the most exciting possibilities. Methods are all based on the fact that small areas of nucleic acid exist within the genome that are unique to an organism and can be used to identify its presence. Once the sequence of these regions is known, synthetic fragments can be made which will bind to the specific areas on the pathogen DNA. These areas can then be made to replicate to produce multiple copies that can then be visualised either by electrophoresis or by fluorescent-based methods. The polymerase chain reaction (PCR) is fundamental to the nucleic acid technologies as it provides a way to amplify selected specific regions of DNA so that they can be visualised and thus provide a measurable signal indicating the presence of the target DNA.

This first edition of *Plant Pathology Techniques and Protocols* seeks to provide workers with both basic methods and more advanced techniques for the diagnosis of plant pathogens. Those with limited experience will find easy to use protocols and those with more experience will find methods that they may wish to use as alternatives to those already in place. Methods cover pathogens which cause major problems in crop plants globally. Issues of crop identity and authenticity have become more important in recent years and two chapters have been included which will allow workers to genotype samples from two major food crops. Authors are all active researchers and the methods are those currently being used in their laboratories.

I thank all of the members of Diagnostics and Molecular Biology at SASA for their encouragement and support throughout this project

Robert Burns

Contents

Contributors

DIRK U. BELLSTEDT • *Stellenbosch University, Matieland, South Africa*
NEIL BOONHAM • *Central Science Laboratories, Sand Hutton, York, UK*
ISLA A. BROWNING • *Virology and Zoology, Scottish Agricultural Science Agency, Edinburgh, UK*
ROBERT BURNS • *DMB, Scottish Agricultural Science Agency, Edinburgh, UK*
CHIN-AN CHANGE • *Graduate Institute of Biochemical Sciences and Technology, chaoyang University of Technology, Wufeng, Taichung 41349, Taiwan*
TSUNG-CHI CHEN • *Department of Biotechnology, Asia University, Wufeng, Taichung 41354, Taiwan*
MARC A. CUBETA • *Department of Plant Pathology, North Carolina State University, Raleigh, NC, USA*
TRIONA DAVEY • *DMB, Scottish Agricultural Science Agency, Edinburgh, UK*
AGNES DELAUNEY • *INRA, Agrocampus Rennes, Le Rheu, France*
MOHAMED HIJRI • *Institut de Recherche en Biologie Végétale, Université de Montréal, Montréal, QC, Canada*
HEI-TI HSU • *Floral and Nursery Plants Research Unit, U.S. National Arboretum, USDA-ARS, Beltsville, MD 20705, USA*
JAAP D. JANSE • *Department Laboratory Methods and Diagnosis, Dutch General Inspection Service (NAK), Emmeloord, The Netherlands*
ELLEN M. KERR • *DMB, Scottish Agricultural Science Agency, Edinburgh, UK*
BLANKA KOKOSKOVA • *Department of Bacteriology, Crop Research Institute, Prague, Czech Republic*
MARGIT LAIMER • *Plant Biotechnology Unit, IAM, Department of Biotechnology, BOKU, Vienna, Austria*
VINCENT MULHOLLAND • *DMB, Scottish Agricultural Science Agency, Edinburgh, UK*
VICENTE PALLAS • *IBMCP, Universidad Politecnica de Valencia-Consejo Superior de Investigaciones Cientificas, CPI, Valencia, Spain*
ALEX REID • *DMB, Scottish Agricultural Science Agency, Roddinglaw Road, Edinburgh, UK*
MATHIEU ROLLAND • *INRA, Agrocampus Rennes, Le Rheu, France*
ALEXANDRA SCHLENZIG • *Plant Health, Scottish Agricultural Science Agency, Edinburgh, UK*
CATHY SOUTHWORTH • *Napier University, Edinburgh, UK*
ANTONET M. SVIRCEV • *Agriculture and Agri-Food Canada, Vineland Station, ON, Canada*
CHRISTOPHER R. THORNTON • *Washington Singer Laboratories, University of Exeter, Exeter, UK*
ELAINE WARD • *Rothamsted Research, Harpenden, Hertfordshire, UK*

Chapter 1

Bioassay for Diagnosis of Plant Viruses

Isla A. Browning

Summary

Bioassay provides a useful means of detecting and identifying plant viruses. The choice of procedure and environmental conditions for the test depends on the virus under investigation.

Key words: Bioassay, Virus, Test plant, Indicator plant, Inoculation, Abrasive, Transmission, Symptoms, Local lesion, Systemic symptom, Inhibitor, Inactivator.

1.1. Introduction

It is often impossible to diagnose plant virus infections merely by observing host symptoms. This is not only because several viruses may induce similar symptoms in the same plant, but virus-like symptoms may also develop for physiological reasons. In addition, symptoms may be very slight and inconclusive, or plants may be infected symptomlessly. Bioassay is a useful procedure for detection and identification of viruses and uses indicator plants which react to infection by showing characteristic symptoms. Over the last few decades, laboratory based virus test methods, such as Enzyme-Linked-Immuno-Sorbent Assay (ELISA) and Polymerase Chain Reaction (PCR), have been developed and are now being used routinely in many laboratories. Automation of these procedures enables a high throughput of samples and provides rapid results. However, in many instances, bioassay remains an indispensable tool. Inoculations can be carried out rapidly with the possibility of detecting a wide range of viruses on a fairly small number of plant species. Verhoeven and Roenhorst *(1)* reported

R. Burns (ed.), *Methods in Molecular Biology, Plant Pathology, vol. 508*

DOI: 10.1007/978-1-59745-062-1_1

that bioassay is a useful tool for post-entry quarantine testing in view of the fact that three indicator species can be used to screen for 30 known viruses whilst it provides an opportunity to detect unexpected or new virus infections.

There are several ways of introducing viruses into test plants: grafting, use of vectors and mechanical transmission. Grafting can be used to transmit viruses between plants that are quite closely related. It is useful for viruses such as Potato Leaf Roll Virus (PLRV) which is translocated in the phloem and can not be transmitted by mechanical inoculation. It is also routinely used for assessing virus resistance. Resistance reactions are frequently more severe than those obtained by vector or mechanical inoculation as large numbers of virus particles are transferred directly into the vascular tissue and in potato this leads to severe systemic necrosis in cultivars with hypersensitivity genes. A more uncommon method, similar to grafting, is the use of dodder (*Cuscuta* spp.), which develops a parasitic union with the host and can facilitate virus transmission between plants that are not closely related. Procedures for virus transmission by grafting and use of dodder are described by Gibbs and Harrison *(2)*.

Many viruses are transmitted by insects, fungi, or nematodes. Insect vectors include aphids, thrips, whitefly, and leafhoppers. Soils may be tested for fungal and nematode vectors by carrying out 'bait-tests'. Generally this involves growing test plants in soil from the sites under investigation. After a period of time, diagnostic symptoms may develop on the test plants. Alternatively, indicator plants may be mechanically inoculated with sap from the test plant roots and observed for symptom development. Detailed procedures for vector transmission of viruses have been described by Dijkstra and de Jager *(3)*.

This chapter concentrates on mechanical inoculation, the simplest and most commonly used bioassay procedure. A perusal of the available literature on bioassay throughout the last century indicates that there is not one method that can be regarded as optimum for all virus/indicator plant combinations. Many factors affect the success of mechanical inoculation and suitable conditions must be identified.

1.2. Materials

1. Inoculum (*see* **Subheading 1.3.2.1**).
2. Indicator (test) plants (*see* **Note 1** for test plant species, **Note 2** for test plant numbers and **Note 3** for test plant age).
3. Pestle and mortar or other device for sap extraction (*see* **Note 4**).

4. Water (tap or distilled).
5. Abrasive (e.g. carborundum) (*see* **Subheading 1.3.2.2**).
6. Disposable gloves.
7. Glasshouse.

1.3. Methods

An example of a basic bioassay procedure is described in **Subheading 1.3.1**. However there are many procedural variations depending on the material and virus under test. **Subheading 1.3.2.1** provides information on inoculum diluents and **Subheading 1.3.2.2** on different abrasives and inoculum application procedures. **Subheading 1.3.3** details pre- and post-inoculation environmental conditions that may influence bioassay results. **Subheading 1.3.4** describes precautions that must be taken when carrying out the bioassay procedure whilst **Subheading 1.3.5** provides information on virus symptom diagnosis.

1.3.1. Example of a Basic Bioassay Method

1. Place test plants on the bench
2. Wearing a suitable protective mask sprinkle carborundum mesh size 400 on the leaves to be inoculated (**Fig. 1.1**)
3. Place a leaf or several leaves showing symptoms in a sterile mortar with a small quantity of water. Unless carrying out

Fig. 1.1. Sprinkling carborundum onto *Nicotiana tabacum* cv White Burley.

experimental work, the volume of tissue is not critical (it is possible to transmit virus from a single local lesion). To minimise the effect of inhibitors, sap may be diluted with water (a dilution of up to 1/50 is suitable for most purposes)

4. Grind sample using a pestle (**Fig. 1.2**)
5. Wearing gloves, apply the inoculum with a finger to the leaves in one movement from the base of the leaf to the top, supporting the leaf in the other hand (**Fig. 1.3**)

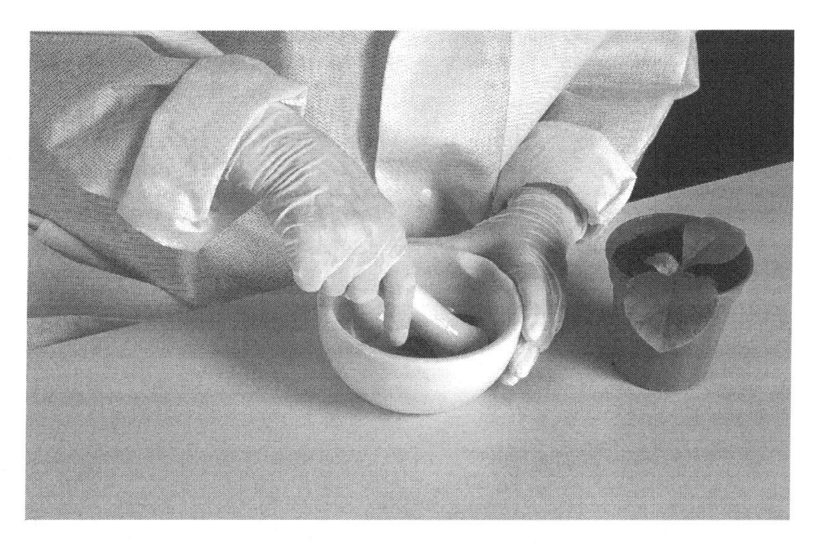

Fig. 1.2. Grinding inoculum using pestle and mortar.

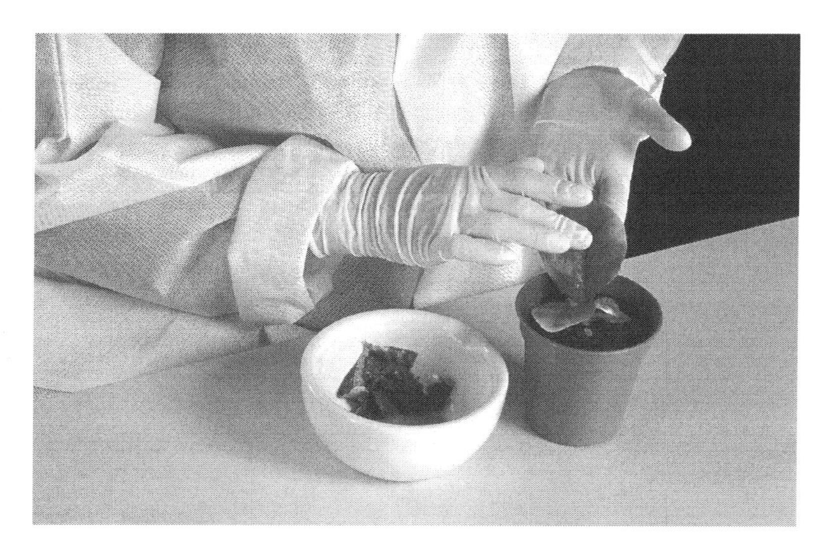

Fig. 1.3. Applying inoculum.

6. After inoculating all the indicator plants for this test allow them to remain on the bench for approximately 2 min
7. Wash plants briefly with a spray of tap or distilled water (*see* **Note 5**)
8. Place plants on bench in glasshouse
9. Observe plants regularly for symptom development which generally takes place between 3 days and 3 weeks

1.3.2. Composition and Application of Inoculum

1.3.2.1. Inoculum Diluents

It is not advisable to use undiluted sap as inoculum as some plants contain substances that inhibit infection or inactivate the virus. Many virus inhibitors are proteins or polysaccharides. When virus is in high concentration inhibitors can be 'diluted out' using water *(4)*. For many years, detection and diagnosis of potato viruses by bioassay at the Scottish Agricultural Science Agency has been successfully carried out using tap water for sap dilution but distilled or deionised water should be used if the quality of the tap water is likely to be poor. With less stable viruses the use of a buffer is advisable.

Virus inactivators include tannins, products of oxidase action and nucleases. Tannins are present in many woody plants, such as those of the family of Rosaceae, and may precipitate viruses, rendering them non-infective. They can be inactivated by adding to the inoculum substances that produce alkaline conditions, such as nicotine or caffeine. Martin and Converse *(5)* compared a range of buffer formulations containing different additives for transmission of Raspberry Bushy Dwarf from *Rubus* and Tobacco Streak and Tomato Ringspot viruses from *Fragaria* to *Chenopodium quinoa* and obtained best results with 0.05 M phosphate buffer (pH 7.0) + 2% polyvinylpyrrolidone (MW 10,000).

Inactivation of virus resulting from the products of oxidase activity may be overcome by the use of reducing agents (such as dithiothreitol or thioglycollic acid) or chelating agents (such as sodium diethyldithiocarbamate). Defective virus particles, lacking a protective protein coat, such as those of the NM type of Tobacco Rattle Virus (TRV) are susceptible to degradation by nucleases. Addition of substances such as bentonite clay can be added to adsorb the nucleases whilst alkaline buffer may act as an inhibitor. The use of phenol has been particularly successful for the transmission of NM type TRV *(6)* but because of the health and safety risks involved with handling this chemical its use should be avoided if at all possible.

Many workers have found improved results for transmission of some viruses when salts have been added to the inoculum. Thornberry *(7)* tested a wide range of salts and two phosphate buffer solutions on the transmission of Tobacco Mosaic Virus (TMV) to *Phoseolus vulgaris* and discovered that at a concentration of 0.1 M nearly all salts tested increased infectivity, whilst concentrations of above 1.0 M reduced infectivity. In particular, dibasic phosphate salts at a concentration of 0.1 M greatly enhanced virus infectivity. This result corresponded with that of Beraha et al. *(8)* who tested a range of concentrations of phosphate buffer (0–1.0 M) at pH 8.5 for transmission of TMV to *P. vulgaris* and found the optimum buffer concentration to be 0.1 M. The effect of phosphate was also studied by Yarwood *(9)* who reported that 1% K_2HPO_4 added to the inoculum dramatically increased the number of local lesions produced by nine viruses on bean. This study, involving several virus/host plant combinations, indicated that phosphate was acting on the host rather than on the virus. However some viruses are adversely affected by phosphate and require the use of other buffers, e.g. borate, citrate, or Tris *(3)*.

1.3.2.2. Application of Inoculum

In order to introduce virus into host cells it is necessary to damage the cells by some means. In the early days of bioassay, Bean Mosaic Virus was successfully transmitted by rubbing the undersides of leaves with infected leaves rolled together and crushed with the fingers *(10)*. Subsequently Priode *(11)* discovered that local lesions were produced prior to development of systemic symptoms when tobacco leaves were pricked with a pin dipped in inoculum containing Tobacco Ringspot nepovirus. Viruses that are naturally transmitted by insect vectors probing into the vascular tissue can be successfully introduced to indicator plants using a pin to prick deeply into the vascular tissues *(2)*.

It was soon discovered that the use of abrasives resulted in improved virus transmission and the first abrasive to be used was sand *(12)*. Subsequently, charcoal and talc *(13)*, carborundum (silicon carbide) *(14)* aluminium oxide *(15)* and Celite *(16)* were reported. The most commonly used abrasives are carborundum, which is available in a range of mesh sizes, or Celite, a diatomaceous earth. By dusting the leaf surfaces of indicator plants with carborundum and subsequently applying the inoculum with a ball of absorbent cotton that had been dipped in extracted infective sap, Rawlins and Tomkins *(17)* successfully transmitted five viruses that had proved difficult to transmit by other means. Lamborn et al. *(18)* found carborundum to be significantly better than Celite for assaying TMV on *P. vulgaris*.

Abrasives can be incorporated in the inoculum, dusted onto leaves, or fired at leaves using a pressurised dispenser or gun prior to inoculation. Costa *(15)* compared three methods of application of carborundum (dusting or sprinkling onto leaves prior to inoculation or added to the inoculum) and observed no difference in results for infection of *Nicotiana tabacum* × *N. glutinosa* with TMV. However, using Celite, McKlusky and Stobbs *(19)* obtained widely variable local lesion counts for TMV when the abrasive was added to the inoculum and improved results were obtained when the abrasive was lightly dusted and inoculum subsequently applied with a brush.

Laidlaw *(20, 21)* described the development of two compressed air guns, a carborundum gun used to puncture cell walls and an inoculation gun used to spray the inoculum onto the damaged leaf surface. A mesh size of 400 was found to be optimum for inoculation of leaves of *N. tabacum* with TBRV. Particles smaller than mesh size 400, fired at leaves in a stream of compressed air, possessed insufficient momentum to puncture leaf cells whilst larger particles damaged some cells so severely that they were killed. The author provided evidence that cytoplasmic extrusions emerge from cells within several minutes and are gradually withdrawn. A peak of susceptibility to virus occurring between 1 and 5 min after cell puncture coincided with both the peak in number and surface area of cytoplasmic extrusions. This finding is somewhat similar to that obtained by Allington and Laird *(22)* who reported a gradual reduction in leaf susceptibility within 5 min of rubbing with carborundum.

When Laidlaw *(20)* compared the use of carborundum (mesh sizes 240–1,200) sprinkled onto leaves before manual inoculation, local lesion numbers obtained were similar for mesh sizes 240–800 whilst a reduction was obtained for mesh sizes 1,000 and 1,200. However, results of inoculations carried out in a similar manner by Kalmus and Kassanis *(16)* and using carborundum (mesh sizes 100–600) indicated that, as for the carborundum gun, mesh size 400 was optimum for manual transmission of TMV to *Nicotiana glutinosa*. Using an ultrasonic inoculation method, Lamborn et al. *(18)* reported that a carborundum mesh size of 1,000 gave optimum results. The procedure involved dusting leaves with carborundum prior to inoculation and they found that infectivity increased the greater the amount of carborundum applied. Beraha et al. *(8)* reported similar results with 600 mesh carborundum added to the inoculum. However by carrying out a comparison of infectivity using quantities of 240 and 800 mesh carborundum composed of equal numbers of particles they provided evidence to suggest that infectivity is related to the number of abrasive particles and not to the size of the particles.

Inoculum can be applied to leaves using a finger, brush *(23)*, pad of gauze *(7)*, cotton bud *(24)* or using a sap spray device. In comparative infection studies with four viruses, Laidlaw *(21)* found gun inoculation to be much more sensitive than manual inoculation. Inoculation guns, however, are not readily available and for most purposes, manual inoculation provides an adequate degree of sensitivity.

For manual inoculation it is important that the pressure applied to the leaf is sufficient to cause adequate wounding of the leaf surface and not so great as to result in cell death. Yarwood *(23)*, studied the effect of extra brushing with carborundum before or after inoculation and found that wet brushing caused a reduction in infection compared with controls not given the extra brushing treatment. This suggests that when carrying out manual inoculations it may be advisable to restrict the number of passes over the same area of leaf surface to one.

1.3.3. Environmental Conditions

1.3.3.1. Seasonal Variation

The environmental conditions under which plants are grown greatly affect symptom expression and seasonal variation has been clearly evident for some viruses *(25)*.

1.3.3.2. Light

Bawden and Roberts *(26)* demonstrated the importance of shading during the summer months in increasing susceptibility to infection by four viruses: reducing the light intensity increased both the numbers of local lesions produced and the virus titre. Subsequent work showed that pre-inoculation conditions were more important than post-inoculation lighting conditions and that maintaining indicator plants in the dark for 1–5 days increased susceptibility to infection *(27)*. It is important that growing conditions of indicator plants are conducive to production of plants that have 'soft' leaves and high light intensities generally give rise to 'hard' plants.

1.3.3.3. Pre-inoculation Temperature

Kassanis *(28)* showed that maintaining bean plants at 36°C for up to 2 days prior to inoculation with Tobacco Necrosis Virus (TNV) increased susceptibility to infection. This effect was also demonstrated by Laidlaw *(21)* who found that maintaining indicator plants at 35°C for 1 h prior to inoculation increased the number of local lesions of Tomato Black Ring Virus (TBRV) on *N. tabacum* cv White Burley (WB) whilst maintaining them at 5°C reduced the number compared with controls kept at 20°C. In this experiment, leaves were damaged prior to inoculation using a carborundum gun and then inoculated using a sap spray gun. The results obtained in this study may have been due to the fact that at the lower temperature the cytoplasmic extrusions that emanated from injured cells were smaller than those produced at the higher temperature thus presenting a lower surface area for 'capture' of the virus particles.

1.3.3.4. Post-inoculation Temperature

At temperatures of 30–40°C virus replication ceases, a feature that has led to the use of high temperature treatment for eradication of virus from infected plants. High temperatures can reduce virus symptoms or reduce a plant's ability to resist infection. Thus, in general, test plants should be grown at temperatures of 15–24°C. A day temperature of around 24°C alternating with a night temperature of 15°C is likely to produce satisfactory results.

1.3.3.5. Leaf Water Potential

The importance of water potential has been demonstrated by several authors. Laidlaw *(21)* compared local lesion production by TBRV on *N. tabacum* cv WB for plants given adequate water, plants that had been deprived of water for 24 h and plants that showed signs of wilting and obtained a mean of 41, 12, and 0 lesions respectively. Thus plants with a reduced water potential prior to inoculation produced fewer cytoplasmic extrusions and developed fewer local lesions. Surprisingly, McKlusky and Stobbs *(19)* reported the opposite effect for inoculation of detached leaves of *N. tabacum* cv. Xanthi-NC with TMV following dusting with Celite. In this study more local lesions developed on leaves that were wilted at the time of inoculation compared with those that were unwilted. These apparently conflicting results may be accounted for the results obtained by Yarwood *(23)* who studied the effect of inoculating detached leaves of bean (*Phaseolus vulgaris*) at varying degrees of water deficit with TMV: infection was increased when leaves were wilted to 90–95% but reduced when wilted to 72–80% of their original weight.

When abrasives have been used, inoculated leaves tend to wilt and so it is advisable to maintain plants in a humid and draught free environment to allow them to recover.

1.3.3.6. Nutrition

Nutritional conditions that favour plant growth also favour virus multiplication and healthy plants grown in a glasshouse with adequate nutrition and watering are likely to provide best results.

1.3.3.7. Endogenous Rhythms

A diurnal cycle that operates irrespective of any recent changes that have been made to environmental conditions has also been found to affect sensitivity to infection for four host/virus combinations *(29, 30)* and is an important factor to consider.

1.3.4. Precautions

Some viruses, such as TMV or Potato Virus X (PVX), which can be present in high concentration, are readily contact transmitted, with infection occurring through damaged leaf hairs. Thus care must be taken not to brush leaves of infected plants against those of healthy plants. As a general rule indicator plants should not be touched by hand following inoculation to avoid transmission of such viruses. Todd *(31)* clearly demonstrated the risk of contamination by PVX when he reported that the virus could be transmitted on trousers several weeks after they had been worn in an infected crop.

When carrying out the 'finger rubbing' technique, the risk of contamination from a previous inoculation can be reduced by thorough hand washing between samples. Ideally disposable gloves should be worn and changed between samples. Care must be taken not to contaminate items used in the procedure. Pestles and mortars should be boiled for 15 min or autoclaved following use. Glasshouses containing indicator plants must be sealed to avoid intrusion of insect vectors and vents covered with either thrip- or aphid-proof mesh as appropriate.

1.3.5. Symptom Diagnosis

Considerable experience is required to identify symptoms in test plants and to differentiate non-virus lesions from virus symptoms. Some test plant species can produce misleading symptoms for physiological reasons thus illustrating the importance of ensuring that control plants inoculated with healthy sap of the species under test are included.

Local lesions may develop within 3 days or more depending on the plant/virus combination. They occur at the point of entry of the virus and appear as chlorotic or necrotic spots or as ringspots (**Fig. 1.4**).

Systemic symptoms (**Fig. 1.5**) develop as a result of virus spreading throughout the plant and may or may not appear after local lesion development. They may be expressed in a variety of ways such as stunting, mosaics, chlorosis, necrosis, and rugosity (puckering of leaves). Several of these symptoms may be expressed sequentially in the same plant. The virus titre in the inoculum can affect the severity of symptom expression. Mosaics consist of varying patterns of light and dark green patches. They may appear as chlorotic areas varying in shape. Sometimes leaf veins appear to stand out, a feature often described as vein clearing. Another type of systemic symptom is vein banding when bands of dark green areas appear adjacent to the veins. Systemic necrosis may result in death of the plant whilst in some instances plants can 'grow away' from the initial symptom, e.g. Potato Virus Y^N and TBRV in *N. tabacum* cv. WB.

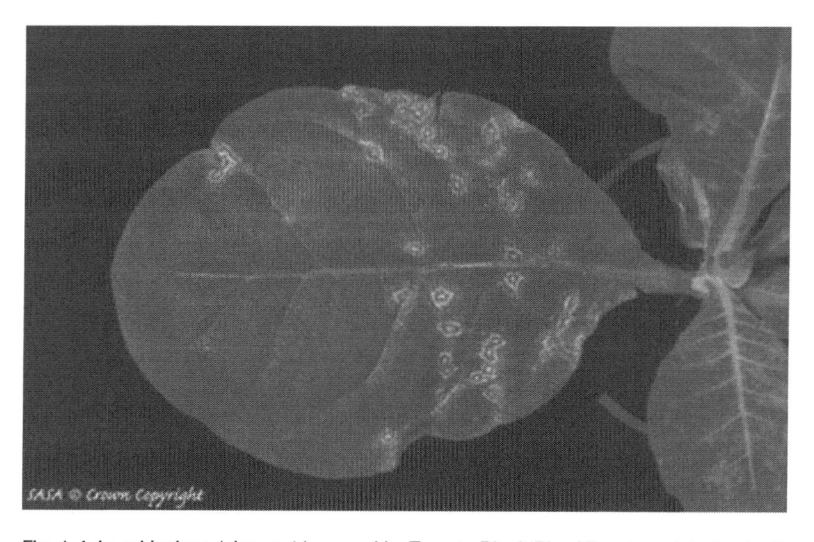

Fig. 1.4. Local lesions (*ringspots*) caused by Tomato Black Ring Virus inoculated onto *N. tabacum* cv White Burley.

Fig. 1.5. Systemic necrosis caused by Potato Virus Y^N inoculated onto *N. tabacum* cv White Burley.

1.4. Notes

1. For information on suitable indicator plant species to use for detection or diagnosis of viruses the VIDE database *(32)* can be viewed online at http://www.image.fs.uidaho.edu/vide. This database provides information on most species of virus known to infect plants. Another useful site (http://www.dpvweb.net/dpv/index.php) is supported by the Association of Applied Biologists and the Zhejiang Academy of Agricultural Sciences, Hangzhou, Peoples Republic of China and contains information on a wide range of viruses. In addition, a list of 32 indicator plants and their symptoms following inoculation by 37 potato viruses is found in a FAO/IPGRI publication by Jeffries *(33)* and similar publications are available for viruses of other crops.
2. The number of plants to inoculate depends on the virus in question. Viruses that have a high titre in the inoculum, such as is regularly obtained with PVX will readily transmit to one test plant. However as a routine at least two plants should be used and for viruses that are difficult to transmit at least six plants should be inoculated.
3. Generally it is best to inoculate young plants. Nicotiana tabacum should be inoculated when plants measure approximately 10 cm across, bearing 2–4 leaves. However inoculation of *Chenopodium quinoa*, can be successfully carried out when

plants are fairly large with around 10–12 leaves. The stage of development suitable for inoculation of a range of commonly used test plants is provided by Dijkstra and de Jager *(3)*.

4. Infective sap for inoculation may be obtained by grinding leaves using a pestle and mortar. It is not essential, as described by many authors, to express sap produced in this way through muslin, unless of course a dilution series is required. Sap may also be expressed through muslin using pliers. Alternatively, a sap press such as that supplied by Erich Pollahne GmbH may be used. Modification of a press to provide a flow of hot water at 80–85°C to sterilise the rollers between samples has been described by Laidlaw *(20)*.
5. It was reported by Holmes *(34)* that washing with water immediately after 'pin' inoculation of Nicotiana glutinosa with TMV increased the number of local lesions produced, and today probably the most common procedure following inoculation is to wash inoculated leaves with water. In many laboratories leaves are washed immediately with tap water as a routine procedure *(35)*. Following inoculation of *P. vulgaris* with TMV, Yarwood *(23)* studied the effect of varying periods of washing leaves with water from a compressed air atomizer. It was found that washing for up to 10 s increased infection but that washing beyond 10 s reduced infection. In another study, Yarwood *(36)* compared washing in slowly flowing water for 2 s and quick drying (using a jet of compressed air at 35 psi for 4 s) for their effect on transmission success for several viruses. Washing was found to be consistently less effective than quick drying. Washing followed by quick drying was less effective than quick drying alone. Kahn and Schachtner *(37)* has also reported an increase in infection following quick drying of leaves inoculated with TMV, Tobacco Necrosis Virus, and Southern Bean Mosaic Virus.

References

1. Verhoeven, J.T.J. & Roenhorst, J.W. (2003) Detection of a broad range of potato viruses in a single assay by mechanical inoculation of herbaceous test plants. *OEPP/EPPO Bull.* 33, 305–311.
2. Gibbs, A. & Harrison, B. (1976) *Plant Virology: The Principles.* Edward Arnold, London.
3. Dijkstra, J. & de Jager, C.P. (1998) *Practical Plant Virology: Protocols and Exercises.* Springer, Berlin.
4. Gendron, Y. & Kassanis, B. (1954) The importance of the host species in determining the action of virus inhibitors. *Ann. Appl. Biol.* 41, 183–188.
5. Martin, R.R. & Converse, R.H. (1982) An improved buffer for mechanical transmission of viruses from *Fragaria* and *Rubus. Acta Hortic.* 129, 69–74.
6. Sanger, H.L. & Brandenburg, E. (1961) Uber die gewinnung von infektiosem presssaft aus 'Wintertyp'-pflanzen des Tabak-Rattle-Virus durch phenolextraktion. *Naturwissenschaften* 48, 391.
7. Thornberry, H.H. (1935) Effect of phosphate buffers on infectivity of tobacco mosaic virus. *Phytopathology* 25, 618–627.

8. Beraha, L., Varzandeh, M. & Thornberry, H.H. (1955) Mechanism of the action of abrasives on infection by Tobacco Mosaic Virus. *Virology* 1, 141–151.
9. Yarwood, C.E. (1952) The phosphate effect in plant virus inoculations. *Phytopathology* 42, 137–143.
10. Reddick, D., & Stewart, V.B. (1919) Transmission of the virus of bean mosaic in seed and observations on thermal death-point of seed and virus. *Phytopathology* 9, 445–450.
11. Priode, C.N. (1928) Further studies in the ring-spot disease of tobacco. *Am. J. Bot.* 15, 88–93.
12. Fajardo, T.G. (1930) Studies on the mosaic disease of the bean (*Phaseolus vulgaris* L.). *Phytopathology* 20, 469–494.
13. Vinson, C.G. & Petrie, A.W. (1931) Mosaic disease of tobacco. II. Activity of the virus precipitated by lead acetate. *Contrib. Boyce Thompson Inst.* 3, 131–145.
14. Rawlins, R.E., & Tompkins, C.M. (1934) The use of carborundum as an abrasive in plant-virus inoculations. *Phytopathology* 24, 1147.
15. Costa, A.S. (1944) Quantitative studies with carborundum and its use in local-lesion tests. *Phytopathology* 34, 288–300.
16. Kalmus, H. & Kassanis, B. (1945) The use of abrasives in the transmission of plant viruses. *Ann. Appl. Biol.* 32, 230–234.
17. Rawlins, R.E., & Tompkins, C.M. (1936) Studies on the effect of carborundum as an abrasive in plant virus inoculations. *Phytopathology* 26, 578–587.
18. Lamborn, C.R., Cochran, G.W. & Chidester, J.L. (1971) An ultrasonic inoculation method for Tobacco Mosaic Virus bioassay. *Phytopathology* 61, 1015–1019.
19. McKlusky, D.J. & Stobbs, L.W. (1985) A modified local lesion assay procedure with improved sensitivity and reproducibility. *Can. J. Plant Pathol.* 7, 347–350.
20. Laidlaw, W.M.R. (1986) Mechanical aids to improve the speed and sensitivity of plant virus diagnosis by the biological test method. *Ann. Appl. Biol.* 108, 309–318.
21. Laidlaw, W.M.R. (1987) A new method for mechanical virus transmission and factors affecting its sensitivity. *OEPP/EPPO Bull.* 17, 81–89.
22. Allington, W.B. & Laird, E.F. (1954) The inhibitive effect of water on infection by Tobacco Mosaic Virus. *Phytopathology* 44, 546–548.
23. Yarwood, C.E. (1955). Deleterious effects of water in plant virus inoculations. *Virology* 1, 268–285.
24. Harris, P.S. & Browning, I.A. (1980) The effects of temperature and light on the symptom expression and viroid concentration in tomato of a severe strain of potato spindle tuber viroid. *Potato Res.* 23, 85–93.
25. Smith, K.M. & Bald, J.G. (1935) A description of a necrotic virus disease affecting tobacco and other plants. *Parasitology* 27, 231–245.
26. Bawden, F.C. & Roberts, F.M. (1947) The influence of light intensity on the susceptibility of plants to certain viruses. *Ann. Appl. Biol.* 34, 286–296.
27. Bawden, F.C. & Roberts, F.M. (1948) Photosynthesis and predisposition of plants to infection with certain viruses. *Ann. Appl. Biol.* 35, 418–428.
28. Kassanis, B. (1952) Some effects of high temperature on the susceptibility of plants to infection with viruses. *Ann. Appl. Biol.* 39, 358–369.
29. Matthews, R.E.F. (1953) Factors affecting the production of local lesions by plant viruses. I. The effect of time of day of inoculation. *Ann. Appl. Biol.* 40, 377–383.
30. Matthews, R.E.F. (1953) Factors affecting the production of local lesions by plant viruses. II. Some effects of light, darkness and temperature. *Ann. Appl. Biol.* 40, 556–565.
31. Todd, J.M. (1957) Spread of Potato Virus X over a distance. *Proceedings of the Third Conference on Potato Virus Diseases*, Lisse-Wageningen, 24–28 June, 1957.
32. Brunt, A., Crabtree, K., Dallwitz, M., Gibbs, A. & Watson, L. (1996) *Viruses of Plants: Descriptions and Lists from the VIDE Database.* 1484 pp. C.A.B. International, UK
33. Jeffries, C.J. (1998) *FAO/IPGRI Technical Guidelines for the Safe Movement of Germplasm.* No. 19. Potato. Food and Agriculture Organization of the United Nations,Rome/ International Plant Genetic Resources Institute, Rome.
34. Holmes, F.O. (1929) Inoculating methods in Tobacco Mosaic studies. *Bot. Gaz. (Chicago)* 87, 56–63.
35. Walkey, D.G.A. (1991) *Applied Plant Virology*, 2nd edition, Chapman & Hall, London.
36. Yarwood, C.E. (1973) Quick drying versus washing in virus inoculations. *Phytopathology* 63, 72–76.
37. Kahn, R.P. & Schachtner, N.D. (1954) The relative effect of rate of drying and potassium phosphate in the inoculum on plant virus infections (Abstract) *Phytopathology* 44, 389.

Chapter 2

Development of Enzyme Linked, Tissue Blot and Dot Blot Immunoassays for Plant Virus Detection

Hei-ti Hsu

Summary

Immunoassays are among the most powerful and useful techniques for analysis of biological materials. There are numerous variations in which immunoassays can be performed. Coupled with enzyme, using chromogenic substrates, the enzyme immunoassay technique is used to trace the target antigen in tissues. The technique also is used to measure the concentration of antigen in tissue extracts. This chapter provides fundamental information that is needed to carry out the routinely used procedures in plant virus research.

Key words: ELISA, Dot blot, Tissue blot, $F(ab')_2$, Protein A, Enzyme-labeling, Biotinylation.

2.1. Introduction

Enzyme Immunossay (EIA) is a simple, effective and powerful tool for identification and analysis of biological materials. The method is based on the principle of specificity of antibody–antigen interaction. Coupled with enzyme using chromogenic substrates, the technique is used to trace target antigen in tissue sections or tissue blots. Upon hydrolysis of the substrate, the presence of antigen is visualized by an insoluble color precipitate. The technique also is used in quantitative measurement of antibodies in body fluids (*see* **Note 1**) or antigens when extracted from tissues. In this procedure, a color soluble substance is produced by the enzyme. The concentration of the antigen present in the extract is determined spectrophotometrically.

R. Burns (ed.), *Methods in Molecular Biology, Plant Pathology, vol. 508*

DOI: 10.1007/978-1-59745-062-1_2

Before getting into the details of basic EIA, a couple of terms commonly used in serological procedures need to be clarified. Direct EIA utilizes enzyme molecules that are covalently linked to antibodies specific to the target antigen in question. In indirect EIA, an unlabeled target-specific antibody is first reacted with antigens. These antibodies are called primary antibodies. In second step, this globulin–antigen complex is reacted with enzyme-labeled anti-globulin conjugate (secondary antibody). If the enzyme activity is detected following the addition of enzyme substrate, we have indirect evidence of the presence of the specific antigen–antibody complex formed in the first step.

Indirect EIAs are versatile. The extension of the indirect procedure is the use of Protein A- or protein G-labeled antibody conjugate and biotinylated antibody. The ability of protein A molecules to bind specifically to the carbohydrate moiety of many immunoglobulins makes enzyme-labeled protein A or protein G conjugate a useful detection reagent *(1, 2)*. In procedures where biotinylated primary or secondary antibodies are used, avidin–enzyme conjugate is employed as a detection reagent *(3)*. Both enzyme-labeled protein A and protein G and biotinylated species specific anti-globulin antibody conjugates are readily available from a number of reliable commercial sources. Its cost effectiveness and consistency in quality make the modified procedures very attractive to investigators.

Direct detection is the simplest, most reliable and highly specific of all EIA procedures but is somewhat a little less sensitive than the indirect method. One drawback of the direct method is that each virus specific antibody reagent must be labeled with the enzyme.

The merit of the indirect method lies on the amplification of primary and secondary antibodies. Maximal amount of the target specific antibody reacts with the antigen followed by the excess amount of labeled secondary antibody reacting with the primary antibody yielding stronger signal than the direct method. With the increase of reacting reagents, the specificity of detection is slightly compromised.

This chapter intends to provide the fundamental information needed to execute the commonly used technology in biological sciences, and is written for plant pathologists having little or no previous exposure with EIA. It is almost impossible that the chapter be exhaustive and extensive to cover every possible variation and modification because of versatility and applicability of the EIA to almost every discipline in biological sciences. I shall focus on a few EIAs routinely used in plant virus laboratories. These shall include basic Enzyme Linked Immunosorbent Assay (ELISA) *(4)*, Tissue Blot Immunoassay (TBIA) *(5)* and Dot Blot Immunoassay (DBIA) *(6)*. Useful variations and modifications will also be discussed. Step-by-step basic procedures shall be presented. It is hoped that the technique will extend its application to various fields in plant pathology.

2.2. Materials

2.2.1. Preparation of Enzyme–Antibody Conjugates

1. Alkaline phosphatase, Type VII (Sigma).
2. 0.01 M Phosphate buffered saline solution (PBS) pH 7.5.
3. 10% Glutaraldehyde, reagent grade (freshly opened vial).
4. 1% Bovine serum albumin (BSA) in PBS.
5. 10% Sodium azide (*see* **Note 2**).
6. Centrifuge.

2.2.2. Preparation of $F(ab')_2$ Antibody Fragments

1. Protein-A Sepharose.
2. 0.5 M Sodium acetate, pH 4.5.
3. Pepsin solution: 1 mg pepsin in 1 mL warm 0.1 M sodium acetate buffer, pH 4.5.
4. Tris–HCl salt.
5. 0.1 M Glycine, pH 3.2.
6. Saturated ammonium sulfate, pH 7.8.
7. Normal saline solution: 0.85% sodium chloride.

2.2.3. Preparation of Biotinylated Antibody

1. 0.2 M Borate buffer,\ pH 8.5.
2. *N*-Hydroxy succinimidobiotin.
3. Dimethylsulfoxide (DMSO).
4. 0.01 M Phosphate buffered saline (PBS), pH 7.5.

2.2.4. ELISA Procedure

1. ELISA plates.
2. Coating-buffer: (0.05 M carbonate, pH 9.6). Dissolve 1.59 g sodium carbonate, and 2.93 g sodium bicarbonate in 1,000 mL distilled water.
3. 0.01 M Phosphate buffered saline (PBS), pH 7.5.
4. PBS-Tween: 0.05% (v/v) Tween-20 in PBS.
5. Blocking solution: 1% nonfat dry milk, 0.5% BSA in PBS-Tween.
6. Alkaline phosphatase–antibody conjugates (either virus-specific antibody or species-specific anti-globulin antibody) in PBS-Tween.
7. Substrate buffer: 97 mL diethanolamine, 0.1 g $MgCl_2 \cdot 6H_2O$, pH 9.8. Add distilled water to 1,000 mL.

8. Enzyme substrate: Dissolve 25 mg *p*-nitrophenyl phosphate in 25 mL substrate buffer.
9. Shaker.
10. ELISA Reader.

2.2.5. TBIA and DBIA Procedures

1. 0.2- or 0.45-µm nitrocellulose membrane.
2. Razor blades.
3. Gloves or forceps.
4. Small (about 10 × 12 cm) glass tray.
5. Phosphate suffered saline solution (PBS): pH 7.5.
6. PBS-Tween: 0.05% Tween-20 in PBS.
7. Blocking solution: 1% nonfat dry milk, 0.5% BSA in PBS-Tween.
8. Alkaline phosphatase-labeled antibody conjugate.
9. Substrate buffer: Solution A: 35.5 mL distilled water, 2.0 mL 2.5 M Tris–HCl, pH 9.5, 2.0 mL 11% sodium chloride, 0.2 mL 1 M magnesium chloride; Solution B: 14 mg nitro tetrazolium in 300 µL methanol; Solution C: 7 mg 5-bromo-4-chloro-3-indolyl *p*-toluidine salt in 50 µL dimethylsulfoxide. Prior to use, add B to A, mix well. Add C dropwise with shaking, mix completely. Stop solution. 0.01 M Tris-HCl, 1 mM EDTA (pH 7.5).
10. Shaker.

2.3. Methods

2.3.1. Preparation of Enzyme–Antibody Conjugates

During early years when the ELISA technique was developed, scientists had to prepare enzyme–antibody conjugates for their own research *(7)*. In recent years with the exception of a few viral pathogens and previously unreported new viral causal agents, enzyme–antibody conjugates to many viruses are now readily available from a number of commercial suppliers (*see* **Note 3**). The great benefits of obtaining conjugates from commercial sources are that they come in either alkaline phosphatase form or horseradish peroxidase form and that they are consistent in quality. However, the following illustrates the procedures used to prepare reagents for use in EIA.

This procedure is useful for either alkaline phosphates or horseradish peroxidase.

1. Centrifuge alkaline phosphatase, type VII, which comes in ammonium sulfate suspension. Dissolve 2 mg alkaline

phosphates in 0.5 mL of PBS containing 1 mg specific immunoglobulin. Dialyze against PBS.

2. Add gluteraldehyde to the enzyme–antibody mixture to a final concentration of 0.2%. Stir at room temperature for 3 h. Dialyze overnight against PBS.
3. Dilute the conjugate solution with PBS containing 1.0% bovine serum albumin (or egg albumin) to make a 10% solution (0.5 mL of enzyme–antibody + 4.5 mL PBS + 0.01 mL 10% azide sodium). Store at 4°C. This is stock solution.

2.3.2. Preparation of $F(ab')_2$ Antibody Fragment

In the double antibody sandwich ELISA (DAS-ELISA) procedure, use of $F(ab')_2$ antibody fragments in coating microtiter plates eliminates the necessity of preparing various enzyme–antibody conjugates for each of individual different viruses. Enzyme-labeled protein A is used as a universal signal generating reagent.

1. Take 5 mg antibody which has been passed through protein A column, and 0.5 M sodium acetate buffer, pH 4.5, to give 0.1 M sodium acetate, pH 4.5 (5 mg antibody in 0.4 mL saline + 0.1 mL 0.5 M acetate).
2. Prepare pepsin solution: 1 mg pepsin in 1 mL warm 0.1 M sodium acetate, pH 4.5. Add 50 μL pepsin solution (i.e. 50 μg) to antibody.
3. Incubate at 37°C for 20–24 h.
4. Neutralize by adding few Tris salt crystals to pH 8.0 to pH 8.4.
5. Run digested antibody through protein A column. The first peak is $F(ab')_2$. Then elute column with 0.1 M glycine, pH 3.0 to remove any undigested antibody.
6. Concentrate $F(ab')_2$ fraction by adding one-half volume saturated ammonium sulfate.
7. Dialyze against several changes of saline. Centrifuge at low speed to remove any insoluble material.

2.3.3. Preparation of Biotinylated Antibody

Avidin is an egg white protein which is a tetramer with four identical subunits. Each contains a high affinity binding site for biotin. The biotin molecule can be coupled to antibodies as illustrated.

1. Dilute antibody to 1 mg/mL in 0.2 M borate buffer, pH 8.5.
2. Prepare 1 mg/mL biotinyl-*N*-hydroxysuccinimide in DMSO.
3. Add 100 μL of biotin solution to 1 mL antibody dropwise while slowly stirring.
4. Continue stirring 5–10 min. Let sit for 4 h at room temperature.
5. Dialyze against PBS, with several changes at 4°C overnight.
6. Aliquot and store at 20°C below zero.

2.3.4. ELISA Procedure

Either antigen or antibody can be attached to the surface of microtiter plates made from polystyrene or polyvinyl. Although round bottom well plates are acceptable, flat bottom well plates are recommended if results are to be read spectrophotometrically (*see* **Note 4**). The optimal conditions for running the test are pre-determined by checker-board titration method using reference reagents. These include the concentration of reactants, time and temperature of incubation. Once determined, these conditions are to be followed in subsequent tests if reproducible results are to be anticipated.

For most viral investigation, protein (viruses, viral proteins, and antibodies) at 1–10 μg/mL in coating buffer at pH 9.6, are used for coating microtiter plates. Coating are usually carried out at room temperature (some people prefer 37°C) for 1–2 h. For convenience, incubation at 4°C overnight gives satisfactory results. Blocking is an essential step in EIA. Egg albumin, BSA and/or nonfat dry milk are used following the step of coating. Blocking or quenching is a process in which unoccupied protein binding sites on the surface of microtiter wells are saturated so that detecting antibody molecules in direct procedure (also primary and secondary antibodies in indirect methods) do not bind non-specifically resulting in background formation. Washing is the separation of bound and free reactant after an appropriate time of reaction. It is usually accomplished on a shaker for 10–20 min with washing solution.

2.3.4.1. ELISA Using Antibody-Sensitized Plates, Direct Procedure

The following assay system illustrates the most commonly used DAS-ELISA direct procedure in plant virus research. Antigen in tissue extract is specifically reacted with and bound to antibody that is already attached to microtiter wells. Unless otherwise noted, incubations are at room temperature for 1–2 h or 4°C overnight.

1. Coat plates by adding 100–150 μL antibody-coating solution to each well and incubate.
2. Discard the antibody solution and wash the plate once.
3. Add 200 μL blocking solution and incubate.
4. Invert and tap plates to remove blocking solution and wash once.
5. Add samples and incubate.
6. Discard samples and wash once.
7. Add enzyme-labeled virus-specific antibody and incubate (*see* **Note 5**).
8. Empty enzyme–antibody conjugate.
9. Wash 3–4 times, 3–5 min each! This is an important step. Be sure that no residual unbound enzyme conjugate is left behind.

10. Add enzyme–substrate solution and incubate for color to develop.
11. Read spectrophotometrically (A_{405}) or record visually.

2.3.4.2. ELISA Using Antigen-Sensitized Plates, Indirect Procedure

Antigen coating procedure in indirect ELISA offers an advantage that one single, species-specific enzyme-labeled anti-globulin antibody conjugate can be used in assays of various different viruses. It eliminates the necessity of preparing various enzyme–antibody conjugates for individual viruses.

1. Dilute infected tissue extract in coating buffer.
2. Coat plate by adding 100–150 μL antigen-coating solution to each well.
3. Discard the coating solution and wash the plate once.
4. Add 200 μL blocking solution and incubate.
5. Discard blocking solution and wash once.
6. Add 150 μL virus-specific mouse monoclonal antibody or dilute rabbit antiserum and incubate.
7. Empty plate and wash once.
8. Add 150 μL mouse specific alkaline phosphatase-labeled anti-globulin antibody (or rabbit specific anti-globulin antibody if rabbit anti-virus antibody is used), and incubate (*see* **Note 5**).
9. Discard the conjugate solution and wash the plate 3–4 times, 5–10 min each. Be sure that no residual unbound enzyme-conjugate is left behind.
10. Add enzyme-substrate solution and incubate for color development.
11. Read spectrophotometrically (A_{405}) or record visually.

2.3.5. TBIA and DBIA Procedures

Similar to coating procedure in ELISA, proteins can be immobilized on nitrocellulose membranes. The superior binding capacity of at least 80 μg of protein per square centimeter of nitrocellulose membrane makes it an excellent solid substrate in blot immunoassay of viral antigens *(8)*. Membranes 0.45 μm are suitable for TBIA or DBIA *(5, 6)*. Use gloves or forceps while handling the membrane.

Tissue blot is a process of transfer of protein antigens from a freshly cut tissue surface to nitrocellulose membrane. It is achieved simply by bringing a freshly cut tissue surface in direct contact with a dry nitrocellulose membrane. The tissue imprint is made by application of a slight pressure of the cut tissue surface while it is in contact with the nitrocellulose membrane. Do not squeeze juice out from tissues. This will smear the imprint.

In the Dot Blot procedure, the samples are applied to nitrocellulose membrane while it is attached and clamped into a manifold. Add 0.25–100 μL tissue extract containing virus or viral protein to each well. Membranes are removed. From here on both membranes from tissue blot or dot blot are processed similarly as microtiter plates in ELISA. Either direct or indirect procedure can be used.

Detection enzymes for use in both tissue blot and blot immunoassay procedures should be carefully evaluated before the system is established. Horseradish peroxidase is commonly employed in enzyme-linked immunoassay for detection of a number of plant pathogens. It may not be suitable for use in tissue blot immunoassay on nitrocellulose membranes since endogenous peroxidase from plant tissue interferes with the assay. Alkaline phosphatase-antibody (both primary virus-specific and secondary immunoglobulin-specific) conjugates are generally available in plant pathology laboratories and are convenient to investigators. Enzyme-labeled protein A or avidin–biotin system works equally well in membrane blot assays. Substrates that produce soluble colored products are used in ELISA. With the same enzyme conjugate in TBIA, the substrates that yield insoluble colored products precipitating at the site of enzyme reaction should be the choice. When chemiluminescent substrates are utilized in tissue blot immunoassay, the presence of antigens can be recognized by the image registered on a light-sensitive X-ray film *(9)*. This is especially useful when colored pigments of plant tissue origin interfere with the results of analysis using chromogenic substrates.

2.3.5.1. TBIA Direct Procedure

1. Excise tissues (leaves, petioles, stems, flower buds, emerging shoots, bulb, etc.).
2. For thin tissue such as leaves, roll them into a tight core.
3. Hold tissue in one hand and cut with a new razor blade in steady motion with other hand to obtain a single-plane cut surface (*see* **Note 6**).Press for about a full second the newly cut surface onto a nitrocellulose membrane to obtain a tissue blot. Use a firm but gentle force (*see* **Note 7**).
4. Block tissue blots by immersing nitrocellulose membranes in the blocking solution for 30–60 min with occasional shaking at room temperature (*see* **Note 8**).
5. Wash blots once with washing solution for about 1 min with gentle shaking (about one rotation per second on a mechanical shaker).
6. Incubate tissue blots in a glass dish with alkaline phosphatase-labeled virus-specific antibodies diluted in PBS-Tween for 60 min at room temperature. Be sure that the reagent solution covers the blot (*see* **Note 9**).
7. Wash blots 4 times in washing solution for 30 min with shaking.
8. Soak blots in substrate solution 2–5 min at room temperature to detect enzyme activity.

9. Rinse blots for a few seconds in distilled water.
10. Stop reaction by immersing blots in stopping solution, two to three changes, 10 min each.
11. Dry nitrocellulose membranes on two to three layers of tissue wipes in a dust-free area.

2.3.5.2. DBIA Indirect Procedure

1. Grind 0.2 g of liquid nitrogen frozen midrib excised from phytoplasma infected plants in 3.0 mL PBS buffer in a mortar and pestle. This constitutes 1:15 dilution of antigen preparation. Transfer the extract to a glass test tub and let the extract undisturbed for 5 min at room temperature. Leaving a sample suspension undisturbed for a few minutes allows particulates, which may interfere with the sample application onto the membrane, to settle by gravity.
2. Wet nitrocellulose membrane in PBS for 1–2 min.
3. Place wet nitrocellulose onto manifold and clamp into place following manufacturer's instructions.
4. Turn on vacuum; add 50 μL antigen to well.
5. Let antigen filter through membrane completely under vacuum. Wash each well with 100 μL PBS.
6. Turn off vacuum and disassemble apparatus.
7. Block tissue blots by immersing nitrocellulose membranes in the blocking solution for 30–60 min with occasional shaking at room temperature.
8. Wash blots once with PBS-Tween for about 1 min with gentle shaking.
9. Incubate tissue blots in a glass dish with virus-specific mouse monoclonal antibodies diluted in PBS-Tween for 60 min at room temperature. Be sure that the reagent solution covers the blot.
10. Incubate tissue blots with alkaline phosphatase-labeled mouse-specific antibodies diluted in PBS-Tween for 60 min at room temperature. Be sure that the reagent solution covers the blot (*see* **Note 8**).
11. From here on, follow **step 8** described in **Subheading 2.3.5.1**.

2.4. Notes

1. Labeled antigen is routinely utilized in EIA in medical diagnosis of infectious diseases to measure the level of specific antibodies after viral infection. This format is rarely applicable in plant sciences.

2. Sodium azide inhibits the horseradish peroxide activity. Instead, use merthiolate as preservative.
3. Companies that offer services and/or antisera for plant viral disease diagnosis are: Adgen Ltd, http://www.adgen.co.uk; Agdia, Inc., http://www.agdia.com; American Type Culture Collection, http://www.atcc.com; Bactochem Ltd, Fax 972-8-9401439; Bio-Rad Laboratories, http://www.bio-rad.com; Bioreba AG., http://www.bioreba.ch; Durviz, Inc., http://www.durviz.com; German Collection of Microorganisms and Cell Cultures, http://www.dsmz.de; LCA BIOTEST, http://perso.wanadoo.fr/lcab/; LOEWE Biochemica GmbH, http://www.loewe-info.com; and Plant Research international, http://www.plant.wageningen-ru.nl/
4. Round bottom well plates are manufactured specifically for hemaglutination, complement fixation and other serological tests.
5. Horseradish peroxidase-labeled antibody conjugate is suitable for use in ELISA, but is not recommended due to that its substrate, *ortho*-phenylene diamine, is possibly carcinogenic.
6. In tissues that contain a high concentration of latex, the cut surface is first drained on tissue paper to remove excess exudate before blotting onto nitrocellulose membranes.
7. Membrane blots, after preparation, can be stored for a period of time up to 4 weeks in a dust-free environment or mailed to a processing laboratory. This allows membrane blots to be prepared in one location and sent to another location for diagnosis.
8. Blocking should only be done immediately before incubation with antibodies.
9. For economical reasons, alkaline phosphatase labeled antibody conjugate can be recovered and used repeatedly for a few times before the solution shows losses of enzyme activity.

References

1. Goding, J.W. (1978). Use of staphylococcal protein A as an immunological reagent. *J. Immunol. Methods* 20:241–253.
2. Langone, J.J. (1982). Protein A of *Staphylococcus aureus* and related immunoglobulin receptors produced by streptococci and pneumonococci. *Adv. Immunol.* 32:157–251.
3. Guesdon, J.L., Ternynck, T., and Avrameas, S. (1979). The use of avidin–biotin interaction of immunoenzymatic techniques. *J. Histochem. Cytochem.* 27:1131–1139.
4. Clark, M.F., and Adam, A.N. (1977). Characterization of the microplate method of enzyme-linked immunosorbent assay for the detection of plant viruses. *J. Gen. Virol.* 34: 475–483.
5. Lin, N.S., Hsu, Y.H., and Hsu, H.T. (1990). Immunological detection of plant viruses and a mycoplasma like organism by direct tissue blotting on nitrocellulose membranes. *Phytopathlogy* 80:824–828.
6. Hsu, H.T., Lee, I.M., Davis, R.E., and Wang, Y.C. (1990). Immunization for generation of hybridoma antibodies specifically reacting with plants infected with a mycoplasma like organism (MLO) and their use in detection of MLO antigens. *Phytopathology* 80:946–950.

7. Avrameas, S. (1969). Coupling of enzymes to proteins with glutaraldehyde. Use of the conjugates for the detection of antigens and antibodies. *Immunochemistry* 6:43–52.
8. Gershoni, J.M., and Palade, G.E. (1982). Electrophoretic transfer of proteins from sodium dodecyl sulphate–polyacrylamide gels to a positively charged membrane filter. *Anal. Biochem.* 12: 396–405.
9. Makkouk, K.M., Hsu, H.T., and Kumari, S.G. (1993). Detection of three plant viruses by dot-blot and tissue-blot immunoassays using chemiluminescent and chromogenic substrates. *J. Phytopathol.* 139:97–102.

Chapter 3

Immunisation Strategies for Antibody Production

Robert Burns

Summary

A range of immunisation techniques can be used for the successful production of antibodies. The choice of method used is dependent on the nature of the antigen and the type of antibody required by the user.

Key words: Immunisation, Antibody, Antigen, Immunogen, Immunity, Epitope, Monoclonal, Polyclonal, Immunoglobulin.

3.1. Introduction

Vertebrate immune systems are capable of producing antibodies to a greater or lesser degree in response to the presence of a foreign protein within the tissues of the animal. The presence of the foreign protein initiates a sequence of events, mediated by the cells of the immune system that leads to the release of antibody molecules in blood and some body secretions.

Antibodies produced by vertebrate immune systems bind strongly to the protein that elicited their formation and it is this unique ability which is harnessed in all branches of immunochemistry.

Vertebrates have evolved this immunological strategy to help them to combat pathogens of viral, bacterial and fungal origin; however, almost all foreign substances regardless of source can induce antibody responses. The immunological response in mammals is particularly well developed and it is this group of vertebrates that are normally used for antibody production.

There are two main approaches for antibody production in vertebrates each having strengths and weaknesses depending upon

R. Burns (ed.), *Methods in Molecular Biology, Plant Pathology, vol. 508*

DOI: 10.1007/978-1-59745-062-1_3

intended application. Prior to discussing these two approaches it would be beneficial to describe some background to immunology.

Mammalian embryos are extremely tolerant of foreign proteins while still in utero, and all substances within the developing organism are accepted as "self". This is essential during development to ensure that immune responses are not raised to proteins and peptides produced during this time. Any immunological response to developmental proteins, hormones and growth factors would have disastrous results.

Shortly before or immediately after birth the neonatal immune system matures and learns to differentiate between "self" and "non-self" *(1)*. The immature immune system contains millions of cells within the bone marrow capable of producing antibodies (B-lymphocytes). This cell population is in effect a "starter pack" containing cells which will be capable of responding to a huge number of target proteins. These neonatal lymphocytes are produced by random re-assortment of the antibody genes and because of this many of them will recognise and respond to proteins within the developing individual. A process of clonal deletion takes place and any lymphocytes, which recognise proteins within the developing organism, are killed. As the young mammal matures it is incapable of mounting an immunological response to "self" antigens as a result of the process of clonal deletion.

The remaining B-lymphocytes in the bone marrow have the potential to respond to an enormous number of foreign substances (antigens). Once exposed to antigens the cells which have the best fit antibody to the target undergo clonal expansion to increase the cell numbers and affinity maturation to increase the specificity (fit) of the antibody molecules produced.

The ability of the mammalian immune system to respond to foreign substances is based on the molecular shape of antigen fragments produced as a result of digestion by cells called macrophages. The antigen fragments produced by this process are generally about the size that one antibody binding site can physically adhere to. These fragments are known as epitopes and although there will be many on any target substance, a single antibody will only be able to recognise and bind to one of them.

The response to any antigen will involve the recruitment of many B-lymphocytes, each making antibodies to an individual epitope on the target molecule. These lymphocytes, which have responded to the epitopes on the antigen, undergo clonal expansion so that many descendant B-lymphocytes are produced from each of the original cells. This process gives rise to a population of cells descended from single progenitors (clones) each with their own specific antibody to epitopes on the antigen. These clones of cells are resident in lymphoid tissue and are particularly concentrated in the marrow, spleen and gut associated lymphoid tissue (GALT). The resulting pool of antibody molecules produced by these cells is know as polyclonal antibody as it is derived from multiple clones each with unique specificity to single epitopes.

Monoclonal antibodies are produced as a result of immortalising and expanding the individual antibody secreting cells artificially in tissue culture *(2)*. Cells grown in this way all have identical epitope specificity and as they are derived from single clones their product is known as monoclonal antibody. Cells that secrete monoclonal antibodies are known as hybridomas and are typically derived by the fusion of two cell types. B-lymphocytes, which have the capacity to make antibody, are obtained from a donor spleen and are physically fused to a tumour cell line, which is immortal. The resulting hybridomas are immortal and produce antibody into the synthetic medium in which they are growing.

Exposure to antigens can be through a variety of routes but ultimately the cellular changes leading to B lymphocyte activation are blood borne. Natural immunisation takes place as a consequence of infection through respiratory, digestive, urogenital and skin surfaces. Medical immunisation to prevent infection by pathogens is normally carried out by intramuscular injection although other routes such as oral dosing for poliomyelitis and intradermal injection for tuberculosis may be used. The main feature, which characterises immunisation, is the presentation of antigens to the cells of the immune system, which induces B lymphocyte priming. These primed B cells will undergo clonal expansion and will secrete antibodies until the antigen has been destroyed. As soon as the antigen has been removed, the B cell lineage making antibodies to it will become quiescent and will form a stable population within the tissues of the immune system (memory cells). If the antigen is encountered again by the organism these quiescent B cells can undergo rapid clonal expansion and can mount an antibody response much faster than during the primary challenge. Each time that the immune B cells are exposed to the antigen, the affinity (fit) of the antibody produced will be improved and the number of quiescent B cells after each challenge increases. Each challenge also increases the amount of antibody produced in the blood and after three or four immunisations the individual reaches a status of hyperimmunity. This is characterised by high levels of circulating specific antibodies typically in the range of 10–20 mg/ml of serum. Hyperimmunity is rarely ever seen as a result of natural immunisation but is commonly used for the in vivo production of polyclonal antibody. Risks are associated with hyperimmunity as further exposure to the antigen can lead to an overwhelming immune response known as anaphylaxis which can be rapidly fatal. Paradoxically, repeated exposure to the antigen can lead to immunological tolerance, where the B cells making the antibody are destroyed by the immune system leaving the individual unable to mount an immune response to the antigen.

As previously stated immunisation is a phenomenon mediated by the cells of the immune system and is normally the result of a blood-borne challenge by antigen. The route of introduction can be very important in determining how well the individual will respond to an antigen.

It is extremely important when immunising animals for antibody work to choose the correct approach for the type of antigen to be used. This chapter describes a number of immunisation routes for polyclonal antibody production, monoclonal antibody cell donors and also one method for inducing selective immune tolerance as a preparative method.

Polyclonal and monoclonal antibodies should be seen as complimentary in their use. Each has strengths and weaknesses and the choice of which to use should be carefully evaluated prior to embarking on antibody production. In general, polyclonal antibodies have a much broader specificity as the antiserum pool comprises many species of antibody molecule each with different target epitopes on the antigen. This lack of specificity is advantageous in situations where variation in the target substance is known and polyclonal antibodies may provide a more robust test. Monoclonal antibodies are derived from clonal cell lines and their specificity is directed to a single epitope on the antigen. The highly specific nature of the monoclonal antibody allows the development of assays where two very closely related substances can be differentiated from each other. Examples of these highly specific tests are found in virus testing for strain differentiation and in clinical assays where levels of a synthetic hormone may be detected in spite of the presence of its naturally occurring counterpart.

3.1.1. Legislation

There are strict regulations governing the welfare of laboratory animals used for antibody production in most countries. Before deciding on a particular approach for antibody production it is important that the appropriate authorities are contacted with a project proposal to ensure that the methods to be used are permissible. Local ethical review committees may also have inputs into project design to ensure that numbers of animals used are appropriate and that other diagnostic alternatives have been investigated. Legislation and ethical review typically covers animal species, numbers to be used, immunisation route, bleeding regimes, and welfare issues such as project duration.

3.2. Materials

1. Balb-c mice are the preferred laboratory strain used as cell donors in monoclonal antibody work (*see* **Note 1**).

2. New Zealand rabbits are the preferred laboratory strain used for polyclonal antiserum production (*see* **Note 2**).
3. Suitable adjuvant for addition to antigen (*see* **Note 3**).
4. Animal house facilities licensed for the specific required procedures under animal welfare legislation.
5. Parenteral anaesthetic agents as prescribed by veterinary surgeon (*see* **Note 4**) (Hypnorm + Hypnovel).
6. Diethyl ether.

3.3. Methods

3.3.1. Immunising Rabbits for Polyclonal Antiserum Production

1. Mix 0.5 ml of 1 mg/ml solution of antigen with 0.5 ml appropriate adjuvant.
2. Inject into muscle of hind leg or subcutaneously into neck scruff.
3. Repeat on days 14 and 44.
4. Test bleed (1–2 ml) on day 54 and assess for antibody activity (ELISA, etc.).
5. Bleed from marginal ear vein on day 60 and then every 28 days until the antibody titre drops.
6. Give boost dose and either commence bleeding regime 10 days later or perform terminal exsanguination under anaesthesia.

3.3.2. Immunising Mice for Monoclonal Antibody Production

1. Mix 0.15 ml of 0.5 mg/ml solution of antigen with 0.15 of appropriate adjuvant and mix (volumes based on group of three mice).
2. Inject 0.1 ml of adjuvant/antigen mixture per dose intraperitoneally (IP).
3. Repeat on days 14 and 44.
4. Obtain test bleeds by tail tip amputation under light anaesthesia on day 54 (*see* **Note 5**) and assess for antibody titre. Mice should be marked by ear punching or tattooing to allow subsequent identification.
5. Rest the mice for a period of 60 days or more to allow the B cells to become quiescent.
6. Inject (boost) the best responding mouse with an IP injection of 0.05 ml antigen (0.5 mg/ml solution) without adjuvant.
7. Kill the mouse 3 days later by cervical dislocation and remove its spleen aseptically for cell harvesting.

3.3.3. Modified Immunisation Protocol for Non-Anamnestic Antigens

When it is known from other work that the antigen is highly glycosylated or comes from a source known to be rich in polysaccharides (bacterial cell walls, etc.) it is highly unlikely that the animal will ever produce a classic full immune response. Usually these antigens do not produce quiescent B cells following the rest period and each immunisation is seen as a primary challenge. Substances which invoke this incomplete response are known as non-anamnestic antigens and immunoglobulins produced to them are always class M. There is no point in carrying out a full immunisation protocol over several months and a shortened one is recommended.

1. Mix 0.15 ml of 0.5 mg/ml solution of antigen with 0.15 of appropriate adjuvant and mix (volumes based on group of three mice).
2. Inject 0.1 ml of adjuvant/antigen mixture per dose intraperitoneally (IP). Repeat on day 14.
3. Take a test bleed by tail tip amputation under light anaesthesia on day 21 and assess for circulating antibodies. All antibodies produced will be IgM and so the assessment method (ELISA, etc.) must take this into account. Mark the mice by ear punch or tattoo for subsequent identification.
4. Give the mouse exhibiting the highest titre of circulating antibody an IP injection of 0.1 ml (0.5 mg/ml solution) of antigen on day 23.
5. Harvest the spleen and perform the cell fusion on day 30.

3.3.4. Induction of Immune Tolerance in Neonatal Mice (3, 4)

As previously mentioned foetal and new-born mammals have immature immune systems which do not yet have the capacity to differentiate self and non-self. This lack of maturity can be harnessed to our advantage when working with antigens, which are naturally found along with closely related (cross-reacting) substances. This technique does not guarantee success but can swing the odds in favour of the researcher producing hybridomas with the desired specificity. Situations where this methodology is used include work on viruses, bacteria and fungi where there may be many shared epitopes between the organism of interest and closely related species. The cross-reacting antigen used for the technique is a whole preparation of the closely related species. This technique causes the suppression of an immune response to *shared* epitopes which may swamp the immune response and favours specific immunity to epitopes found only on the species of interest.

1. Obtain a "time mated" female Balb-c mouse 10 days into the pregnancy and maintain in standard cage used for mouse breeding.
2. Observe until litter are born.

3. Inject the neonates daily on days 1–5 after birth with 0.025 mg of *cross-reacting* antigen into the neck scruff. (The neonatal immune systems are maturing at this time and because the cross-reacting antigen is present they will adopt these proteins as "self" and lose the ability to mount an immune response to them.)
4. Immunise three of the animals with specific antigen of interest when 6 weeks old using the standard monoclonal immunisation protocol given above.

3.3.5. Intrasplenic Immunisation

Intrasplenic immunisation is used for the production of hybridomas in situations where only very small quantities of the antigen are available. Typically it lends itself extremely well to producing antibodies to proteins that have been purified by electrophoresis and subsequent blotting onto nitrocellulose. Antibodies produced by this route are always immunoglobulin class M as only one immunisation is used. This method is covered by Animal Procedures legislation in most countries as it is an invasive surgical procedure and welfare issues must be addressed.

The antigen must be in a highly aggregated or immobilised form. This technique is frequently used to produce antibodies to proteins separated by electrophoresis. The gel is stained in the normal manner and the proteins blotted over onto nitrocellulose. The band containing the protein of interest is then excised from the blot and used as the antigen. The presence of protein stains and other reagents makes little difference and as the "in vivo" part of the protocol is very short there are no long term animal welfare issues.

1. Induce and maintain anaesthesia using a mixture of fentanyl/fluanisone with midazolam (Hypnorm/Hypnovel) (*see* **Note 4**). The dosage for these agents is 0.25 ml of each active ingredient plus 0.5 ml water given IP at a rate of 0.1 ml per mouse.
2. Shave the hair along the mid-scapular line above the position of the spleen and make a 1 cm incision made through the skin. Cut through the muscle wall to expose the spleen, and then deliver it through the opening complete with its pedicle.
3. Introduce the antigen into a pocket made in the spleen through the capsule and return the organ to the body cavity.
4. Close the muscle layer with three or four sutures and then close the skin likewise.
5. Keep the mice warm and close to a source of drinking water until they recover from the anaesthetic which takes about 1–2 h. Generally the mice suffer no side effects and will be feeding, grooming and showing no signs of discomfort soon after recovery from anaesthesia.

6. Use the mice as donors for cell fusion 7 days after the intrasplenic immunisation.

3.3.6. In Vitro Immunisation

This technique is performed in tissue culture and works well when only small quantities of antigen (1–2 μg) are available. It also allows the production of antibodies to substances that are toxic in whole animals. There is no immunological processing of antigens so only soluble, simple antigens such as peptides can be used for this approach.

A source of interleukins 4 and 5 is required for the method to work and the easiest way of obtaining them is from thymocyte conditioned medium.

3.3.6.1. Thymocyte Conditioned Medium

1. Kill two 6-week-old Balb-c mice and remove their thymus glands aseptically.
2. Homogenise the tissue to produce single cells and resuspend in 10 ml RPMI 1640 medium containing 15% FBS.
3. Incubate the medium at 37°C/ 5% CO_2 for 24 h (*see* **Note 6**) and then harvest the supernatant by centrifugation (700 × *g*). Store the conditioned medium at –20°C until required.

3.3.6.2. Immunisation

1. Kill a non-immunised Balb-c mouse and remove its spleen aseptically.
2. Homogenise the spleen to produce single cells, and then resuspend them in 10 ml of the thymocyte-conditioned medium.
3. Add 1–2 μg of antigen to the cell suspensions and incubate at 37°C/ 5% CO_2 for 72 h (*see* **Note 7**).
4. Harvest the spleen cell by centrifugation (500 × *g*) and use immediately for a cell fusion.

3.4. Notes

1. Balb-c mice are an inbred strain ideally suited to monoclonal antibody work. Females are usually used, as they do not fight when housed together in project groups of 3–5 individuals. This mouse is also known as the "barber" strain, as the dominant female will remove the whiskers from the others in the group.
2. New Zealand rabbits are normally used for serum production; they are easily handled and adapt well to individual cages or group floor pens. This strain has half-lop ears, which make blood collection from marginal veins a fairly straightforward procedure.

3. Most antigens require an adjuvant to increase their immunogenicity and a number of formulations can be used. Regulations on their use should be consulted prior to embarking on a course of immunisations. For many years Freund's complete and incomplete adjuvants were the formulation of choice for all immunisation work. In recent years welfare issues have been raised over the use of these adjuvants and a number of alternatives based on water soluble bacterial cell wall components have become available.
4. Parenteral anaesthetic agents are preferable to gaseous ones as the size of the mouse creates problems when using standard anaesthetic machines.
5. Tail tip amputation is normally used to obtain test bleeds from mice and this is normally carried out under light anaesthesia induced with diethyl ether.
6. It is extremely important that the medium is harvested after 24 h of incubation. Longer incubation periods may induce the formation of suppressing cytokines, which will block the desired cell stimulation.
7. It is important that the thymocytes are not disturbed for the full 72-h incubation and that they are rapidly harvested and fused after this period of time.

References

1. Roitt, I., Brostoff, J. and Male, D. (1996) *Immunology*. Mosby, London, pp. 8.13–8.16.
2. Kennet, R. H., Denis, K. A., Tung, A. S. and Klinman, N. R. (1978) Hybrid plasmocytoma production: fusions with adult spleen cells, monoclonal spleen fragments, neonatal spleen cells and human spleen cells. *Current Topics in Microbiological Immunology* 81, 495.
3. Hsu, H. T., Lee, I., Davis, R. E. and Wang, Y. C. (1990) Immunisation for generation of hybridoma antibodies specifically reacting with plants infected with a mycoplasma like organism (MLO) and their use in detection of MLO antigens. *Phytopathology* 80, 946–950.
4. Hsu, H. T., Wang, Y. C., Lawson, R. H., Wang, M. and Gonzalves, D. (1990) Spenocytes of mice with induced immunological tolerance to plant antigens for construction of hybridomas secreting tomato spotted wilt virus-specific antibodies *Phytopathology* 80, 158–162.

Chapter 4

Preparation of Immunogens and Production of Antibodies

Hei-ti Hsu, Tsung-chi Chen, Chin-an Chang, and Shyi-dong Yeh

Summary

The quality of reagents greatly affects the interpretation of serological tests. Methods used in conventional viral purification and molecular cloning and expression of target viral proteins to obtain antigens for immunization are presented. Immunization of rabbits, mice and chickens and isolation of immunoglobulin from immunized animals also are described.

Key words: Antibody, Antigen, IgG, IgY, Immunization, Immunoglobulin isolation, Protein expression.

4.1. Introduction

Immunology, a more commonly used term than serology, has had an important role in the advancement of biological science. Interest in diagnosing causal agents has provided the greatest impact to the development of immunodiagnostic techniques. In plant virology the primary concern is the detection of viral agents that causes plant diseases. The basis of the technique lies in the specific interaction of antigen (viruses) with immunoglobulins (antibodies).

When a foreign substance is introduced into an animal it elicits an immune response resulting in humoral antibody production. Plant viruses and viral proteins are immunogenic when used as immunizing agents (immunogen).

Numerous immunization protocols have been used successfully to generate antibodies to a variety of plant viral antigens. No

R. Burns (ed.), *Methods in Molecular Biology, Plant Pathology, vol. 508*

DOI: 10.1007/978-1-59745-062-1_4

single protocol proves to be the best *(1)*. Many of the variables including the choice of animals, the doses of antigen, route of immunization, frequency of injection and choice of adjuvants need to be considered. They all, however, share common features that include a primary immunization followed by a 1- to 2-week resting period before subsequent one or more booster injections, and collection of immune sera about 1–2 weeks after the last injection.

Among the most important considerations for production of antibodies is the selection of a species or strain of animal. Polyclonal antibodies (antisera) have been prepared in numerous animal species. If a large volume of diagnostic reagent or extensive research utilizing the antibody reagent is the primary objective, goats or horses would be the better option. On the other hand, if antibodies are to be used for analysis in a few experiments, the use of mice which produce an adequate amount of antibody-rich ascitic fluids is an appropriate choice. The rabbit is, however, the most common species for polyclonal antibody production. Chicken antibodies (IgYs) are gaining more and more attention in basic research but are not fully utilized in the study of plant viruses *(2)*.

Immunization of animals with highly purified antigens produces antiserum that does not contain significant amount of antibodies to host-plant proteins. The numbers of injections of a given antigen will, however, affect the specificity of an antiserum produced. Antibodies to major antigenic determinants can be produced by just one or two injections of purified virus antigens. The result is a highly specific antiserum. Too many injections of animals with a viral preparation may produce higher titer antisera that contain antibodies to both major and minor antigenic determinants resulting in an antiserum with broad-range reactivity. In addition, virus preparation considered pure may otherwise contain contaminants that also elicit immune responses when injected repeatedly into animals. Such sera react generously with plant tissue antigens and require extensive clean-up before use.

When hybridoma technology producing monoclonal antibodies was introduced, many thought that polyclonal antibody reagents were things of the past. Progresses made in molecular biology have extended the application of polyclonal antibodies into areas previously thought to be exclusive to monoclonal antibodies. It is now possible to produce mono-specific polyclonal antibodies to short, define peptide segments *(3)*.

In this chapter we shall describe preparation of antigens, immunization of animals and isolation of immunoglobulins.

4.2. Materials

4.2.1. Preparation of Antigens

4.2.1.1. Virus Purification

Purified viruses, and recently the isolated capsid proteins, are routinely prepared from infected plants for production of diagnostic polyclonal antibodies. An excellent resource reference is available. Readers are recommended to consult *Plant Virology Protocols* in the Methods in Molecular Biology series *(4)*. We will describe a procedure using *Watermelon silver mottle virus* (WSMoV) as an example for purification of nucleocapsids of tospoviruses *(5)*. The same procedure has been successfully used in the isolation of nucleocapsids of other members in the *Tospovirus* genus.

1. Infected plant tissue: Tospoviruses are readily transmitted by manual inoculation and virions attain high concentration in the infected *Chenopodium quinoa* leaf tissues when chlorotic spots developed. The inoculum is prepared in 0.1 M potassium phosphate, pH 7.0, containing 0.01 M sodium sulfite. *C. quinoa* plants are inoculated at 8–10 leaf stage and kept in a greenhouse at 25–30°C (*see* **Note 1**).
2. TB extraction buffer: 0.01 M Tris–HCl, pH 8.0, containing 0.01 M sodium sulfite and 0.1% cysteine.
3. Cheese cloth or Miracloth.
4. TBG buffer: TB buffer containing 0.01 M glycine.
5. 35% Cesium sulfate in TBG buffer.
6. Triton X-100.
7. Beckman high speed JA41, ultracentrifuge SW 41, 35 Ti and 60 Ti rotors.
8. Protein dissociation buffer.
9. 0.25 M KCl.

4.2.1.2. E. coli Expression

Traditionally, virions purified from infected tissue are used as immunogen to prepare diagnostic polyclonal antiserum. However, not every virus is easily purified from infected tissue. Some viruses accumulate very low concentration in infected tissues, and others are unstable when extracted from plant tissues. Such preparations after purification procedures are unsatisfactory to incite immune responses in injected animals. We have successfully used *E. coli* expressed viral coat proteins as immunogen to produce diagnostic antibodies *(6, 7)*. We will describe a procedure using *Chrysanthemum virus* B (CVB) as an example to illustrate the preparation of *E. coli* expressed CVB coat protein for antiserum production *(8)*. The same protocol has been successfully applied to prepare antisera against many ornamental plant viruses *(9–11)*.

1. Culture of CVB maintained in chrysanthemum plants.
2. pCRII-TOPO vector and cloning kit (Invitrogen, Carlsbad, CA).
3. Sequence analysis ScanDNASIS program (Hitachi Software Engineering America, Ltd, CA, USA).
4. Protein expression vector plasmid pET28b(+) (Novogen, Inc., Madison, WI, USA).
5. Restriction enzymes *Nco*I and *Xho*I.
6. *E. coli* strain Rosetta(DE3).
7. Bacteria culturing M9 medium containing 50 μg/ml kanamicin and 34 μg/ml chloramphenicol.
8. 1 mM of isopropyl β-d-thiogalactopyranoside (IPTG).
9. VCX 600 sonicator (Sonics & Materials Inc., CT, USA).
10. Sodium dodecyl sulfate–polyacrylamide gel electrophoresis (SDS–PAGE) system.
11. Primers T7 promoter (5′-TAATACgACTCACTATAGGG-3′) and T7 terminator (5′-gCTAgTTATTgCTCAgCgg-3′).
12. Protein denaturing solution (0.25 M Tris–HCl, pH 6.8, containing 2% (w/v) of SDS, 4% (v/v) of 2-mercaptoethanol and 10% (w/v) of sucrose).

4.2.1.3. ZYMV Expression

Recent success of plant viral vectors provide a fast and convenient technique for obtaining various proteins in eukaryotic plant cells *(12, 13)*. In this section, we will illustrate a method that use a Zucchini yellow mosaic virus (ZYMV) vector to express the nonstructural NSs protein of WSMoV in zucchini squash plants *(3)*. The NSs open reading frame (ORF) was inserted in between the P1 and HC-Pro cistrons of ZYMV following a hexahistidine tag and an additional NIa protease cleavage sequence for purification and processing of the expressed protein, respectively. The expressed NSs protein was purified from squash tissues for production of antibodies.

1. WSMoV maintained in *Nicotiana benthamiana* plants.
2. ZYMV vector: p35SZYMVGFPhis plasmid.
3. TOPO TA cloning kit (Invitrogen; Carlsbad, CA).
4. Ultraspec RNA isolation system (Biotex Laboratories; Houston, TX).
5. *E. coli* strain DH5α.
6. Oligonucleotide primers.
7. Restriction enzymes *Sph*I and *Kpn*I.
8. T4 DNA ligase.
9. Agarose and DNA gel equipment.

10. Ampicillin.
11. TE buffer: 10 mM Tris–HCl and 1 mM EDTA, pH 8.0.
12. Miracloth (Calbiochem, La Jolla, CA).
13. Buffer A: 50 mM Tris–HCl, pH 8.0, 15 mM $MgCl_2$, 10 mM KCl, 20% glycerol, 0.05% β-mercaptoethanol (β-Me), and 0.1 mM phenylmethylsulphonyl fluoride (PMSF).
14. Buffer B: 50 mM Tris–HCl, pH 8.2, 15 mM $MgCl_2$, 20% glycerol, 0.05% β-Me, and 0.1 mM PMSF.
15. 0.45-μm Filters (Millipore, Billerica, MA).
16. Ni-NTA SUPERFLOW (Qiagen; Germany).
17. Wash buffer: buffer B containing 5 mM imidazole.
18. Elution buffer: buffer B containing 250 mM imidazole.
19. Monoclonal antibody (MAb) against the histidine tag (MAb-His, Amersham Pharmacia Biotech; Buckinghamshire, England).
20. SDS–PAGE equipment.
21. AlphaInnotech IS2000 image system (Alpha Innotech Corporation, San Leandro, CA).

4.2.2. Immunization and Isolation of Immunoglobulins (IgG/IgY)

4.2.2.1. Production and Isolation of Rabbit IgG

1. Two to three New Zealand White rabbits.
2. Adjuvant: Complete Freund's adjuvant (CFA), a water-in-oil emulsifying agent with heat-killed *Mycobacterium tuberculosis* or incomplete Freund's adjuvant (IFA) in which the bacterium is omitted.
3. PBS: Phosphate buffered saline solution, 0.01 M disodium phosphate, pH 7.5, 0.85% sodium chloride.
4. 50-ml Centrifuge tubes, serum vials.
5. Preservative stock solutions: 1% Merthiolate, 10% sodium azide.
6. Ammonium sulfate solution, saturated at room temperature, adjusted to pH 7.8 prior to precipitation of gamma globulin. 2 *N* sodium hydroxide.
7. 10% Barium chloride.

4.2.2.2. Production and Isolation of Mouse IgG

1. Three to five Balb/c mice.
2. Pristane (Sigma-Aldrich, St. Louis).
3. PBS: Phosphate buffered saline solution, 0.01 M disodium phosphate, pH 7.5, 0.85% sodium chloride.
4. Adjuvant: Complete Freund's adjuvant (CFA).
5. Mouse myeloma cells.
6. 70% Ethanol.

7. Protein-A Sehparose.
8. 0.02 M Disodium phosphate, pH 7.3.
9. 0.1 M Glycine, pH 3.0.
10. 2.0 M Tris–HCl, pH 8.3.

4.2.2.3. Production and Isolation of Chicken IgY

1. Two to three egg-laying hens.
2. Adjuvant: Complete Freund's adjuvant (CFA).
3. PBS: Phosphate buffered saline solution, 0.01 M disodium phosphate, pH 7.5, 0.85% sodium chloride.
4. Polyethylene glycol 8000 (PEG, formally PEG 6000).

4.3. Methods

4.3.1. Preparation of Antigens

All purification steps should be carried out in a 0–4°C cold room with pre-cooled buffer, equipment, and rotors.

4.3.1.1. Purification of WSMoV Nucleocapsids

1. Harvest infected *C. quinoa* leaves 4–6 days after inoculation.
2. Homogenize the tissue with a Waring blender in TB extraction buffer (300 ml/100 g tissue).
3. Filter through four layers of cheesecloth.
4. Centrifuge the filtrate at 15,000 × *g* in a Beckman JA 14 rotor for 10 min to remove cell debris.
5. Carefully decant the supernatant into a beaker and discard pellets.
6. Add to supernate 1 to 2% Triton X-100 (final concentration) and stir for 5 min in the cold room.
7. Centrifuge at 80,000 × *g* in a Beckman 35 Ti rotor for 120 min through a 20% sucrose cushion.
8. Resuspend pellets in TBG buffer and further centrifuge in 35% cesium sulfate at 84,000 × *g* in a Beckman SW41 rotor for 16–18 h.
9. Collect opalescent zones containing virus nucleocapsids and centrifuge for 75 min at 160,000 × *g* in a Beckman 60 Ti.
10. Pellets resuspended in TBG buffer (3 ml/100 g initial weight of tissue) constitute the purified nucleocapsids. If not used immediately nucleocapsid preparation may be stored at –20°C. Nucleocapsid proteins are isolated by the following gel electrophoresis procedure.
11. Proteins are dissociated by the addition of 1/3 volume of 4× protein dissociation buffer.

12. The mixture is heated in boiling water for 3 min.
13. Centrifuge at 2,500 × *g* for 5 min to remove insoluble materials.
14. Supernatant is loaded onto stacking gel and electrophoresed at 80 V for 12 h. The NP proteins are visualized by soaking the gel in cold 0.25 M potassium chloride and eluted from the gel using an ISCO electrophoretic concentrator. Purified proteins are measured by absorbance at 280 nm, and store at −20°C.

4.3.1.2. Coat Protein Expression by *E. coli*

A protocol developed for the preparation of *E. coli* expressed CVB coat protein *(9)* is outlined as follows:

Construction of CVB Coat Protein Expression Vector

1. Total RNA was extracted from CVB-infected chrysanthemum tissue by the use of QIAamp total plant RNA mini kit (Qiagen, Hilden, Germary).
2. Coat protein gene of CVB was amplified using primers CVB-dw (5′-ATCTTCACAATGACATCCAT-3′) and CVB-up (5′-TAGGTTGTGGAGTGGTTACA-3′) by reverse transcription-polymerase chain reaction (RT-PCR) using 3 μl of total RNA extracted from CVB-infected chrysanthemum as template.
3. A PCR product about 1,000 bp was consistently amplified and it was cloned into pCRII-TOPO vector for sequencing studies.
4. The revealed 5′- and 3′-terminal sequences of CVB coat protein open reading frame were used to design PCR primers for the cloning of complete coat protein sequence.
5. Two restriction enzymes, *Nco*I and *Xho*I digestion sites (underlined) were created at the 5′-end of the upstream primer, CVB-up1 (5′-AGTCA<u>CCATGG</u>CTCCCAAA-3′) and downstream primer, CVB-dw1 (5′-ACA<u>CTCGAG</u>CACCACCACCACCACCACTGA-3′), respectively, to facilitate subsequent directional cloning of the coat protein into expression vector plasmid pET28b(+).
6. Using the CVB coat protein inserted pCRII-TOPO vector as template, the complete CVB coat protein open reading frame was amplified by PCR with CVB-up1 and CVB-dw1 as primers.
7. The amplified PCR product was digested with *Nco*I and *Xho*I and then ligated with a *Nco*I–*Xho*I cleaved pET28b(+).
8. The recombinant pET28b(+) vector was subsequently transformed into *E. coli* strain Rosetta(DE3) for protein expression.

9. Bacterial clones containing recombinant pET DNA were identified by PCR using primer pairs T7 promoter (5′-TAATACgACTCACTATAGGG-3′) and T7 terminator (5′-gCTAgTTATTgCTCAgCgg-3′).
10. Bacteria clones identified to have CVB CP inserted pET28b(+) were grown overnight at 37°C in M9 medium containing 50 μg/ml kanamicin and 34 μg/ml chloramphenicol.
11. When optical density (OD600) of bacteria culture filtrate reached 0.6–0.7, 1 mM of isopropyl β-d-thiogalactopyranoside (IPTG) was added to the medium to induce protein expression.
12. Four hours after IPTG induction, bacteria culture was subjected to centrifugation at 8,000 × g for 10 min. Bacteria cells were resuspended in TE buffer (10 mM Tris–HCl, pH 8.0, 0.5 mM EDTA) and frozen at -20°C overnight.
13. After thawing, the bacteria cells were further disrupted by a VCX 600 sonicator (Sonics & Materials Inc., CT, USA) followed by centrifugation at 3,000 × g for 10 min to remove cell debris.
14. Supernatants were treated with an equal volume of protein denaturing solution (0.25 M Tris–HCl, pH 6.8, containing 2% (w/v) of SDS, 4% (v/v) of 2-mercaptoethanol and 10% (w/v) of sucrose and heated in boiling water bath for 3 min. The sample was then analyzed in SDS–PAGE *(14)*. Size and expression level of viral CP was identified by western blotting analysis with a CVB antiserum purchased from Agdia Inc. (Elkhart, IN, USA).
15. Bacteria clones with satisfactory CP expression ability were selected and grown in a 1,000-ml flask of M9 medium containing 50 μg/ml kanamicin and 34 μg/ml chloramphenicol. Protein expression was induced and processed similarly as in **steps11–14**. Bacteria expressed CP was further purified by a preparative SDS–PAGE protocol as described previously *(7, 9)*.
16. Purified protein was always adjusted to a concentration of 1.0 OD_{280} per ml, divided into 1.0 ml aliquot and preserved at –80 C for subsequent immunization.

4.3.1.3. NSs Protein Expression by ZYMV

An affinity chromatography *(13, 15)* was modified for purification of the ZYMV-expressed NSs protein from infected zucchini squash plants *(3)*.

Construction of ZYMV Vector

1. The full-length NSs ORF was amplified from WSMoV S RNA using primers WNSs67KS (5′-GGGTACCGCATGCATGTCTACTGCAAAGAATGCTGCT-3′) and WNSs1383cK

(5′-GGGTACCTTCTGCTTTCACAACAAAGTGCTG-3′) by RT-PCR.

2. The PCR product was cloned into pCR2.1-TOPO by TOPO TA cloning kit (Invitrogen, Carlsbad, CA) to generate pTOPO-WNSs.
3. The DNA fragment corresponding to NSs ORF was released from pTOPO-WNSs using restriction enzymes *Sph*I and *Kpn*I, and then ligated with the *Sph*I/*Kpn*I-digested ZYMV vector p35SZYMVGFPhis.
4. The plasmid of the ZYMV recombinant carrying NSs ORF was isolated by the mini-prep method, dissolved in TE buffer, and mechanically introduced with a glass spatula on *C. quinoa* leaves (10 μg in 10 μl per leaf) dusted with 600 mesh carborundum.
5. Local lesions developed were individually transferred to cotyledons of single zucchini squash plants.
6. Total RNAs extracted from symptomatic squash leaves using the Ultraspec RNA isolation system and primers WNSs67KS and WNSs1383cK were used to check the presence of the insert in the recombinant by RT-PCR.
7. PCR products were analyzed in 1.0% agarose gels by electrophoresis.

Purification of ZYMV-Expressed NSs Protein

1. 50 g Infected zucchini squash leaves were ground in 100 ml buffer A with a blender.
2. Extracts were clarified by centrifugation at 3,000 × *g* for 10 min, and supernatants were filtered through Miracloth.
3. The filtrates were treated with 1% Triton X-100 at 4°C for 30 min and then centrifuged at 30,000 × *g* for 30 min.
4. The supernatants were filtered through 0.45-μm filters.
5. Approximately 1 ml of Ni^{2+}-NTA resins, pre-equilibrated in buffer B, was added.
6. The mixtures were gently shaken for 1 h at 4°C and loaded onto a column.
7. After allowing the resins to settle, the unbound materials were discarded and the resins were washed with twofold bed volume of wash buffer.
8. The proteins bound to the resins were eluted with 10 ml elution buffer.
9. The ZYMV-expressed NSs protein was further purified by gel electrophoresis method.
10. Each fraction of the purification steps was monitored by western blot using MAb-His.

11. The amount of purified NSs protein was estimated with standardized histidine-tagged GFP using MAb-His in western blotting or by comparison with bovine serum albumin (BSA) in SDS–PAGE, and estimated by the software Spot Density of AlphaInnotech IS2000.

4.3.2 .Immunization and Immunoglobulin Isolation

Although marking experimental animals is of primary importance in laboratory studies, it is generally not required for plant virology investigations. Unless a mass production of antisera for a variety of viruses is a major responsibility of the facility, temporary cage marking for identification is sufficient since it is less stressful to the animals through the period of production. Procedures minimizing stress and pain and proper handling and care of animals should be observed all the time.

Both ammonium sulfate precipitation and Protein A-Sepharose chromatography are preferred methods for isolation of mammalian immunoglobulin (IgG) from immune sera and ascitic fluids. Protein A chromatography is not applicable to isolation of chicken immunoglobulin (IgY) since IgY does not bind with mammalian Fc-receptors nor with protein A or G. Isolation of chicken antibodies employs polyethylene glycol fractionation and precipitation.

4.3.2.1. Rabbits Immunization

1. Day 0: Primary, intramuscular injection with 0.5–1.0 mg protein in 0.5 ml Tris–acetate buffer (containing 0.002 M EDTA) emulsified with 0.5 ml CFA equally divided into two hind legs (*see* **Notes 2** and **3**).
2. Day 7: Booster injection with 0.5–1.0 mg protein in a 0.5 ml Tris–acetate buffer (containing 0.002 M EDTA) emulsified with 0.5 ml ICA subcutaneously behind the neck or equally divided into two hind legs (*see* **Note 3**).
3. Day 14: Repeat **step 2**.
4. Day 21: Repeat **step 2**.
5. Day 28: Repeat **step 2**.
6. Day 35 and weekly thereafter: Serum collection (*see* **Note 4**).
7. Allow the blood to clot at room temperature for 2–4 h.
8. Dislodge the clot by rimming the tube with a wood applicator.
9. Centrifuge at 10,000 × g for 5 min and carefully transfer the serum to a new tube.
10. Sterile immune serum can be held at 4°C for a long time. If in doubt, add merthiolate (0.01% final concentration) or sodium azide (0.1% final concentration). Freezing at –20°C or lower and lyophilization will preserve sera for a long time, however.

Isolation of Rabbit IgG

Rabbit serum volume 50 ml or less is convenient and easy to manipulate. Using a 50 ml serum sample, ammonium sulfate precipitation procedure is presented.

1. At room temperature, while stirring, add, drop-by-drop, 25 ml saturated ammonium sulfate (adjust to pH 7.8 prior to use) to 50 ml serum sample in a 150-ml beaker. Continue stirring the suspension for an additional 2–3 h.
2. Centrifuge at 1,400 × *g* for 30 min in a clinical centrifuge.
3. Discard supernatant, and dissolve pellets in PBS to restore the original serum volume.
4. Repeat **steps 1–3**.
5. Repeat **steps 1** and **2**.
6. Dissolve precipitate in one half or less (25 ml or less) volume of PBS.
7. From hereon all steps are performed at 4°C. Dialyze against several changes of PBS overnight at 4°C and test the presence or absence of sulfate ions by adding a few drops pf 10% barium chloride to a small volume of dialysate.
8. Remove the solution from the dialysis tubing and centrifuge the solution at 4°C for 30 min at 1,400 × *g*.
9. Determine the immunoglobulin concentration by measuring the absorbance at 280 nm.
10. Purified immunoglobulin can be stored frozen.

4.3.2.2. Mice (Balb/c)

Immunization

When limited amounts of antigens are available for immunization, mice are excellent substitutes for larger animals. Antibody rich ascitic fluids can be induced in immunized mice by injection of FCA or implantation of mouse myeloma cells in the peritoneal cavity. Rouse sarcoma virus and the virus transformed tumor cell lines although serve the same purpose, are not recommended for plant pathologists who are not trained to handle animal pathogens.

1. Mice are primed with an intraperitoneal injection of Pristane (0.5 ml/mouse) 1–2 weeks before immunization (*see* **Note 5**).
2. Day 0: Intraperitoneal injection of 10–50 μg proteins in PBS (*see***Notes 3** and **5**).
3. Day 7: Intraperitoneal injection of 10–50 μg proteins in PBS with CFA.
4. Day 21: Repeat **step 2** and allow 3–5 weeks for development of ascites. Alternatively inject peritoneally (10^6) myeloma cells in PBS, and allow 2–3 weeks for ascitic fluids to develop.
5. Tap mice when abdomen becomes distended. Surfaces sterilize the abdomen with 70% ethanol lightly and, with a finger, insert an 18 gauge needle. The fluid will drip out. Gentle

massaging of the abdomen also increases the yield of ascites during tapping. Tapping can be repeated two to three times whenever abdomen becomes distended (*see* **Note 6**).

Isolation of Mouse IgG

Mouse immunoglobulins are generally purified by the method employing Protein A or Protein G column chromatography because of its small volume. The details of the isolation procedure can be found with accompanied literature from Protein A suppliers. Ammonium sulfate precipitation technique outlined in the rabbit immunoglobulin purification also can be applied in isolation of murine immunoglobulins.

1. Rehydrate 0.5 or 1 g Protein A-Sepharose in 0.02 M sodium phosphate, pH 7.3. Pack into a small column by running several bed volumes of buffer through column no faster than 1.0 ml/min.
2. Load sample on top and continue adding buffer. If the sample is serum, then dilute it with an equal volume of buffer before loading. Continue running buffer through until absorbance returns to zero.
3. Elute antibody off column with 0.1 M glycine, pH 3.0. Collect 1.0 ml fractions and immediately neutralize with 40 µl 2.0 M Tris–HCl, pH 8.5.
4. Concentrate antibody by ammonium sulfate precipitation (using 0.9 volumes saturated ammonium sulfate) and followed by dialysis against several changes of PBS.
5. Stabilize column with 0.02 M sodium phosphate, pH 7.3. Add a drop of sodium azide solution and store at 4°C.
6. For the next use, warm the column and all buffers to room temperature. De-gas the buffers before running through the column to eliminate air bubbles.

4.3.2.3. Chicken

Use of chickens for antibody production has been reported but not fully utilized in plant virology *(1, 16)*. The concentration of chicken immunoglobulin Y (IgY, which is equivalent to mammalian IgG) in the yolk is essentially similar to (in some cases much higher than) the concentration of IgY in the serum. IgY is transferred from plasma through egg follicle to the yolk. Daily collection of chicken egg eliminates bleeding steps that every animal worker tries to avoid. A single chicken egg may yield as much as 100–200 mg of IgY.

Immunization

1. Day 0: Primary intramuscular injection with 0.1–0.5 mg protein in CFA.
2. Day 14: Booster injection with similar amount of protein in CFA.
3. Start collecting eggs 2–3 weeks after the beginning of immunization.

Isolation of Chicken IgY

1. Separate yolk and egg white.
2. Measure the volume of the yolk.
3. Add to 1 part egg yolk + 2 parts PBS, while stirring, PEG 8000 to 3.5% (w/v). Incubate at room temperature for 30 min.
4. Centrifuge the mixture at 14,000 × *g* for 10 min.
5. Carefully decant the liquid phase onto absorbent cotton in a funnel to remove fatty materials.
6. Add PEG 8000 to a final concentration of 12% (remember there was 3.5% before) while stirring and incubate at room temperature for 30 min.
7. Centrifuge at 14,000 × *g* for 10 min.
8. Dissolve the precipitate in PBS to the original yolk volume.
9. Add PEG 8000 to 12% (w/v) as in **step 6** and sit at room temperature for 30 min.
10. Centrifuge at 14,000 × *g* for 10 min.
11. Remove as much as possible the supernatant containing PEG.
12. Dissolve the precipitate in one half the original yolk volume.
13. Aliquot the IgY preparation and store at –20°C.

4.4. Notes

1. Other susceptible species *Nicotiana benthamiana* and *N. tabacum*, etc., have been used as propagation hosts for tospovirus purification. Virion concentration, however, decreases before leaf necrosis occurs in these hosts.
2. Hyperimmunized animals may become very sensitive and thus susceptible to anaphylaxis. It is suggested that lower dose of antigen be used.
3. For most soluble antigens prepared with adjuvant, subcutaneous and intramuscular injections are preferred in rabbits, and intraperitoneal is more practical for mice. Interdermal and footpad injections are not recommended since injection via this route often causes animals discomfort due to local swelling, inflammation, and infection.
4. For a successful bleeding, ear vein should be dilated by rubbing the ear with your hand/or a gauze soaked in hot water (that your hand can tolerate). Coat the ear with a thin layer of Vaseline. With a quick stroke of a brand new razor nick the vein diagonally. Allow the blood to drip into a centrifuge tube. Care must be taken not to cut through the vein.

5. Injections should be made in the lower abdomen. To prevent the injected materials leak out of the abdomen. Injections work best when the needle first pierces the skin followed by inserting needle, parallel to and under the skin, about one cm before puncturing the abdominal wall.
6. Lipids in the fluid can be removed by passing through glass wool placed in a funnel.

References

1. Van Regenmortel, M.H.V. (1982). Serology and Immunochemistry of Plant Viruses. 302 pp. Academic Press, New York.
2. Schade, R., Behn, I., Erhard, M., Hlinak, A., and Staak, C. (2001). Chiken Egg Yolk Antibodies, Production, and Application: IgY-Technology. 255 pp. Springer, Berlin, Heidelberg, New York.
3. Chen, T.C., Huang, C.W., Kuo, Y.W., Liu, F.L., Hsuan Yuan, C.H., Hsu, H.T., and Yeh, S.D. (2006). Identification of common epitopes on a conserved region of NSs proteins among tospoviruses of *Watermelon silver mottle virus* serogroup. *Phytopathology* 96:1296–1304.
4. Foster, G.D., and Taylor, S.C. (1998). Plant Virology Protocols. Humana Press, Totowa, NJ.
5. Yeh, S.D., Chao, C.H., Cheng, Y.H., and Chen, C.C. (1996). Serological comparison of four distinct tospoviruses by polyclonal antibodies to purified nucleocapsid proteins. *Acta Hort.* 431:122–134.
6. Li, R.H., Zettler, F.W., Purcifull, D.E., and Hiebert, E. (1998). The nucleotide sequence of the 3 -terminal region of *Dasheen mosaic virus* (Caladium isolate) and expression of its coat protein in *Escherichia coli* for antiserum production. *Arch. Virol.* 143:2461–2469.
7. Chen, C.C., Hsiang, T., Chiang, F.L., and Chang, C.A. (2002). Molecular characterization of Tuberose mild mosaic virus and preparation of its antiserum to the coat protein expressed in bacteria. *Bot. Bull. Acad. Sin.* 43:13–20.
8. Lin, M.J., Chang, C.A., Chen, C.C., and Cheng, Y.H. (2005). Occurrence of *Chrysanthemum* mosaic virus B in Taiwan and preparation of its antibody against bacteria-expressed coat protein. *Plant Pathol. Bull.* 14:181–192.
9. Lin, M.J., Chang, C.A., Chen, C.C., and Cheng, Y.H. (2004). Molecular and serological detection of *Tomato aspermy virus* infecting chrysanthemum in Taiwan. *Plant Pathol. Bull.* 13:291–298.
10. Chen, C.C., Hsu, H.T., Cheng, Y.H., Huang, C.H., Liao, J.Y., Tsai, H.T., and Chang, C.A. (2006). Molecular and serological characterization of a distinct potyvirus causing latent infection in calla lilies. *Bot. Stud.* 47: 369–378.
11. Chen, C.C., Hsu, H.T., Chiang, F.L., and Chang, C.A. (2006). Serological and molecular properties of five potyviruses infecting calla lily. *Acta Hort.* 722:259–270.
12. Gleba, Y., Marillonnet, S., and Klimyuk, V. (2004). Engineering viral expression vectors for plants: the 'full virus' and the 'deconstructed virus' strategies. *Curr. Opin. Plant Biol.* 7:182–188.
13. Chen, T.C., Hsu, H.T., Jain, R.K., Huang, C.W., Lin, C.H., Liu, F.L., and Yeh, S.D. (2005). Purification and serological analyses of tospoviral nucleocapsid proteins expressed by *Zucchini yellow mosaic virus* vector in squash. *J. Virol. Methods* 129:113–124.
14. Laemmli, U.K. (1970). Cleavage of structural proteins during the assembly of the head of bacteriophage T-4. *Nature* 227:680–685.
15. Gal-On, A., Canto, T., and Palukaitis, P. (2000). Characterization of genetically modified *Cucumber mosaic virus* expressing histidine-tagged 1a and 2a proteins. *Arch. Virol.* 145:37–50.
16. Hsu, H.T., and Lawson, R.H. (1985). Comparison of mouse monoclonal antibodies and polyclonal antibodies of chicken egg yolk and rabbit for assay of *carnation etched ring virus. Phytopathology* 75:778–783.

Chapter 5

Enzyme-Linked Immunosorbent Assay Detection of *Ralstonia solanacearum* in Potatoes: The South African Experience

Dirk U. Bellstedt

Summary

Outbreaks of bacterial wilt caused by *Ralstonia solanacearum* in South Africa in the 1980s necessitated the development of a sensitive assay for its detection. In this chapter, the development of an enzyme-linked immunosorbent assay (ELISA) for the detection of *Ralstonia solanacearum* including antibody production, the ELISA method itself, potato sample preparation for testing, ELISA result validation and interpretation, and confirmation of infection are discussed. Data showing the drop in the *Ralstonia solanacearum* infection rate, after testing was implemented, are presented and shows how the use of this method has brought the disease under control in South Africa.

Key words: *Ralstonia solanacearum*, ELISA, South Africa.

5.1. Introduction

In South Africa, enzyme-linked immunosorbent assay (ELISA) testing for *Ralstonia solanacearum* has been implemented for the past 11 years *(1)*, and the valuable experiences gained through this testing will be described in this chapter. Other ELISAs for the detection of *Ralstonia solanacearum* have been described *(2)*; this bacterium can also be detected by immunofluorescence *(3, 4)* and PCR *(5)*. *Ralstonia solanacearum* is the causative organism of bacterial wilt in many crops including potatoes, resulting in total crop failure. It is prevalent in the tropical areas of the world, rather than in the more temperate

R. Burns (ed.), *Methods in Molecular Biology, Plant Pathology, vol. 508*

DOI: 10.1007/978-1-59745-062-1_5

zones. The pathogen tends to be latent under low temperature conditions, but becomes pathogenic when temperatures are higher. South Africa has both temperate and more tropical areas in which potatoes are cultivated, as a result of which conditions for outbreaks of the disease always exist. The first outbreaks of the disease occurred in South Africa in the late 1970s. As long-lasting soil contamination occurs as a result of infection, the disease has very severe consequences and effectively means that infected soils are rendered unsuitable for potato production for many years. The detection of infections in potatoes in South Africa was therefore viewed to be very important and this lead to the development of an ELISA for the detection of *Ralstonia solanacearum* in 1989 *(1)*. After extensive specificity and sensitivity testing as well as laboratory staff training, testing was implemented as part of the Potato Certification Scheme in South Africa in 1996. The ELISA test is used as a first screen to detect infections after which confirmation is obtained using standard bacteriological identification methods. These may soon be replaced by PCR detection methods. Although concerns have been expressed about the specificity of polyclonal antibodies for the detection of *Ralstonia solanacearum (6–8)*, our experience does not indicate this to be a problem. ELISA for the detection of *Ralstonia solanacearum* has therefore played a major role in bringing potato bacterial wilt infections under control in South Africa.

5.2. Materials

1. All chemicals should be Analar grade, only the suppliers of specialized products will be indicated.
2. *Ralstonia solanacearum* cultures should be obtained from contaminated potato sources *(9)* (*see* **Notes 1** and **2**).
3. Rabbits are used for antibody production.
4. Microtiter plates (Nunc Maxisorb).
5. Positive control suspensions of *Ralstonia solanacearum* are prepared by suspending freeze-dried bacteria in sterile water to an OD of 0.2 at 660 nm.
6. Carbonate buffer: Make a stock solution of 1 M $NaHCO_3$ (84.01 g/l) in distilled water. Dilute 1 in 20 (5 ml plus 95 ml distilled water) and adjust pH to 9.6 with 0.1 M NaOH. This diluted buffer (0.05 M) is then used in the assay. This buffer should be stored at 4°C but tends to be unstable if stored for longer than 2–3 weeks.

7. Casein buffer *(10)*: 4.5 g NaCl; 2.5 g casein (BDH, UK); 0.788 g Tris–HCl; 0.1 g thiomersal. Dissolve all the components except the casein in 500 ml of distilled water, adjust pH to 7.6, add casein and allow to dissolve overnight at room temperature with gentle stirring (sprinkle the casein on the buffer and take care that no foam is allowed to form). Store at 4°C (*see* **Note 3**).
8. Casein buffer containing 0.1% Tween 20: Add 0.1 ml of Tween 20 to 100 ml of casein buffer. The same conditions with regard to stability apply as for casein buffer. Store at 4°C (*see* **Note 3**).
9. Citrate buffer: 7.35 g citric acid monohydrate; 19.12 g sodium citrate dehydrate and make up to 1 L with distilled water and adjust to pH 5. This buffer can be stored at 4°C for up to 4 weeks.
10. PBS: 80 g NaCl; 2 g KCl; 11.5 g Na_2HPO_4, 2 g KH_2PO_4 Make up to 1 L with distilled water. Dilute 1 in 10 before use and adjust pH to 7.2.
11. PBS containing 0.1% Tween 20: Add 0.1 ml of Tween 20 to 100 ml of PBS.
12. Biotinamidecaproate-*N*-hyroxysuccinimide ester (Sigma).
13. Avidin-horse radish peroxidase conjugate (Zymed): Diluted and stored as specified by the supplier.
14. Substrate buffer for ELISA: add 0.05% 2,2′-azino-di-[3-ethylbenzothiazoline-6-sulfonic acid] (ABTS) plus 0.015% H_2O_2 in 0.1 M citrate buffer, pH 5.
15. TZC medium: prepare the two following solutions: Solution A: Bactopeptone, 10 g; casein hydrolysate, 1 g; Agar, 18 g; glycerol 5 ml; make up to 975 ml with distilled water; autoclave for 15 min at 121°C. Solution B: 2,3,5-triphenyltetrazoliumchloride, 0.05 g; make up to 25 ml with distilled water; autoclave for 7 min at 121°C. Add solution A and solution B to each other and mix well. Fill Petri dishes with this solution and allow to cool before use.
16. SMSA medium *(11)*: Prepare the following solution: bactopeptone, 10 g; glycerol 5 ml, casein hydrolysate, 1 g; Agar, 15 g; make up to 1,000 ml with distilled water; autoclave for 15 min at 121°C. Allow to cool to 50°C and then add the following solutions: Crystal violet (1%, m/v), 0.5 ml; Polymyxin B sulfate (1%, m/v), 10 ml; 2,3,5-triphenyltetrazoliumchloride (1%, m/v), 5 ml; Bacitracin (1%), 2.5 ml; penicillin (0.1%), 0.05 ml; chloramphenicol (1%), 0.5 ml. Mix these solutions and then fill Petri dishes with this solution and allow to cool before use.

5.3. Methods

5.3.1. Cultivation of Ralstonia solanacearum

1. Obtain potato tubers from potato plants showing symptoms of bacterial wilt. In our methodology tubers were obtained from different localities in South Africa *(9)*.
2. Wash the tubers and allow them to air dry, peel and cut into thin slices, after which obtain sap from them with a suitable plant tissue press.
3. Plate the sap onto TZC medium *(12)* and incubate at 30°C for approximately 48 h until colonies of 5 mm in diameter are apparent.
4. Harvest the bacteria from the TZC plates by flooding with small volumes of sterilized water, agitate and then pellet by centrifugation (3,000 × *g* for 30 s). After centrifugation, resuspend the bacteria in water and re-centrifuge. After a second cycle of resuspension and centrifugation, freeze-dry the bacteria for later use. Store freeze dried cultures at 4°C until used for immunization purposes or for the preparation of positive control suspensions for ELISA.

5.3.2. Antibody Production in Rabbits

The method employed for the production of antibodies against *Ralstonia solanacearum* was based on that developed for the immunization of protein antigens adsorbed to acid-treated *Salmonella minnesota* R595 bacteria *(13)*.

1. Immunize rabbits intravenously in the peripheral ear vein with 200 μg of *Ralstonia solanacearum* bacteria in 0.5 ml of PBS on days 1, 4, 7, 14, 17, and 21, and again at 28, 31, and 35 days in order to achieve high antibody titers (*see* **Note 4**).
2. Take blood samples from the rabbits for use as pre-immune controls prior to the start of immunizations. Seven days after the immunization series take blood samples to test specific activity of antiserum.
3. Test serum by direct Double Antibody Sandwich (DAS) ELISA.
   (a) Coat plates with *Ralstonia solanacearum*, incubate and wash.
   (b) Add test serum (doubling dilutions) incubate and then wash.
   (c) Add secondary antibody conjugate at recommended dilution, *(13)* incubate and wash.
   (d) Add substrate solution and read plates at 405 nm.

5.3.3. Enzyme-Linked Immunosorbent Assay for Ralstonia solanacearum Detection

The ELISA for *Ralstonia solanacearum* detection is based on the biotin/avidin system. Antibodies against *Ralstonia solanacearum* are used both as capture antibodies and in biotinylated form as detection antibodies. Streptavidin horse radish peroxidase conjugates are subsequently used for detection of the biotinylated antibodies, ABTS is used as the substrate.

5.3.3.1. Isolation of Immunoglobulin from Rabbit Serum

1. Mix 1ml antiserum thoroughly with 2 ml PBS then add 3 ml saturated ammonium sulfate solution to aid precipitation of the antibody fraction cool for 20 min at 4°C.
2. Isolate the resulting precipitate by centrifugation at 15,000 × *g* for 20 min.
3. Dissolve the precipitate in 2 ml PBS, precipitate again with 2 ml saturated ammonium sulfate and cool for 20 min at 4°C.
4. Isolate the resulting precipitate again by centrifugation as in **step 2**, after which dissolve in 2 ml PBS.
5. Dialyze the redissolved immunoglobulin fraction against three changes of carbonate buffer (0.1 M, pH 8.3).
6. Determine the protein concentration of the immunoglobulin (Ig) by measuring the absorbance at 280 nm (1 mg BSA gives an absorbance of 1.4 at 280 nm).

5.3.3.2. Biotinylation of Ig

1. Dilute the immunoglobulin solution prepared in **step 6** in **Subheading 5.3.3.1** to a concentration of 5 mg/ml solution in carbonate buffer (0.1 M, pH 8.3).
2. Dissolve 2 mg biotinamidecaproate-*N*-hyroxysuccinimide ester in 1 ml *N,N*-dimethylformamide.
3. Add 250 μl of the biotinamidecaproate-*N*-hyroxysuccinimide ester solution to 1 ml of the immunoglobulin solution over 5 min and then stir for 2 h at room temperature.
4. Dialyze the resulting conjugate solution against PBS and store in PBS/50% glycerol at –20°C.

5.3.3.3. Potato Sample Preparation

1. Testing is performed on samples consisting of 4,605 tubers. This sample size ensures that there is a high statistical probability (99.99%) if *Ralstonia solanacearum* bacteria are present in the sample that they will be detected with the ELISA *(14)*. Tubers are combined into sub-samples of 100 tubers each, with the last five of these samples containing 101 tubers (*see* **Note 5**). This therefore gives 46 samples that need to be tested.
2. At the stolon end of each tuber, peel an area of approximately 1 cm^2 in diameter and remove a small conical section (the base of which is approximately 5 mm in diameter) tuber containing vascular material. Place the pieces from the 100 tubers into a Bioreba® extraction bag. This bag has a finely perforated plastic mesh bisecting the bag. Place all of the tuber sections on the one side of the bisecting mesh. Using a flat headed rubber hammer, pulp the tuber sections with a few blows. Add ten ml of distilled sterile water to each of the bags

of homogenized tuber sections and allow to stand for 5 min, after which the tuber sap is removed from the opposite side of the bisecting mesh for ELISA testing.

3. After removal of the tuber section for ELISA testing, cut each tuber in half and visually inspect for ring symptoms. Any tubers exhibiting ring symptoms or showing suspicious symptoms are extracted separately as in **step 2** and the tuber sap of the single tuber tested by ELISA.

5.3.3.4. ELISA Procedure

1. Coat the first eleven rows of wells in Nunc Maxisorb ELISA plates with rabbit antibodies diluted 1 in 800 in 0.05 M carbonate buffer. A total volume of 100 μl per well is used, incubate the plates at 4°C overnight. Decant the coating solution from the plates.
2. Add 200 μl/well casein buffer to all the wells of the plate to block non-specific binding and incubated at 37°C for 1 h. Wash plates three times using PBS containing 0.1% Tween 20 (200 μl/well).
3. Plate layout: the plate layout for the positive control and tuber samples is shown in **Fig. 5.1**.
4. Positive control suspensions are used to make a dilution series on the ELISA plate: Add 100 μl/well casein buffer containing

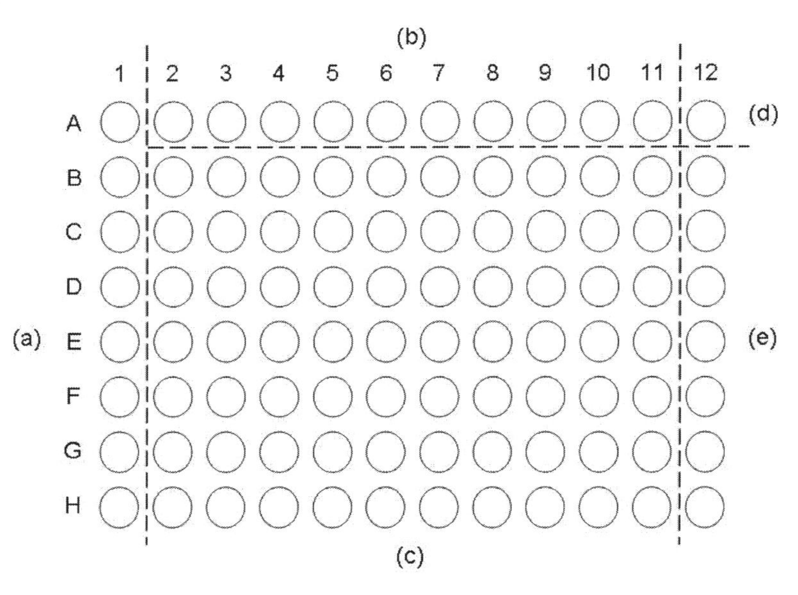

Fig. 5.1. Plate layout for the *Ralstonia solanacearum* ELISA. Wells in rows 1–11 are coated with anti-*Ralstonia solanacearum* antibodies. Wells in row 12 are left uncoated. Different samples are pipetted into demarcated blocks of wells: (**a**) Wells containing buffer only to serve as blanks, (**b**) the dilution series of positive control samples, (**c**) tuber sap samples, (**d**) a single dilution of the positive control sample, (**e**) duplicates of the tuber sap samples pipetted into the wells of row 11.

Tween 20 to wells A3–A11. Dispense 180 μl casein buffer containing Tween 20 in well A2 and add 20 μl of positive control to make a 1:10 dilution. Mix well with a pipette and transfer 100 μl to well A3, mix well again and transfer 100 μl to well A4, etc., so as to establish a twofold dilution series of the positive control. Remove the excess 100 μl from well A11 and discard. Place 100 μl of a 1:10 dilution of positive control suspension in casein buffer containing Tween 20 into well A12 (uncoated) as a non-specific binding control.

5. Place 100 μl tuber sap samples diluted 1:4 in casein buffer containing Tween 20 in the allocated wells of the microtiter plate as shown in the plate layout. The tuber sap samples used in row 11 are also used in row 12 as non-specific binding controls.
6. After the positive control and tuber sap samples have been pipetted in the plate incubate overnight at 37°C. Plates are then washed three times using PBS containing 0.1% Tween 20 (200 μl/well).
7. Add 100 μl/well antibody biotin conjugate diluted 1:100 in Casein-Tween to the wells and incubate at 37°C for 1 h. Plates are subsequently washed three times using PBS containing 0.1% Tween 20 (200 μl/well).
8. Add 100 μl/well streptavidin peroxidase conjugate diluted 1:100 in Casein-Tween solution to each well and incubate at 37°C for 1 h. Plates are subsequently washed three times using PBS containing 0.1% Tween 20 (200 μl/well).
9. Add 100 μl/well substrate solution and incubate at 37°C for 30 min after which the absorbance is measured on a microtiter plate reader at 405 nm.

5.3.3.5. ELISA Result Interpretation

1. Although the plate layout incorporating a positive control suspension dilution series in row A and a non-specific binding control in column 12 may seem cumbersome and labor intensive, experience has shown this to be invaluable in the interpretation of results.
2. The blank controls in column 1 must give absorbance values below 0.15 (*see* **Note 6**).
3. The positive control suspension dilution in well A2 must give an absorbance value higher than 0.7 (*see* **Note 7**). The positive control suspension dilution series must give an absorbance reading of 0.15 or higher at the dilution in well A11. This serves to validate assay performance as opposed to the single positive control well routinely used in other plant pathogen ELISA procedures (*see* **Note 8**).
4. The non-specific binding control of the positive suspension in A12 and the tuber samples added in C12, D12, E12, F12,

G12 and H12 must give absorbance values of 0.1 or lower. This control is very important in order to monitor non-specific binding which can occur in some tuber samples (*see* **Notes 9** and **10**).

5. Tuber sap samples giving values of 0.15 or higher are interpreted as indicative of positive detection of *Ralstonia solanacearum* (*see* **Notes 11** and **12**). The original sap sample is then used for testing to confirm the presence of *Ralstonia solanacearum*. This is performed in one centralized laboratory in South Africa.

5.3.4. Confirmation of Infection

1. Plate out tuber sap samples giving absorbance readings of 0.15 and greater in the ELISA in tenfold dilutions on plates containing TZC medium from undiluted to 10^{-8} dilution and in SMSA medium *(14)* from undiluted to 10^{-3} dilution. Colonies conforming to *Ralstonia solanacearum* colony morphology are then streaked out on individual plates to generate pure cultures. These can then be used in medium tests as described *(15, 16)*. Positive test results for these tests are then interpreted as confirmation of the presence of *Ralstonia solanacearum* (*see* **Notes 13–15**).
2. In instances where tuber sap from individual tubers gave positive ELISA results, the confirmatory tests were performed in the same way.

5.4. Notes

1. *Ralstonia solanacearum* Biovar II (race 3) was isolated from contaminated potato stocks in South Africa, but any *Ralstonia solanacearum* race can be isolated and used for antibody production and be detected using this method.
2. *Ralstonia solanacearum* Biovar III (race 2) was isolated from other contaminated plants (tobacco, tomatoes, green peppers) in South Africa, but only very rarely from potatoes. However, antibodies against *Ralstonia solanacearum* Biovar II do cross-react with *Ralstonia solanacearum* Biovar III and are therefore also detected with this ELISA.
3. In the original description of casein buffer *(10)* it was indicated that thiomersal could be added as a preservative, but in our experience this buffer cannot be stored for more than 2 weeks at 4°C, whether thiomersal has been added or not. As thiomersal is hazardous, we now no longer incorporate this in the buffer.
4. Antibody production in rabbits varies, and although some rabbits produced high levels of antibodies by day 46, our

experience has been that it takes most rabbits 60–70 days, after about four series of booster immunizations as described, to produce high levels of antibodies. Although antibodies against *Ralstonia solanacearum* Biovar II (race 3) were used in our ELISAs, the method was also successful for antibody production against *Ralstonia solanacearum* Biovar III (race 2) bacteria.

5. During the initial phases of ELISA testing, microbiological trials were conducted to assess the number of bacteria present in tubers that were latently infected, i.e., those that exhibited no symptoms. These trials revealed that the vast majority of latently infected tubers contained significant numbers of bacteria and not just low numbers. If sap from a latented infected tuber was diluted as much as 1:100 it still gave a significantly elevated ELISA reading indicative of a positive identification of the *Ralstonia solanacearum* bacteria. On the basis of these findings, the decision was taken that testing in which 100 tubers were combined in a single sample, would not dilute the bacteria from a latently infected tuber below the detection limit of the ELISA.
6. If absorption values of over 0.15 are obtained in the blank controls, this indicates that either the carbonate buffer or the casein buffer is no longer suitable for use.
7. Undiluted positive controls routinely give absorbance values of 2 and higher.
8. As the ELISA described here is supplied in kit form to a number of laboratories in outlying areas of South Africa, the inclusion of a dilution series of the positive control, serves as a very effective tool in monitoring assay performance.
9. The non-specific binding control (in row 12) serves as a very important tool in monitoring elevated background values which (as a result of the wells not being coated with antibodies) are not the result of antibody cross-reactivity. During the many years of testing with the ELISA, it became apparent that the following potato tuber sample factors could lead to non-specific background readings: a. Any tuber with signs of rotting; b. Any tuber sample containing potato skin; c. Any tuber of which the skin was damaged during mechanical harvesting; d. Tubers grown under wet water-logged soil conditions in 3 weeks immediately prior to harvesting.
10. Edge effects: These effects are common in ELISA detection of plant pathogens. It was found that all edge effects could be eliminated if the reagents to be added in **steps 7–9** in the ELISA procedure were prepared 30 min prior to pipetting onto the plate and incubated at 37°C until use.
11. The sensitivity of the ELISA was tested and it was determined that the number of *Ralstonia solanacearum* cells that

gave an absorbance reading of 0.15 in the ELISA was approximately 10,000.

12. During the initial testing phases during which the ELISA was evaluated (1993–1995) attempts were made to increase the numbers of *Ralstonia solanacearum* bacteria in potato samples by incubation of tubers at 37°C for periods up to 2 weeks prior to ELISA testing. It was found that the numbers of bacteria in potato tubers did increase, but at the same time rotting of potato tubers also increased dramatically. This lead to an increase in background values in the ELISA, which therefore effectively invalidated these results. As a result, the practice of warm incubation of potatoes prior to ELISA testing was discontinued.
13. The incidence of infections detected by ELISA and subsequently confirmed since testing was made compulsory, is shown in **Fig. 5.2**. The drop in infection rates from 1996 to 1999 was as a result of testing by means of the ELISA. Subsequent confirmed cases could be attributed to soil contamination, which must have occurred prior to the implementation of mandatory testing.
14. Confirmation of infection: During the cultivation of bacteria from potato samples suspected of containing *Ralstonia solanacearum* it was found that in spite of the use of selective media, these bacteria are easily overgrown by other bacteria. As a result, inspection of bacterial growth at short intervals, the isolation of colonies and recultivation is essential in order to ensure that symbiotic and/or contaminating bacteria do not overgrow the *Ralstonia solanacearum* bacteria. If this is not done, this may lead to a failure of the detection of *Ralstonia solanacearum* bacteria in the potato sample. In many instances the SMSA medium proved to be more selective than the TZC medium, as the growth of symbionts was in all likelihood inhibited better by the medium components of the SMSA medium.

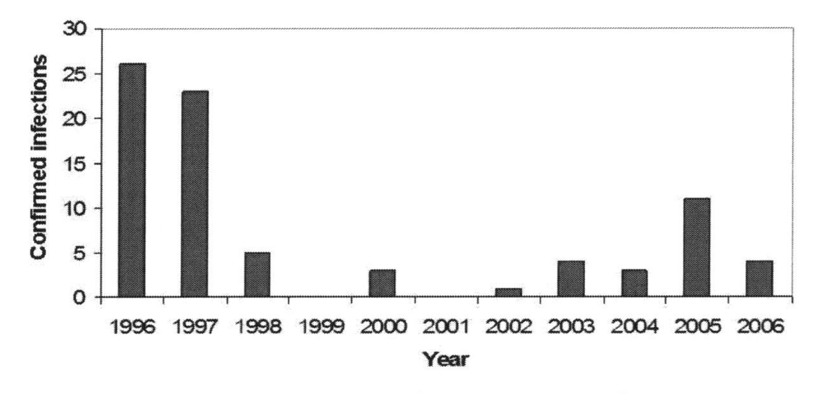

Fig. 5.2. Number of *Ralstonia solanacearum* infections detected by ELISA and subsequently confirmed by selective media usage in South Africa during the period 1996–2006.

15. Time taken for confirmation of infection by *Ralstonia solanacearum*: Although characterization of bacterial isolates by medium testing as described is effective, it can take up to 3 weeks, which has economic implications for producers. As a result, the use of PCR based confirmatory tests is currently being investigated.

Acknowledgments

The expert technical assistance in the development of this ELISA and its supply to the industry by Mrs. Coral de Villiers, is gratefully acknowledged. Dr. Neil Theron, for many years the head of Potatoes SA's laboratory services and in the last few years the CEO of Potatoes SA, played a major role in instating ELISA testing for *Ralstonia solanacearum* in South Africa and his major contribution in implementing this testing procedure is hereby posthumously acknowledged. The assistance of the following persons in the development and application of this ELISA is gratefully acknowledged: Ms. Elisabeth Gustaffson, Ms. Anita Swanepoel, Ms. Marieta Botha, Ms. Anel Espach and Mr. Chris Kleingeld. Dr. Ben Pieterse, the present CEO of Potatoes SA and research manager, is thanked for his continued financial support for research in the improvement of *Ralstonia solanacearum* ELISA testing.

References

1. Bellstedt, D.U., and Van der Merwe, K.J. (1989) The Development of ELISA Kits for the Detection of *Pseudomonas solanacearum* Bacterial Wilt in Potatoes. *Proceedings of the First South African Potato Research Symposium*, Warmbaths, South Africa, pp. 64–69.
2. Robinson-Smith, A., Jones, P., Elphinstone, J. G., and Forde, S. M. D. (1995) Production of antibodies to *Ralstonia solanacearum*, the causative agent of bacterial wilt. *Food Agric. Immunol.* 7:67–79.
3. Anon. (1992) Quarantine procedure no. 26. *Pseudomonas solanacearum. Bull. OEPP* 20:255–262.
4. Wullings, B. A., van Beuningen, A. R., Janse, J. D., and Akkermans, A. D. L. (1998) Detection of *Ralstonia solanacearum*, which causes brown rot of potato, by fluorescent *in situ* hybridization with 23S rRNA-targeted probes *Appl. Environ. Microbiol.* 64:4546–4554.
5. Seal, S. E., Jackson, L. A., Young, J. P. W., and Daniels, M. J. (1993) Differentiation of *Pseudomonas solanacearum*, *Pseudomonas syzygii*, *Pseudomonas pickettii* and blood disease bacterium by partial 16S rRNA sequencing: construction of oligonucleotide primers for sensitive detection by polymerase chain reaction. *J. Gen. Microbiol.* 139:1587–1594.
6. Robinson-Smith, A. (1993) Polyclonal and Monoclonal Antibody-Based Enzyme-Linked Immunosorbent Assay for *Pseudomonas solanacearum*. In: Hardy, B. and French, E. P., Eds. The Integrated Management of Bacterial Wilt. Proceedings of an International Workshop, New Delhi, India, 11–16 October, 1993.
7. Griep, R. A., Charlotte van Twisk, C., José van Beckhoven, J. R. C. M., van der Wolf, J. M., and Schots, A. (1998) Development of specific recombinant monoclonal antibodies against the lipopolysaccharide of *Ralstonia solanacearum* Race 3. *Phytopathology* 88: 975–803.

8. Caruso, P., Gorris, M. T., Cambra, M., Palomo, J. L. Collar, J., and Lòpez, M. M. (2002) Enrichment double-antibody sandwich indirect Enzyme-Linked Immunosorbent Assay that uses a specific monoclonal antibody for sensitive detection of *Ralstonia solanacearum* in asymptomatic potato tubers. *Appl. Environ. Microbiol.* 68:3634–3638.
9. Swanepoel, A. E., and Young, B. W. (1988) Characteristics of South African strains of *Pseudomonas solanacearum. Plant Dis.* 72: 403–405.
10. Kenna, J. G., Major, G. N., and Williams, R. S. (1985) Methods for reducing non-specific antibody binding in enzyme-linked immunosorbent assays. *J. Immunol. Methods* 85:409–419.
11. Engelbrecht, M. C. (1994) Modifications of a semi-selective medium for the isolation of *Pseudomonas solanacearum. Bact. Wilt Newsl.* 10:3–5.
12. Kelman, A. (1953) The bacterial wilt caused by *Pseudomonas solanacearum. N. C. Agric. Exp. Stn. Tech. Bull.* 99:1–193.
13. Bellstedt, D. U., Human, P. A., Rowland, G. F., and Van der Merwe, K. J. (1987) Acid-treated, naked bacteria as immune carriers for protein antigens. *J. Immunol. Methods* 98:249–255.
14. Clayton, M. K., and Slack, S. A. (1988) Sample size determination in zero tolerance circumstances and implications in stepwise sampling: bacterial ring rot as a special case. *Am. Potato J.* 165:711–723.
15. Holt, J. G. (ed.) (1994) Bergey's Manual of Determinative Bacteriology, 9th Edition, Williams & Wilkins, Baltimore, MD.
16. Schaad, N. W. (1994) Laboratory Guide for Identification of Plant Pathogenic Bacteria, 2nd Edition, American Phytopathological Society, St. Paul, MN.

Chapter 6

Production of Monoclonal Antibodies to Plant Pathogens

Christopher R. Thornton

Summary

The use of monoclonal antibodies in plant pathology has improved the quality and specificity of detection methods for diseases. Hybridoma technology allows the limitless production of highly specific antibodies which can be used to identify pathogens to the species or even sub-species level.

Key words: Monoclonal antibody, Hybridoma, Immunogen, Epitope, Antiserum.

6.1. Introduction

With the advent of hybridoma technology, highly specific monoclonal antibodies have been developed for the detection, quantification and visualization of plant pathogens and their antagonists in soil and *in planta (1–3)*. The exquisite specificity of mouse mAbs enables the identification of individual species in mixed populations, a feat that was unattainable using cross-reactive rabbit antisera. This chapter is aimed at the non-specialist. It will not describe immunization and hybridoma fusion procedures per se, since these are specialist techniques that are unlikely to be performed by most plant pathologists, and which are adequately described elsewhere *(4, 5)*. Hybridoma technology is available commercially and so the development of bespoke mAbs is now a reality. However, this still requires the preparation and provision of immunogens, and the supply of antigens for hybridoma screening, by the client. Subsequent characterization of antibodies can be performed by the client, thereby reducing the overall production costs. This chapter will therefore focus on strategies

R. Burns (ed.), *Methods in Molecular Biology, Plant Pathology, vol. 508*

DOI: 10.1007/978-1-59745-062-1_6

for selecting and preparing immunogens, the design and implementation of hybridoma screening procedures and methods for the characterization of mAbs and identification of antigens and epitopes. The development of a mAb (GE6) against *Pythium sylvaticum* will be used to illustrate the procedures involved in the production of mAbs against plant pathogens. While *P. sylvaticum* is an oomycete, the techniques employed in mAb generation and characterization are similar to those used for fungi, and which are described in detail elsewhere *(5)*.

6.2. Materials

1. Bicarbonate buffer, pH 9.6: 15 mM Na_2CO_3, and 35 mM $NaHCO_3$.
2. Phosphate-buffered saline (PBS) buffer, pH 7.4: 137 mM NaCl, 2.7 mM KCl, 8 mM Na_2HPO_4, and 1.5 mM KH_2PO_4.
3. Phosphate buffered saline with Tween-20 (PBST) buffer: PBS buffer containing 0.005% polyoxyethylene(20)sorbitan monolaurate.
4. Microtitre wells arranged in 8-well strips (Thermo Electron Corporation).
5. Flat-bottomed Immunlon II HB microtitre plates (Thermo Labsystems).
6. Anti-mouse horseradish peroxidase conjugate (Sigma).
7. DMSO (dimethyl sulfoxide).
8. TMB (3,3′,5,5′-tetramethyl benzidine): 10 mg in 1 mL DMSO at 4°C.
9. Peroxidase substrate: 5 mL dH_2O, 5 mL 0.2 M sodium acetate, 195 μL 0.2 M citric acid, 5 mL 30% H_2O_2, and 100 μL TMB.
10. Peroxidase substrate stop solution: 3 M H_2SO_4.
11. Microplate reader fitted with 450-nm filter.
12. Sodium acetate buffer, pH 4.5: 37 mL of 50 mM NaOAc, and 63 mL 50 mM HOAc.
13. Periodate solution: 20 mM sodium *m*-periodate in sodium acetate buffer.
14. Pronase solution: 9 mg pronase (protease XIV) in 10 mL PBS.
15. Trypsin solution: trypsin tablets (Sigma) dissolved in dH_2O according to the manufacturer's instructions.

16. Trypsin inhibitor solution: trypsin inhibitor (Sigma) dissolved in PBS to give a 0.1 mg/mL solution.
17. Sodium dodecyl sulfate–polyacrylamide gel electrophoresis (SDS–PAGE) and Western blotting equipment.
18. Laemmli sample buffer for denaturing PAGE: 1 mL 0.5 M Tris–HCl, pH 6.8, 3.8 mL dH_2O, 0.8 mL glycerol, 1.6 mL 10% sodium dodecyl sulfate, 0.4 mL β-mercaptoethanol, and 0.4 mL 0.5% bromophenol blue.
19. PAGE running buffer, pH 8.3: 25 mM Tris, 19 mM glycine, and 3.5 mM sodium dodecyl sulfate.
20. Western blotting buffer, pH 8.3: 20% methanol, 25 mM Tris, and 19 mM glycine.
21. Pre-stained molecular weight markers (BioRad).
22. PVDF (polyvinyladine difluoride) transfer membrane (BioRad).
23. Whatman filter paper #1.
24. Immunoassay grade bovine serum albumin (Sigma).
25. Anti-mouse alkaline phosphatase antibody conjugate (Sigma).
26. DMF (*N,N*-dimethyl formamide).
27. BCIP (5-bromo-4-chloro-3-indoyl phosphate): 25 mg in 500 μL DMF.
28. NBT (nitro blue tetrazolium): 50 mg in 1 mL 70% DMF.
29. Alkaline phosphatase substrate buffer: 100 mM Tris–HCl, pH 9.5, 100 mM NaCl, 5 mM $MgCl_2 \cdot 6H_2O$.
30. Alkaline phosphatase substrate solution: 100 mM Tris–HCl, pH 9.5, 100 mM NaCl, 5 mM $MgCl_2 \cdot 6H_2O$, 33 μL BCIP, and 66 μL NBT.
31. Coomassie Blue R-250 protein staining solution: 1.0 g Coomassie Blue R-250 dissolved in 500 mL dH_2O, 400 mL methanol, and 100 mL acetic acid. Filter through Whatman filter paper.
32. Coomassie Blue R-250 protein de-staining solution: 500 mL dH_2O, 400 mL methanol, and 100 mL acetic acid.

6.3. Methods

6.3.1. Choice of Antigen

The choice of antigen for immunization of mice is governed primarily by the final purpose for which a mAb is destined. Experience dictates that the most useful mAbs for the detection, quantification and visualization of plant pathogens in soil and *in planta* are those that recognize extracellular, constitutively expressed, antigens *(1, 2)*. To improve the likelihood of raising mAbs to such molecules, antigens should be prepared from fungi that are actively growing since extracellular antigens are likely to be secreted at the growing tip of hyphae. For the development of mAbs to *P. sylvaticum*, lyophilized mycelium was prepared from actively growing cultures of the oomycete according to the procedure described in **Subheading 6.3.1.1**, and used as the immunogen for mouse immunizations. This procedure should also be used for the preparation of antigens from fungi.

6.3.1.1. Preparation of Immunogen

1. Grow the oomycete or fungus in shake culture under conditions and in a growth medium that allows active growth of the organism. For fungi, where possible, use spores as the culture inoculant as this allows the generation of germlings.
2. Harvest mycelium (or germlings) from young, actively growing, cultures by pouring the contents of flasks through miracloth. Keep the culture filtrate, add sodium azide as a preservative, and store at –20°C.
3. Gently wash the mycelium with dH_2O. This should not be a thorough wash, but enough to remove excess culture medium.
4. Transfer the mycelium to a fresh flask and plunge in liquid nitrogen. Once frozen, lyophilize the sample and store at –20°C.
5. The immunogen is prepared by reconstituting the mycelium in PBS and mixing with an appropriate adjuvant prior to immunization of mice *(4)*.

6.3.2. Screening of Hybridomas

The preparation of robust and clearly defined selection criteria should be established well in advance of hybridoma screening. This is made easier by the supply of antisera in test bleeds from the tails of the immunized mice, supplied by the hybridoma company. As the client, it is important that you ask the company to supply tail bleeds from the mice during the course of immunizations for use as positive controls in the development of the hybridoma screening strategy. Given the large number of cell lines that can be generated from a single fusion, the most appropriate assay for the identification of hybridomas secreting antigen-positive mAbs in a primary screen is an enzyme-linked immunosorbent assay (ELISA). Using this technique, several hundred hybridoma cell lines can be tested for mAb production in a single day. The first step in an ELISA is to coat microtitre plates with antigens (*see* **Subheading 6.3.2.1**), followed by the ELISA itself (*see* **Subheading 6.3.2.2**). Once positive cell lines have been identified using this technique, antigens from other organisms can be tested to determine mAb specificity in ELISA tests. In addition, techniques such as immunofluorescence can be used to examine the spatial distribution of antigens *(2, 3, 5)*. These additional tests form the basis of a secondary screening strategy. The choice of secondary enzyme conjugate in primary and secondary screens is governed by antibody class and is an important consideration in primary and secondary screening steps (*see* **Note 1**). Monoclonal antibody GE6 was identified in a primary screen of mouse hybridomas using ELISA. Secondary screening of GE6 specificity was carried out using a colony-blotting test described in **Subheading 6.3.2.3**. This enabled the mAb to be screened against fastidious organisms (such as *Aphanomyces*, *Pythium* and *Phytophthora* species) that are difficult to grow in liquid culture. The technique showed that mAb GE6 binds to extracellular antigens from *P. sylvaticum* and certain other related *Pythium* species (*see* **Fig. 6.1**).

6.3.2.1. Coating Microtitre Wells with Antigens

1. Reconstitute 1 mg of lyophilized mycelium in 1 mL of PBS or bicarbonate buffer and mix vigorously for 5 min. Larger volumes can be prepared as required. Pellet the mycelium by centrifugation (12,000 × *g*; 5 min).
2. Pipette 50 μL of antigen extract into the wells of microtitre strips or plates. For PBS extracts, antigens should be used to

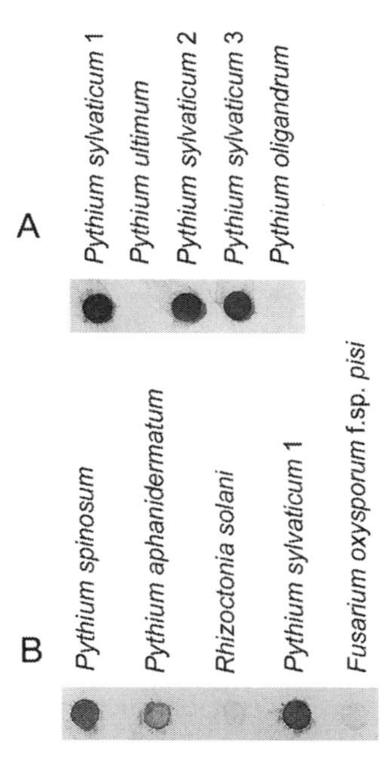

Fig. 6.1. (**a**) Colony blots of mycelium from three isolates of *P. sylvaticum* pathogenic to pea and from the related oomycetes *P. ultimum* and *P. oligandrum*. (**b**) Colony blots of mycelium from an isolate of *P. sylvaticum* pathogenic to pea, other pythiaceous oomycetes and the pea pathogenic fungi *Rhizoctonia solani* and *Fusarium oxysporum* forma specialis *pisi*.

coat microtitre strips. Bicarbonate extracts should be used to coat microtitre plates. Tap the wells gently to ensure even distribution of extracts over well surfaces. Place the wells in a sealed plastic bag and incubate overnight at 4°C to allow immobilization of antigens.

3. Remove antigen extracts from the wells by inverting the strips or plates and gently flicking the wrist. Fill each well to the rim with PBST dispensed from a wash bottle. Leave for 5 min and repeat twice more (5 min each time) with PBST, once with PBS, and once with dH_2O. Remove dH_2O as described and bang the wells dry on paper towelling. Air-dry the wells in a flow bench at 23°C and store the dried wells in a sealed plastic bag at 4°C. Immobilized antigens can be stored in this way for several months without any apparent detrimental effects.

6.3.2.2. Enzyme-Linked Immunosorbent Assay

1. Pipette 50 μL of hybridoma supernatants into antigen-coated wells. Negative control wells should contain an equal volume of mAb-free tissue culture medium. Positive control wells should be incubated with 1,000-fold dilutions of tail bleed sera in PBST. Place the microtitre wells in a sealed plastic bag and incubate for 1 h at 23°C.
2. Remove mAb, TCM or pAb solutions by inversion and flicking and wash the wells three times with PBST (5 min each time) as described in **Subheading 6.3.2.1**.
3. Prepare a 1,000-fold dilution of secondary goat anti-mouse peroxidase conjugate by diluting in PBST. Add 50 μL to each well and incubate for 1 h at 23°C in a sealed plastic bag.
4. Wash the wells three times with PBST and once with PBS as described. Add 50 μL of peroxidase substrate solution to each well and incubate in a sealed plastic bag for 30 min at 23°C. Stop the blue coloured reaction by the addition of 50 μL peroxidase substrate stop solution to each well.
5. Read absorbance values at 450 nm. A positive reaction in ELISA is determined by comparison of absorbance values obtained from mAb supernatants and mAb-free tissue culture medium controls.

6.3.2.3. Colony-Blotting

1. Activate a sheet of PVDF membrane by soaking in methanol for 2 min followed by gentle rinsing in dH_2O. Do not allow the membrane to dry out. To prevent this, place it on a piece of filter paper moistened with PBS contained in a plastic box.
2. Grow the oomycete or fungus on an appropriate solid medium. From the growing edge of colonies remove a plug of mycelium using a cork borer and place it hyphal-face down onto the surface of the membrane. Incubate at 23°C

for 7 h and carefully remove the agar plugs using a scalpel and forceps.

3. Block the membrane for 16 h at 4°C, with gentle mixing, in PBS containing 1% BSA.
4. Transfer the membrane to half-strength GE6 mAb supernatant, diluted in PBS containing 0.5% BSA (PBSA), and incubate for 2 h at 23°C with gentle mixing.
5. Wash the membrane three times (5 min each) in PBS with gentle mixing.
6. Transfer the membrane to PBSA containing a 15,000-fold dilution of goat anti-mouse IgG (whole molecule) alkaline phosphatase conjugate. Incubate with gentle mixing for 1 h at 23°C.
7. Wash the membrane three times in PBS and once in PBST.
8. Transfer the membrane to alkaline phosphatase substrate buffer and equilibrate for 5 min at 23°C.
9. Transfer the membrane to alkaline phosphatase substrate solution and allow the colour reaction to develop. Once the desired intensity has been obtained, wash the membrane thoroughly in dH_2O and dry between sheets of filter paper.

6.3.3. Antibody Characterization

Determining the class of an immunoglobulin is a necessary requirement if the properties of a mAb are to be fully exploited. It allows you to select the most appropriate secondary anti-mouse reporter molecules for ELISA, IF, Western blotting and EM-gold studies and, in the case of IgGs, to readily purify the target antigen and map the antigenic determinant. Immunoglobulin class is most easily determined using the ISO-1 iso-typing kit from Sigma. Full instructions are provided and the four main classes and sub-classes of antibody (IgA, IgG_1, IgG_{2a}, IgG_{2b}, IgG_3, and IgM) can be identified within 1 h. Other assays for immunoglobulin class determination are described elsewhere *(5)*. Using this assay, mAb GE6 was characterized as an immunoglobulin G_1 (IgG_1) antibody.

6.3.4. Epitope Characterization by Chemical Modification

Identification of a mAb with the desired specificity is frequently more of an issue than epitope recognition, but it is often wrongly assumed that selecting mAbs that bind to protein epitopes will improve the likelihood of isolating specific antibodies. A number of mAbs have now been raised to plant pathogens that are highly specific and which bind to carbohydrate epitopes *(6)*. The result of a successful fusion is a panel of cells secreting mAbs of different antibody classes (*see* **Subheading 6.3.3**) that bind to carbohydrate and protein epitopes. If the epitopes are on the same glycoprotein antigen, combinations of antibody classes can be used to expand the repertoire of assays available for pathogen detection.

An example is the development of double-antibody-sandwich-enzyme-linked immunosorbent assays *(5)*. Epitope characterization studies also provided preliminary data on the likelihood that a mAb can be used to identify its antigen in Western blotting studies (*see* **Subheading 6.3.5**). Generally speaking, mAbs that recognize protein epitopes bind to discrete bands in Western blots of SDS–PAGE gels, whereas mAbs that bind to carbohydrate epitopes produce characteristic glycoprotein smears. Selection for mAbs that bind to protein or carbohydrate epitopes can be included in the primary and secondary screening steps described in **Subheading 6.3.2**. Determining the nature of an epitope is therefore a worthwhile and informative step in immunoassay development and was used to identify the proteinaceous nature of the antigenic determinant bound by mAb GE6 (**Tables 6.1** and **6.2**). For an explanation of these results and the interpretation of epitope characterization studies in general (*see* **Note 2**).

Table 6.1
Absorbance values from ELISA tests with protease-treated antigens from *P. sylvaticum* using mAb GE6. Each value represents the mean of replicated values ± SE

	Absorbance (450 nm)			
Temperature (°C)	**Pronase**	**Control**	**Trypsin**	**Control**
4	0.203 ± 0.001	0.744 ± 0.006	0.619 ± 0.004	0.813 ± 0.003
37	0.369 ± 0.003	0.826 ± 0.005	0.163 ± 0.001	0.765 ± 0.004

Table 6.2
Absorbance values from ELISA tests with periodate-treated antigens from *P. sylvaticum* using mAb GE6. Each value represents the mean of replicated values ± SE

	Absorbance (450 nm)	
Time (h)	**Periodate**	**Control**
20	1.109 ± 0.013	0.853 ± 0.015
4	1.127 ± 0.012	0.864 ± 0.009
3	1.119 ± 0.021	0.849 ± 0.018
2	1.101 ± 0.014	0.821 ± 0.013
1	1.116 ± 0.016	0.813 ± 0.009

6.3.4.1. Periodate Oxidation

1. Incubate antigen-coated wells with 50 μL of periodate solution for the appropriate time period at 4°C in sealed plastic bags. Incubate an equivalent number of control wells with acetate buffer only.
2. Wash the wells four times (5 min each) with PBS and assay by ELISA (*see* **Subheading 6.3.2.2**).

6.3.4.2. Protease Digestion

1. Incubate antigen-coated wells with 50 μL of pronase or trypsin solutions for 4 h at 37°C and 4°C in sealed plastic bags. Incubate an equivalent number of control wells with PBS or dH_2O only.
2. Wash the wells four times (3 min each) with PBS. To trypsin treated wells add 50 μL of trypsin inhibitor solution, incubate at 23°C for 10 min, and wash the wells a further three times with PBS. Assay the treated and controls wells by ELISA (*see* **Subheading 6.3.2.2**).

6.3.5. Antigen Characterization by Western Blotting and N-Terminal Sequencing

Added benefits can be reaped from the generation of a mAb if its target antigen can be identified. The whole genomes of a number of plant pathogens have now been sequenced and the genomes of other plant pathogens are likely to be fully sequenced over the forthcoming years. Consequently, if the N-terminal amino acid sequence of a protein antigen can be obtained, the corresponding gene sequence can be identified from the genome and primers designed for gene cloning via polymerase chain reaction. Provided the organism in question is genetically tractable, the function of the gene in the biology of the organism can be determined through targeted gene deletion. The identification of proteins is easily achieved by Western blotting of proteins and N-terminal sequencing, a technique that is now widely available as a chargeable service.

The Western blotting procedure described in **Subheading 6.3.5.1**, is based on the BioRad Mini Protean II gel electrophoresis and electroblotting apparatus and the use of pre-cast 4–20% Tris–HCl gradients gels. Using this procedure, mAb GE6 was found to bind to an extracellular (glyco)protein with a molecular weight of ~M_r 15,000 (*see* **Fig. 6.2a**). Further analysis of the protein band excised from a replica Western blot stained with Coomassie (*see* **Fig. 6.2b**) revealed an N-terminal amino acid sequence of YDHTTTHPFTEFTKLAPVAT. N-terminal sequencing was carried out by Alta Bioscience at the University of Birmingham, UK.

6.3.5.1. SDS–PAGE and Western Blotting

1. Suspend 1 mg of lyophilized mycelium in 1 mL of Laemmli sample buffer. Culture filtrates should be diluted fourfold in buffer. Mix thoroughly by vortexing and heat for 10 min in a boiling water bath. Allow the mixture to cool and centrifuge (12,000 × *g*; 5 min).

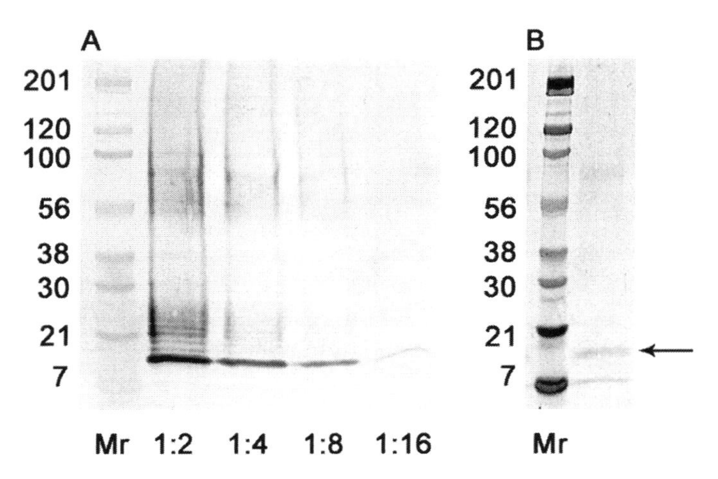

Fig. 6.2. (**a**) Western blot of *P. sylvaticum* extracellular antigens diluted 1:2, 1:4, 1:8, or 1:16 with LaemmLi sample buffer and probed with mAb GE6, as described in **Subheading 6.3.5.1**. (**b**) Replica Western blot using a 1:2 dilution of antigens in LaemmLi sample buffer and stained with Coomassie Blue R-250 for visualization of proteins. The protein indicated by the arrow corresponds to the ~M_r 15,000 protein bound by mAb GE6 (**a**). The protein was excised and the N-terminal amino acid sequence of the antigen determined as described in **Subheading 6.3.5.2**.

2. Load the wells of the SDS–PAGE gel with 40 μL of buffer containing denatured antigens. Load control wells with 5 μL of pre-stained markers for subsequent determinations of molecular weights.
3. Separate the proteins by electrophoresis for approximately 1 h at 165 V.
4. Electroblot the proteins to activated PVDF membrane (*see* **Subheading 6.3.2.3**). This should be conducted at 75 V for 2 h. Successful blotting of proteins is indicated by the transfer of the pre-stained markers from the control wells to the PVDF membrane.
5. Transfer the membrane to PBS containing 1% BSA and block for 16 at 4°C with gentle mixing.
6. Transfer the membrane to half-strength GE6 mAb supernatant, diluted in PBS containing 0.5% BSA (PBSA), and incubate for 2 h at 23°C with gentle mixing.
7. Wash the membrane three times (5 min each) in PBS with gentle mixing.
8. Transfer the membrane to PBSA containing a 15,000-fold dilution of goat anti-mouse. IgG (whole molecule) alkaline phosphatase conjugate. Incubate with gentle mixing for 1 h at 23°C.
9. Wash the membrane three times in PBS and once in PBST.
10. Transfer the membrane to alkaline phosphatase substrate buffer and equilibrate for 5 min at 23°C.

11. Transfer the membrane to alkaline phosphatase substrate solution and allow the colour reaction to develop. Once the desired intensity has been obtained, wash the membrane thoroughly in dH_2O and dry between sheets of filter paper.

6.3.5.2. N-Terminal Sequencing

1. Carry out the western blotting of proteins as described in **Subheading 6.3.5.1**.
2. Transfer the PVDF membrane to Coomassie Blue R-250 protein staining solution and mix gently for 10 min until the membrane is uniformly stained. At this stage protein bands will not be visible against the blue background.
3. Transfer the membrane to de-staining solution and regularly remove with forceps to monitor the appearance of proteins. Once bands are clearly visible and the background has started to fade, allow the membrane to dry between sheets of filter paper. Store the membrane at room temperature in a dry environment.
4. Identify the protein to be sequenced by comparing the results of the Western blot processed using mAb supernatant (*see* **Subheading 6.3.5.1**) with the banding pattern on the replica blot stained with Coomassie. Excise the corresponding band from the Coomassie-stained replica using a pair of clean scissors. Handle the excised band carefully with a pair of clean forceps. Attach the band to a sheet of card by both ends using self-adhesive labels. PVDF blots of proteins are sufficiently stable to allow postage of samples to a third party for N-terminal sequencing, as was the case with the protein bound by mAb GE6.

6.4. Notes

1. The choice of secondary horseradish peroxidase and alkaline phosphatase enzyme conjugates in ELISA, colony-blotting and Western blotting procedures is determined by the immunoglobulin class of the mAb. For primary screening of antibodies, polyvalent conjugates that bind to all classes and sub-classes of mouse antibodies should be used. Once a mAb has been isotyped, secondary conjugates can be used that discriminate between antibody classes. Thus, for IgG immunoglobulins, either anti-mouse IgG (whole molecule) enzyme conjugates or anti-mouse IgG (γ-chain specific) enzyme conjugates can be used. For IgM immunoglobulins, anti-mouse IgM (μ-chain specific) conjugates should be used.

These two classes of immunoglobulin are the most frequently encountered species in mouse hybridoma production.

2. A reduction in mAb binding in ELISA following treatment with pronase, shows that its epitope is proteinaceous. More specifically, sensitivity of a protein epitope to trypsin digestion shows that a mAb binds to an antigen containing positively charged lysine and arginine side chains, as is the case with mAb GE6 (Table 6.1). A reduction in mAb binding following periodate treatment, shows that the epitope contains carbohydrate moieties. Increased signal strength in ELISA after periodate treatment is not uncommon and reflects increased exposure of a protein epitope to mAb binding following removal of glycan moieties from a glycoprotein antigen. Such a response is exhibited by mAb GE6 (Table 6.2). In addition to modification with proteases, antigens can be heated to determine sensitivity of mAb binding to epitopes. Sensitivity indicates that a mAb binds to a protein epitope. Carbohydrate epitopes are generally heat stable and can often withstand autoclaving *(5, 6)*. Other procedures can be used to determine the nature of glycoprotein antigens. Reduction in mAb binding in ELISA following mild alkaline hydrolysis (β-elimination) of antigens *(5)* shows that the antigen is a glycoprotein with glycans O-linked through the β-hydroxy amino acids serine and threonine, but not through tyrosine, hydroxylysine or hydroxyproline, or N-linked through asparagine. Enzymatic treatment of antigens with peptide-*N*-glycosidase (PNGase) or endo-β-*N*-glucosaminidase-H (endo-H) in Western blotting studies *(7)* can show whether antigens are N-linked glycoproteins or whether they contain high-mannose side chains.

References

1. Thornton, C.R., Pitt, D., Wakley, G.E., and Talbot, N.J. (2002) Production of a monoclonal specific to the genus *Trichoderma* and closely related fungi, and its use to detect *Trichoderma* spp. in naturally infested composts. *Microbiology* 148, 1263–1279.
2. Thornton, C.R. (2004) An immunological approach to quantifying the saprotrophic growth dynamics of *Trichoderma* species during antagonistic interactions with *Rhizoctonia solani* in a soil-less mix. *Environ. Microbiol.* 6, 323–334.
3. Thornton, C.R., and Talbot, N.J. (2006) Immunofluorescence microscopy and immunogold EM for investigating fungal infection of plants. *Nat. Protoc.* 1, 2506–2511.
4. Harlow, E., and Lane, D. (1988) *Antibodies, A Laboratory Manual.* Cold Spring Harbor Laboratory Press, New York.
5. Thornton, C.R. (2002) Immunological methods for fungi, in *Molecular and Cellular Biology of Filamentous Fungi* (Talbot, N.J., ed.), Oxford University Press, Oxford, pp. 227–257.
6. Thornton, C.R., Dewey, F.M., and Gilligan, C.A. (1993) Development of monoclonal antibody-based immunological assays for the detection of live propagules of *Rhizoctonia solani* in soil. *Plant Pathol* 42, 763–773.
7. Bleddyn Hughes, H., Carzaniga, R., Rawlings, S.L., Green, J.R., and O'Connell, R.J. (1999) Spore surface glycoproteins of *Colletotrichum lindemuthianum* are recognized by a monoclonal antibody which inhibits adhesion to polystyrene. *Microbiology* 145, 1927–1936.

Chapter 7

Enzyme-Linked Immunosorbent Assay for the Detection and Identification of Plant Pathogenic Bacteria (In Particular for *Erwinia amylovora* and *Clavibacter michiganensis* subsp. *sepedonicus*)

Blanka Kokoskova and Jaap D. Janse

Summary

Enzyme-linked immunosorbent assay (ELISA) is the most commonly used serological diagnostic technique. A number of different ELISA formats can be used for the detection of bacterial plant pathogens and in particular *Erwinia amylovora* and *Clavibacter michiganensis* subsp. *sepedonicus*.

Key words: ELISA, antibodies, PTA ELISA, DAS ELISA, microplate, monoclonal antibodies, polyclonal antibodies.

7.1. Introduction

The principal aim of an immunodiagnostic assay is to detect and/or quantify the binding of the diagnostic antibody with the target antigen. Enzyme-linked immunosorbent assay (ELISA), a technology developed in the 1970s, is the most commonly used diagnostic technique that uses antibodies *(1)*. It involves an enzyme-mediated colour change reaction to detect antibody binding. This is usually done in an appropriate plastic plate where the degree of colour change, usually measured in a computer-controlled plate reader, can be used to determine the amount of pathogen present *(2, 3)*.

The enzyme-linked antiserum is added in wells of microplates that were previously coated with antigen. After the excess antiserum is washed off with phosphate buffered saline, the substrate

R. Burns (ed.), *Methods in Molecular Biology, Plant Pathology, vol. 508*

DOI: 10.1007/978-1-59745-062-1_7

(e.g. *p*-nitrophenylphosphate) of the enzyme is added to the well. The substrate is hydrolysed by the enzyme (usually alkaline phosphatase) bound to antiserum and releases *p*-nitrophenol, changing the colour of the reaction mixture (from colourless to yellow).

ELISA uses polyclonal (PAbs) and monoclonal antibodies (MAbs), the latter are usually more sensitive and also more specific *(4–7)*. The engagement of PAbs enables the detection of a broader spectrum of pathogen strains; however, cross-reactions can occur *(8–13)*. MAbs prepared to one determinant are more specific than PAbs, but they can be too specific and do not detect all strains in the populations of the target bacterium *(14, 15)*. Usually, a mixture of several MAbs is necessary to allow for recognition of all representative strains *(16–18)*.

ELISA techniques have the advantages of being simple, cheap and suitable for the processing of many samples. Commercial ELISA kits are available for many bacteria. The typical sensitivity of ELISA for bacteria is about $10^{6\cdot5}$ cfu/ml, at the lower concentration, the target pathogens might be more difficult to detect *(19–22)*. The technique works best for diagnosis when samples consist of fresh lesions containing very high pathogen titres.

The simplest is the plate-trapped antigen ELISA (PTA-ELISA). In this assay, the microtitre plate wells are directly coated with the test sample. This is followed by incubation with specific antibody (usually PAb), which binds to the target antigen. In direct detection of antigen, the specific antibody is conjugated to the enzyme, in indirect detection the specific antibody is detected by a second antibody, e.g. anti-rabbit (PAb) or anti-mouse (MAb), which is conjugated to the enzyme.

A more commonly used format is the double antibody sandwich enzyme-linked immunosorbent assay (DAS-ELISA). In DAS-ELISA, firstly, the specific antibody is used to coat the microtitre plate, which then traps the target antigen from the test sample. An enzyme-labelled specific antibody conjugate is subsequently used for the detection of bacteria in sample.

DAS-ELISA technique was successfully used for detection of the causal organism of bacterial wilt of potato in soil (See Note 1). The modified DAS-ELISA technique was used for reliable detection of the causal agent of fire blight disease (*See* **Note 2**). ELISA technique has also certain disadvantages and in some cases detecting the plant pathogen could fail (*See* **Note 3**).

In the paragraphs below a short outline of PTA- and DAS-ELISA is presented.

7.1.1. PTA-ELISA

In PTA-ELISA (*see* **Fig. 7.1**) (here introduced for the fire blight pathogen *Erwinia amylovora* – using the NeogenEurope kit), an extract of the plant sample is coated on to the microtitre plate. After a period of incubation, the unbound sample is removed by washing and a blocking buffer with a high proportion of protein is added. After a period of incubation, the blocking buffer is removed by washing and an antibody, which has been produced against the antigen, is added to the microplate. After washing, a

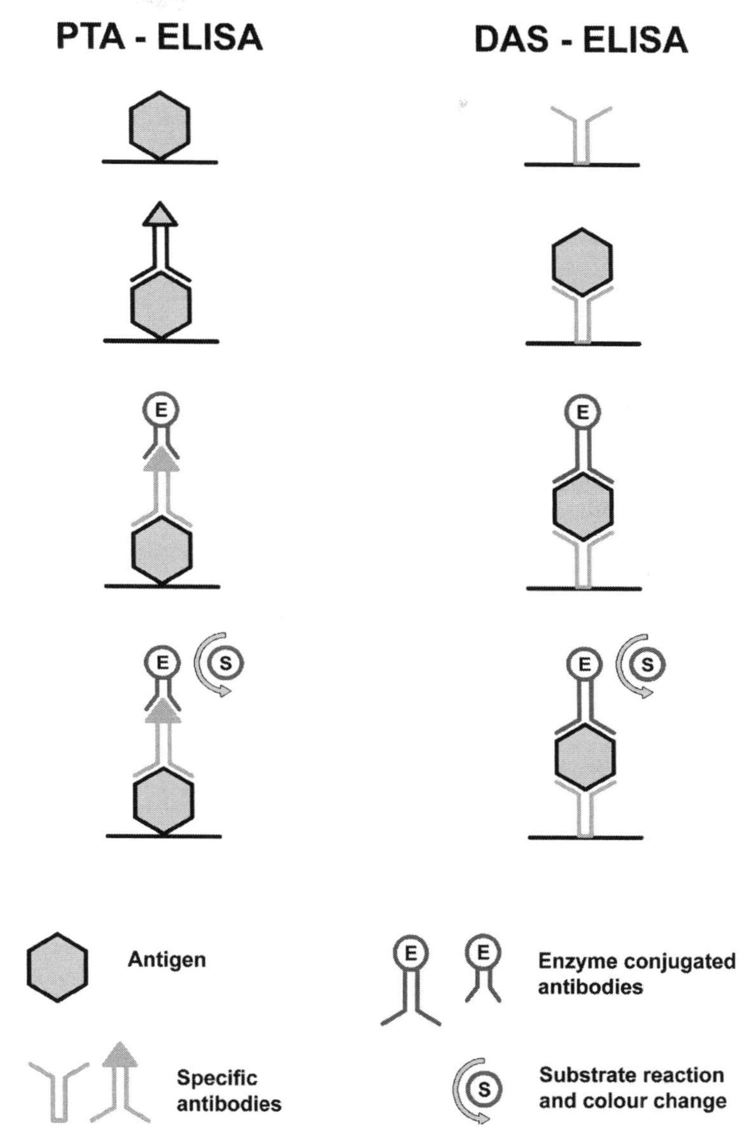

Fig. 7.1. Schematic illustration of indirect PTA-ELISA (plate-trapped antigen) and DAS-ELISA (double antibody sandwich) formats. (Redrawn according to Ward et al. *(3)*).

second antibody, which is conjugated to the enzyme (alkaline phosphatase) is added to the microplate and incubated. The second antibody is specific for the species/pathovar to which the first antibody has been developed. The unbound antibody–enzyme conjugate is washed from the microplate and the substrate for the enzyme is added. If there is any bound antibody–enzyme conjugate, the substrate will change colour.

7.1.2. DAS-ELISA

In DAS-ELISA (here introduced for the causal agent of bacterial ring rot of potato, *Clavibacter michiganensis* subsp. *sepedonicus* – using the Agdia kit), firstly a primary species/pathovar specific antibody is coated on the microplate. After a period of incubation and removing unbound antibody by washing, extract of the plant sample is added to the microplate. After a period of incubation and removing unbound sample by washing, a secondary pathogen-specific antibody (conjugate), which is labelled with enzyme (alkaline phosphatase) is added to the microplate and incubated. The unbound antibody–enzyme conjugate is washed from the microplate and the substrate for the enzyme is added. The result of the hydrolysis of the substrate is a yellow coloured product *(15)*.

7.2. Materials

7.2.1. PTA-ELISA (*Erwinia amylovora* – NeogenEurope Ltd)

1. Antibody (commercially available polyclonal antibody) against the target plant pathogenic bacteria, e.g. *Erwinia amylovora* (*See* **Note 4**).
2. Coating buffer: 1.59 g sodium carbonate, 2.93 g sodium hydrogen carbonate. Make up to 1 L with distilled (d) H_2O. The pH of this buffer is 9.6 and does not need to be adjusted. This buffer is used for dilution of the sample.
3. Phosphate buffered saline PBS: 8 g sodium chloride, 0.2 g potassium dihydrogen orthophosphate, 1.15 g disodium hydrogen orthophosphate, 0.2 g potassium chloride. Make up to 1 L with dH_2O. The pH of this buffer should be 7.2. This buffer is used for dilution of the antibody.
4. Wash buffer (PBS + Tween 20; PBST): 1 L phosphate buffered saline, 0.5 ml Tween 20. This buffer is used for washing of the microplates.
5. Blocking Buffer: 5 g non-fat dried milk powder, PBST 100 ml.
6. Conjugate buffer: 0.2 g bovine serum albumin, 100 ml PBST. This buffer is used for dilution of the second antibody (conjugate).
7. Substrate buffer: 90.39 g diethanolamine, 19.82 g diethanolamine–HCl, 0.1 g magnesium chloride. Make up to 1 L with dH_2O. The pH of this buffer is 9.8 and it does not need to be adjusted. (The diethanolamine and diethanolamine–HCl are liquids; however, it is easier to weigh them than to measure their volumes, as they are extremely viscous.) *p*-Nitrophenylphosphate (pNPP) is added to this buffer at 1 mg/ml to make up the substrate for alkaline phosphatase.
8. Suspension of 24-h culture of a virulent strain of the target bacterium, here *Erwinia amylovora* (e.g. strain NCPPB 683) is the coating buffer with a cell density of approximately 10^8 cfu/ml (turbid suspension equivalent to an optical density of 0.1 at 600 nm) (*See* **Note 5**).

9. Microplates with 96 wells (*See* **Note 6**).
10. ELISA reader.
11. Eight-channel pipette and a pipette with adjustable volume of up to 200 μl, sterile pipette tips.
12. Filter paper.

7.2.2. DAS-ELISA (Clavibacter michiganensis subsp. sepedonicus – Agdia)

1. Antibody (commercially available monoclonal and polyclonal antibody) against *Clavibacter michiganensis* subsp. *sepedonicus* (*See* **Note 4**).
2. Coating buffer (PBS): 1.59 g sodium carbonate, 2.93 g sodium hydrogen carbonate, 0.2 g sodium azide. Make up to 1 L with dH_2O. The pH of this buffer should be 7.4. Store at 3–6°C. Buffer is used for dilution of a first antibody.
3. General extract buffer: Dissolve 1.25 g Tween 20 and 1.0 g non-fat dried milk in 250 ml PBST. Stir for 30 min. This buffer contains non-fat dried milk as a source of protein for blocking. Prepare only as much of this buffer as you will need for 1 day. The pH of this buffer should be 7.4. Store at 3–6°C. Buffer is used for dilution of samples and bacteria.
4. Wash buffer (PBST): 8.0 g sodium chloride, 1.15 g sodium phosphate, dibasic (anhydrous), 0.2 g potassium phosphate, monobasic (anhydrous), 0.2 g potassium chloride, 0.5 g Tween 20. The pH of this buffer should be 7.4. Buffer is used for washing the microplate.
5. Conjugate buffer: Dissolve 0.4 g non-fat dried milk in 100 ml PBST. Stir for 30 min. This buffer contains non-fat dried milk as a source of protein for blocking. Prepare only as much of this buffer as you will need for 1 day. Store at 3–6°C. This buffer is used for dilution of a second antibody (conjugate).
6. Substrate buffer: Dissolve 0.1 g magnesium chloride hexahydrate, 0.2 g sodium azide, 97.0 ml diethanolamine in 800 ml dH_2O. The pH of this buffer should be adjusted with hydrochloric acid to 9.8. Adjust final volume to 1 L with dH_2O. Store at 3–6°C. *p*-Nitrophenylphosphate (pNPP) is added to this buffer at 1 mg/ml to make up the substrate for alkaline phosphatase.
7. Suspension of 4- to 6-day culture of a virulent strain of *Clavibacter michiganensis* subsp. *sepedonicus* (e.g. strain NCPPB 3467) in general extract buffer with cell density of approximately 10^7 cfu/ml (turbid suspension equivalent to an optical density of 0.1 at 600 nm) (*See* **Note 5**).
8. 8–11. (*see* **Subheading 7.2.1.** 9–12)

7.3. Methods

7.3.1. PTA-ELISA

The method consists of the preparation and application of samples including the controls (*See* **Subheading 7.3.1.1.**) and the preparation and application of antibodies (*See* **Subheading 7.3.1.2.**).

7.3.1.1. Preparation and Application of Plant Samples Including the Positive and Negative Controls

1. Macerate the samples to be tested by grinding 0.5 g of tissue with 2 ml of coating buffer in a mortar using a pestle. In some cases the recommended amount of buffer may have to be increased if the extract is too viscous for pipetting. The extract is then used directly in the assay and/or diluted. Use a pure culture of a known *Erwinia amylovora* strain (e.g. NCPPB 683) as a positive control and a strain of other bacterium as negative control (*See* **Note 5**).
2. Pipette a measured standard volume of 100 µl of each test sample extract (diluted 0.1, 0.01) and of the positive and negative control sample solution (diluted 0.1, 0.01) to the wells of the microtitre plate. Each sample, including controls, should be tested in duplicate. Close the microplate with a cover, place it in a plastic bag and incubate the plate overnight at 3–6°C in a refrigerator.
3. Wash the plate three times with PBST by means of filling the wells of the plate with PBST using an 8-channel pipette (200 µl to each well) and invert to remove the buffer, pat the microplate dry with filter paper.
4. Subsequently add 200 µl of blocking buffer to each of the test wells. Wrap the plate as described in **step 2** and incubate at 37°C for 1 h. Wash the plate as described in **step 3**.

7.3.1.2. Preparation and Application of Antibodies

1. Dilute the probe antibody as recommended by the manufacturer for commercial antibody (for *Erwinia amylovora* 1:8,000) in PBS buffer and add 100 µl to each test well (*See* **Note 7**). Wrap the plate as described in **Subheading 7.3.1.1**, **step 2** and incubate at 37°C for 2 h.
2. Wash the plate as described in **Subheading 7.3.1.1**, **step 3**.
3. Dilute the antibody–enzyme conjugate as recommended by the manufacturer on the bottle label (for *Erwinia amylovora* 1:4,000) in conjugate buffer and add 100 µl to each test well. Wrap the plate as described in **Subheading 7.3.1.1**, **step 2** and incubate at 37°C for 1 h.
4. Wash the plate four times as described in **Subheading 7.3.1.1**, **step 3**. An extra wash stage is included at this step to ensure that all the unbound antibody–enzyme conjugate is removed from the wells.
5. Prepare the substrate just before use – add pNPP at 1 mg/ml to the substrate buffer (one 5 mg tablet in 5 ml of buffer).

Alternatively, use one of the NeogenEurope Ltd. liquid substrates. All of these substrates may change colour when exposed to light and should be protected from light to prevent this from occurring. Subsequently add 100 μl of prepared substrate to each test well. Wrap the plate as described in **Subheading 7.3.1.1**, **step 2** and incubate in the dark at room temperature for 1 h.

7.3.2. DAS-ELISA

The method includes preparation and application of the first antibody (*See* **Subheading 7.3.2.1.**), preparation and application of plant samples inclu-ding controls (*See* **Subheading 7.3.2.2.**) and preparation and application of the second antibody (conjugate) (*See* **Subheading 7.3.2.3.**).

7.3.2.1. Preparation and Application of the First Antibody

1. Dilute the captured antibody as recommended by the manufacturer for commercial antibody (for *Clavibacter michiganensis* subsp. *sepedonicus* 1:250) in the coating buffer. Mix the prepared antibody thoroughly and add immediately 100 μl of diluted antibody to each test well. Estimate the volume needed as 10 ml coating buffer for each a full microplate used. Wrap the plate as described above and incubate at room temperature for 4 h or overnight in the refrigerator (3–6°C).
2. Wash the plate. Empty the wells into a sink or waste container. Wash the plate three more times with PBST by each time filling the wells of the plate with PBST using an 8-channel pipette (200 μl to each well) and invert to remove the buffer, pat the plate dry on filter paper.

7.3.2.2. Preparation and Application of Plant Samples Including the Positive and Negative Control

1. Grind and dilute samples. Tuber extracts of potato are prepared according to EU Directive No. 2006/56/EC. Concerning other plants, extract the samples to be tested by grinding the tissue (stem, sprout, seed or leaves preferably with symptoms) in sample buffer using a mortar and a pestle that should be washed and rinsed thoroughly between samples. The extract is then used directly and diluted in the assay. Use pure cultures of positive (*Clavibacter michiganensis* subsp. *sepedonicus*, e.g. strain NCPPB 3467) and a negative control strain (another bacterium) (*See* **Note 5**).
2. Pipette a measured standard volume of 100 μl of each sample extract (diluted 0.5, 0.1 and 0.01), positive and negative control solutions (diluted 0.1 and 0.01) to the wells of the microtitre plate. Each sample including control should be done in duplicate. Close a microplate with cover, place in a plastic bag and incubate it for 2 h at room temperature or overnight at 3–6°C in a refrigerator.
3. Wash the plate five times with PBST by each time filling the wells of the plate with PBST using an 8-channel pipette (200 μl to each well) and invert to remove the buffer, pat the plate dry with filter paper.

7.3.2.3. Preparation and Application of the Second Antibody (Conjugate)

1. A few minutes before the incubation is complete, prepare the enzyme conjugate. Dilute the second antibody as recommended by the manufacturer on the bottle label (for *Clavibacter michiganensis* subsp. *sepedonicus* 1:250) in the conjugate dilution buffer. Mix the prepared antibody thoroughly and add immediately 100 µl of diluted conjugate to each test well. Wrap the plate as described in **Subheading 7.3.2.2, step 2** and incubate at room temperature for 2 h.
2. Wash the plate as described in **Subheading 7.3.2.2, step 3**.
3. Prepare the PNP solution just before use – add pNPP at 1 mg/ml to the substrate buffer (one 5 mg tablet in 5 ml of buffer). Do not touch the PNP tablets. Substrate should be protected from light to prevent spontaneous oxidation showing a change of colour. Light or contamination could cause background colour in negative wells. Subsequently add 100 µl of prepared substrate to each test well. Wrap the plate as described in **Subheading 7.3.2.2, step 2** and incubate in the dark at room temperature for 30–60 min.

7.3.3. Evaluating of Results and Reading of Microplates in the ELISA Reader

1. Evaluate results visually. Results are interpreted usually after 30–60 min of incubation, but they may be interpreted after more than 60 min of incubation as long as negative wells remain virtually clear. Adding 50 µl of 3 M sodium hydroxide to each well might stop the reaction. Sodium hydroxide stops colour development, but the microplate can be interpreted visually or with a plate reader without this step. Wells in which colour develops indicate positive results. Wells in which there is no significant colour development indicate negative results. Test results are valid only if positive control wells give a positive result and buffer wells remain colourless.
2. Read the absorbance levels using a spectrophotometer or specially designed ELISA reader at 405 nm, e.g. the ELISA MRX microplate reader (Dynex Technologies) using its Endpoint Program). The ELISA test is negative if the average optical density (OD) reading from duplicate sample wells is <2× OD of that in the negative sample control wells (provided the OD for the positive controls are above 1.0 after 60-min incubation and are greater than twice the OD obtained for negative samples). The ELISA test is positive if the average OD reading from the duplicate sample wells is >2× OD in the negative sample wells provided that 2× average OD readings in all negative control wells are lower than those in the positive control wells *(24)*.
3. Possible problems could appear related with conducting ELISA test and evaluating ELISA microplates (*See* **Note 8**).

7.4. Notes

1. A reliable, sensitive, low-cost and easy-to-use technique is described for the detection of *Ralstonia solanacearum* (the causal organism of bacterial wilt, BW) in soil. A total of 273 potato isolates belonging to five different biovars (Bv), originating from 33 countries worldwide, were tested and successfully detected by antibodies produced at the International Potato Center (CIP). Isolates of *R. solanacearum* belonging to Bv1 and Bv2A were successfully detected by double antibody sandwich-enzyme-linked immunosorbent assay (DAS-ELISA) at low population levels after incubation of soil suspensions for 48 h at 30°C in a new semiselective broth containing a potato tuber infusion. Detection thresholds of 20 and 200 cfu/g inoculated soil were obtained for Bv1 and Bv2A, respectively. Sensitivity of detection of Bv2A was similar or even higher in five different inoculated soil types. No cross-reactions were obtained in DAS-ELISA after enrichment of soil suspensions (a) prepared from 23 different soils sampled in BW-free areas in six departments of Peru and (b) inoculated with 10 identified bacteria and 136 unknown isolates of soil microbiota isolated from eight different locations. Only the blood disease bacterium gave a low-level reaction after enrichment. In naturally infested soils, average sensitivities of 97.6 (SE 14.8) and 100.9 (SE 22.6) cfu/g were obtained for biovars 1 and 2A, respectively. By making serial dilutions of the soil suspension before enrichment, densities of *R. solanacearum* could be determined in a semi-quantitative way. Results also showed that composite samples of five soils could be analysed to assess field soil populations without reducing detection sensitivity *(23)*.
2. A reliable and sensitive technique, double antibody sandwich indirect (DASI-ELISA), is also described for the detection of *Erwinia amylovora* from symptom-problematic or symptomless plant samples. Monoclonal antibodies for *Erwinia amylovora* are used in DASI-ELISA with native or boiled antigens. A mixture of monoclonal antibodies has been validated in a ring test. The technique was further proposed in association with enrichment on suitable media for a specific and sensitive detection of *E. amylovora (24)*.
3. All available antibodies for use in ELISA have been produced against fluidal strains of *Clavibacter michiganensis* subsp. *sepedonicus*. Extracellular polysaccharides (EPS) situated in surface layers of cells contain antigen determinants utilized

in ELISA *(5)*. That is why probably rough and intermediate strains of ring rot pathogen, having less or no EPS, are less reliably identified in ELISA than fluidal strains *(21, 25–27)*.

4. Validated or commercially available source of antibodies should be used for *Erwinia amylovora*, *Clavibacter michiganensis* subsp. *sepedonicus* or other plant pathogenic bacteria. The antibodies and conjugates should be stored at 4–6°C in a refrigerator upon receiving. Diagnostic products manufacturers (NeogenEurope Ltd., UK; Loewe Biochemica GmbH, Germany; Agdia Intercorporated, USA) and others) supply reagent sets for ELISA for different plant pathogenic bacteria, which include probe antibody and anti-species conjugate, but also offer buffer solutions, pNPP tablets and liquid substrates. Additionally, positive and negative controls are available from these producers.
5. Reference bacterial strains are recommended for use as positive and negative controls in all tests to ensure that all of the test components are working properly.
6. Microtitre plates (Nunc-Polysorp plates or those of the same quality) should be used immediately after freshly coated with antibodies. Capture antibody should be prepared in a container made from polyethylene or similar material that does not readily bind antibodies. Enzyme-conjugate should always be prepared not longer than within 10 min before use. The PNP tablets should not be touched and the PNP solution should not to be exposed strong light, because light and contamination could cause colour change in negative wells.
7. The titre is usually determined for each new batch of antibodies. The titre is the antiserum dilution at which an optimal reaction occurs when tested with the homologous reference strain of the target bacterium used in a concentration of ca 10^6 cfu/ml and using an appropriate labelled-enzyme conjugate according to the manufacturer's recommendation.

Preparation of positive and negative controls:

Produce a 1-day culture of a virulent strain of *Erwinia amylovora* (e.g. strain NCPPB 683) on one of the following media:

King's B medium: 20.0 g proteose peptone N° 3, 10.0 ml glycerol, 1.5 g K_2HPO_4, 1.5 g $MgSO_4 \cdot 7H_2O$, 15.0 g agar, distilled water to 1 L. Adjust pH to 7.0–7.2. Sterilize by autoclaving at 120°C for 15 min *(24, 28)*.

Levan medium: 2.0 g yeast extract, 5.0 g Bacto peptone, 5.0 g NaCl, 50.0 g sucrose, 20.0 g agar, distilled water to 1 L. Adjust pH to 7.0–7.2. Sterilize by autoclaving at 120°C for 15 min *(24)*.

Produce a 4- to 6-day culture of virulent strain of *Clavibacter michiganensis* subsp. *sepedonicus* (e.g. strain NCPPB 3467) on one of the following media:

NCP-88 medium: 23.0 g nutrient agar (Difco), 2.0 g yeast extract (Difco), 5.0 g d-mannitol, 2.0 g K_2HPO_4, 0.5 g KH_2PO_4, 0.25 g $MgSO_4{\cdot}7H_2O$, distilled water to 1.0 L. Dissolve ingredients and adjust pH to 7.2. After autoclaving and cooling down to 50°C add the following antibiotics: 0.003 g polymyxin B sulphate (Sigma), 0.008 g nalidic acid (Sigma), 0.2 g cycloheximide (Sigma). Dissolve antibiotics in stock solutions as follows: nalidic acid in 0.01 M NaOH, cycloheximide in 50% ethanol, polymyxin B sulphate in distilled water. Stock solutions are filter-sterilized. Durability of basal medium is 3 months. After antibiotics are added, durability is 1 month when stored refrigerated.

NBY medium: 23.0 g nutrient agar, 2.0 g yeast extract, 0.5 g KH_2PO_4, 2.0 g K_2HPO_4, 0.25 g $MgSO_4{\cdot}7H_2O$, 5.0 g d-mannitol, distilled water to 1.0 L. Sterilize by autoclaving at 115°C for 20 min *(29, 30)*.

Reference strains are commercially available in plant pathogenic bacteria collections such as National Collection of Plant Pathogenic Bacteria (NCPPB), Central Science Laboratory, York, GB and Culture Collection of the Plant Protection Service (PD), Wageningen, the Netherlands, Collection Française de Bactéries Phytopathogènes (CFBP), INRA Station Phytobactériologie, Angers, France and others.

8. Possible troubles during testing (according to NeogenEurope Ltd.)
   - *A uniform high level of colour appears in all wells.* The microplate may not have been properly washed. The substrate solution may have been contaminated by touching pNPP tablets or exposed to bright light. One of the buffers may have been contaminated by a conjugate.
   - *All wells are coloured but to different degrees.* The microplate may not have been properly washed and/or has been exposed to bright light.
   - *Colour appears in all of the outer wells of the plate.* The outer wells may not have been properly washed, giving rise to so-called 'edge effects'. Use high-quality microplates to avoid this problem.
   - *Colour appears in some negative control well.* A positive sample or positive control may have been added or has been accidentally washed into the negative control well. Contamination of the negative control well may have occurred if the same pipette tip was used. New tips must be used for each sample.
   - *All of the wells, including the positive control, are clear.* The antibody–enzyme conjugate may not have been added to

the conjugate buffer. The positive control may have gone off. The pNPP substrate tablet may not have been added to the substrate buffer.

- *The colour in the positive control wells in very low.* One of the buffers that have been used may be too old. The positive control may be starting to go off. The substrate buffer may have been diluted with PBST and not water as recommended. The phosphate in PBST reduces the amount of colour produced.

References

1. Clark, M. F., and Adams, A. N. (1977) Characteristics of the microplate method of enzyme-linked immuno sorbent assay for the detection of plant viruses. *J Gen Virol* 34, 475–483.
2. Paulin, J. P. (2000) *Erwinia amylovora*: general characteristics, biochemistry and serology, in *Fireblight, the Disease and Its Causative Agent, Erwinia amylovora* (Vanneste, J.L., ed.), CABI Wallingford, UK, pp. 87–115.
3. Ward, E., Foster, S. J., Fraaije, B. A., and McCartney, H. A. (2004) Plant pathogen diagnostics: immunochemical and nucleic-based approaches. *Ann Appl Biol* 145, 1–16.
4. De Boer, S. H., and Wieczorek, A. (1984) Production of monoclonal antibodies to *Corynebacterim sepedonicum*. *Phytopathology* 74, 1431–1434.
5. De Boer, S. H., Wieczorek, A., and Kummer, A. (1988) An ELISA test for bacterial ring rot of potato with a new monoclonal antibody. *Plant Dis* 72, 874–878.
6. Pánková, I., and Kokošková, B. (2002) Sensitivity and specificity of monoclonal antibody Mn-Cs1 for detection and determination of *Clavibacter michiganensis* subsp. *sepedonicus*, the causal agent of bacterial ring rot of potato. *Plant Protect Sci* 38, 17–124.
7. Westra, A. A. G., Slack, S. A., and Drennan, J. L. (1994) Comparison of some diagnostic assays for bacterial ring rot of potato: a case study. *Am Potato J* 71, 557–565.
8. De Boer, S. H. (1982) Cross-reaction of *Corynebacterium sepedonicum* antisera with *C. insidiosum*, *C. michiganense*, and an unidentified *Corynebacterium* bacterium. *Phytopathology* 72, 1474–1478.
9. Kokošková, B., and Mráz, I. (2005) Reliability of diagnostic techniques for *Erwinia amylovora*, the causative agent of fire blight disease. *Folia Microbiol* 50 (3), 217–221.
10. Miller, H. J. (1984) Cross-reactions of *Corynebacterium sepedonicum* antisera with soil bacteria associated with potato tubers. *Neth J Plant Pathol* 90, 23–28.
11. Slack, S. A., Drennan, J. L., Westra, A. A. G., Gudmestad, N. C., and Oleson, A. E. (1996) Comparison of PCR, ELISA, and DNA hybridization for the detection of *Clavibacter michiganensis* subsp. *sepedonicus* in field-grown potatoes. *Plant Dis* 80, 519–524.
12. Sobiczewski, P., Deckers, T., and Pulawska, J. (1997) *Fire blight (Erwinia amylovora). Some aspects of epidemiology and control.* Research Institute of Pomology and Floriculture, Skierniewice, Poland.
13. Zielke, R., Schmidt, A., and Naumann, K. (1993) Comparison of different serological methods for the detection of the fire blight pathogen, *Erwinia amylovora* (Burrill) Winslow et al. *Zentralbl Mikrobiol* 148, 379–391.
14. Lin, C. P., Chen, T. A., Wells, J. M., and van der Zwet, T. (1987) Identification and detection of *Erwinia amylovora* with monoclonal antibodies. *Phytopathology* 77, 376–380.
15. Schaad, N. W., Süle, S., van Vuurde, J. W. L., Vruggink, H., Alvarez, A. M., Benedict, A. A., de Wael, L., and van Laere, O. (1990) Serology, in *Methods in Phytobacteriology* (Klement, F., Rudolf, K., Sands, D.C., eds.), Akademiai Kiado, Budapest, pp. 153–190.
16. Gorris, M. T., Camarasa, E., Lopez, M. M., Cambra, M., Paulin, J. P., and Chartier, R. (1996) Production and characterization of monoclonal antibodies specific for *Erwinia amylovora* and their use in different serological techniques. *Acta Hortic* 411, 47–52.
17. Gorris, M. T., Cambra, M., Llop, P., López, M. M., Lecomte, P., Chartier, R., and Paulin, J. P. (1996) A sensitive and specific detection of *Erwinia amylovora* based on the ELISA-

DASI enrichment method with monoclonal antibodies. *Acta Hortic* 411, 41–45.

18. McLaughlin, R. J., Chen, T. A., and Wells, J. M. (1989) Monoclonal antibodies against *Erwinia amylovora*: characterization and evaluation of a mixture for detection by Enzyme-linked Immunosorbent Assay. *Phytopathology* 79, 610–613.
19. De Boer, S. H., Boucher, A., and De Haan, T. L. (1996) Validation of thresholds for serological tests that detect *Clavibacter michiganensis* subsp. *sepedonicus in potato tuber tissue*. *Bull OEPP/EPPO Bull* 26, 391–398.
20. Gudmestad, N. C., Baer, D., and Kurowski, C. J. (1991) Validating immunoassay test performance in the detection of *Corynebacterium sepedonicum* during the growing season. *Phytopathology* 81, 475–480.
21. Kokošková, B., Mráz, I., Janse, J. D., Fousek, J., and Jeřabková, R. (2005) Reliability of diagnostic techniques for determination of *Clavibacter michiganensis* subsp. *sepedonicus*. *Pfl Krankh* 112 (1), 1–16.
22. Lopez, M. M., Bertolini, E., Olmos, A., Caruso, P., Gorris, M. T., Llop, P., Penyalver, R., and Cambra, M. (2003) Inovative tools for detection of plant pathogenic viruses and bacteria. *Int. Microbiol* 6, 233–243.
23. Priou, S., Gutarra, L., and Aley, P. (2006) An improved enrichment broth for the sensitive detection of *Ralstonia solanacearum* (biovars 1 and 2A) in soil using DAS-ELISA. *Plant Pathol* 55, 1–36.
24. Oepp/Eppo (2004) Diagnostic protocols for regulated pests 7/20. *Erwinia amylovora*. *Bull OEPP/Eppo Bull* 34, 159–171.
25. Bishop, A. L., Clarke R. G., and Slack S. A. (1988) Antigenic anomaly in naturally occurring nonfluidal strain of *Corynebacterium sepedonicum*. *Am Potato J* 65, 237–244.
26. Baer, D., and Gudmestad, N. C. (1993) Serological detection of nonmucoid strains of *Clavibacter michiganensis* subsp. *sepedonicus* in potato. *Phytopathology* 83, 157–163.
27. Henningson, P. J., and Gudmestad, N. C. (1993) Comparison of exopolysaccharides from mucoid and nonmucoid strains of *Clavibacter michiganensis* subsp. *sepedonicus*. *Can J Microbiol* 39, 291–296.
28. Oepp/Eppo (1992) Quarantite procedures. 40. *Erwinia amylovora* – sampling and test methods. *Bull OEPP/EPPO Bull* 22, 225–232.
29. Anon. (2006) Commission Directive 2006/56/EC of 12 June 2006 amending the Annexes to Council Directive 93/85/EC on the control of potato ring rot. *Official J Eur Comm* L182, 1–43.
30. Oepp/Eppo (2006) Diagnostic protocols for regulated pests 7/59. *Clavibacter michiganensis* subsp. *sepedonicus*. *Bull OEPP/EPPO Bull* 36, 99–109.

Chapter 8

Indirect Immunofluorescence Microscopy for the Detection and Identification of Plant Pathogenic Bacteria (In Particular for *Ralstonia solanacearum*)

Jaap D. Janse and Blanka Kokoskova

Summary

Immunofluorescence microscopy is a very sensitive serological test which harnesses both the power of antibodies to bind to targets along with the use of the fluorescence microscope to visualise the structures to which they bind. Antibody binding is visualised by the fluorescent emission from a marker molecule bound to the antibody. The technique is of particular use in bacteriology and pathology where the location and morphology of the bacterial cells can be viewed along with the location of the fluorescently labelled antibodies.

Key words: Antibody, Fluorochrome, Direct IF, FITC, Indirect IF, Polyclonal antiserum.

8.1. Introduction

Immunofluorescence (IF) microscopy was developed by J.S. Ploem in the late 1960s for medical purposes *(1)*. Its principles as well as those of fluorescence microscopy are explained in Herman *(2)* and have been used in plant pathology/bacteriology since the late 1970s *(3)*. It proved to be a very sensitive and specific serological test (detection level of c. 10^3–10^4 cells/ml of plant extract) because the primary reaction of antigen and antibody is made visible. Binding reactions can be observed at very high titres (titre = highest dilution of the crude antiserum where a clear reaction is still visible) of antiserum. Many polyclonal antisera produced against unwashed, whole bacterial cells also contain antibod-

R. Burns (ed.), *Methods in Molecular Biology, Plant Pathology, vol. 508*

DOI: 10.1007/978-1-59745-062-1_8

ies against flagella and extracellular polysaccharide and these can therefore also be made visible under the microscope. Immunological differences in cell wall and flagellar antigens as revealed in IF-microscopy can be used in taxonomy *(4–6)*.

The method has been specifically developed for *Ralstonia solanacearum* in the laboratory of the first author *(7)* and was further refined in an SMT EU project (SMT-4-CT97-2719) *(8)* and included in the EU Control Directive for *Ralstonia solanacearum* *(9, 10)*. It also proved to be a robust and relatively cheap detection test and has been used in our and other EU statutory and private laboratories over the years for literally hundreds of thousands of samples of different (quarantine) organisms with excellent results. Immunofluorescence staining of *R. solanacearum* cells has also been used to label whole colonies on agar plates *(11)*.

In the IF test antibodies are marked (labelled) with a chemical fluorescent dye. For IF a light microscope fitted for epifluorescent light is necessary with the suitable excitation and barrier filters for the selected fluorescent dye. The dyes used in immunofluorescence are excited by light of one wavelength, usually blue or green, (produced by a high pressure mercury or xenon lamp) and emit light of a different (longer) wavelength in the visible spectrum. In plant bacteriology most commonly used fluorescent dyes are fluorescein isothiocyanate (FITC), which emits green/yellow light, rhodamine and Cy3 both of which emit orange/red light. By selective filters in the fluorescence microscope, only the light emitted from the dye or fluorochrome is detected in the fluorescence microscope. When different dyes are used to different antibodies, the distribution of two or more antibodies can be determined in the same cell or two or more different bacterial species can be detected in plant tissue extract. Monoclonal antibodies can also be applied, although fluorescence of bacterial cells is usually somewhat weaker, due to the fact that only one type of antigen (epitope) in the cell wall will react *(12, 13)*.

In so-called *direct IF* antiserum against a certain plant pathogenic bacterium is already labelled with the fluorescent dye. In indirect IF the bacteria are first treated with a pathogen-specific rabbit or mouse antiserum (against the target bacterium, usually produced in a rabbit, goat, or in the case of monoclonal antibodies, in a mouse). After incubation and washing, a second, labelled anti-rabbit, anti goat or anti-mouse serum, prepared in another animal genus than the one used to produce the pathogen-specific antibodies, is applied. This anti-rabbit, anti-goat or anti-mouse serum is labelled with the fluorescent dye (usually FITC) and is called the *conjugate*. Only the antibodies bound to the bacteria will fluoresce, while the others are removed by washing. Indirect IF is usually slightly more sensitive and less specific than direct IF.

The method described in this chapter and developed by Janse *(7)* is by and large, with some minor modifications, presently included in an updated form as Annex II in the EU Commission Directive 2006/63/CE of 14 July 2006 amending Annexes II to VII to Council Directive 98/57/EC on the control of *Ralstonia solanacearum* (Smith) Yabuuchi et al. *(10)*. It is specifically designed for *Ralstonia solanacearum*, but can equally be used for *Clavibacter michiganensis* subsp. *sepedonicus*, *Erwinia amylovora* and actually for any other bacterial pathogen by just changing the specific antiserum and determining the titre of the antiserum used *(14, 15)*.

8.2. Materials

1. Antiserum (validated) against *Ralstonia solanacearum* or other plant pathogenic bacterium appropriate (FITC labelled) conjugate, directed against the animal used to produce the pathogen specific antiserum (*see* **Note 1** and **2**).
2. Pellet buffer: 10 mM phosphate buffer, 2.7 g $Na_2HPO_4 \cdot 12H_2O$, 0.4 g $NaH_2PO_4 \cdot 2H_2O$, 1.0 L distilled water. The pH of this buffer should be 7.2. Dissolve ingredients, check pH and sterilise by autoclaving at 121°C for 15 min. Used for resuspension and dilution of potato tuber heel-end core extracts (or other plant extracts) following concentration to a pellet by centrifugation or resuspension of bacterial cells from pure culture (*see* **Note 3**).
3. IF-buffer: 10 mM phosphate buffered saline (PBS), 2.7 g $Na_2HPO_4 \cdot 12H_2O$, 0.4 g $NaH_2PO_4 \cdot 2H_2O$, 8.0 g NaCl, 1.0 L distilled water. The pH of this buffer should be 7.2. Dissolve ingredients, check pH and sterilise by autoclaving at 121°C for 15 min. Used for dilution of antibodies and as wash buffer.
4. IF-buffer-Tween: IF buffer with 0.1% Tween 20. Used as wash buffer.
5. Phosphate buffered glycerol: 3.2 g $Na_2HPO_4 \cdot 12H_2O$, 0.15 g $NaH_2PO_4 \cdot 2H_2O$, 50 ml (autofluorescence-free) glycerol, and 100 ml distilled water. The pH of this buffer should be 7.6. Mountant fluid on IF slide windows to enhance fluorescence.
6. Anti-fading mountant, e.g. Vectashield® (Vector Laboratories, Burlingame, CA, USA) or Citifluor® (Leica, Wetzlar, Germany).
7. Suspension of a 48-h culture of a virulent strain of *R. solanacearum* race3/biovar2 (e.g. strain NCPPB 4156 = PD 2762 = CFBP 3857) in 10 mM phosphate with a cell density of approximately 2 × 10^8 cfu/ml (faintly turbid suspension

equivalent to an optical density of 0.15 at 600 nm). Other standardised positive control material and negative control materials (*see* **Note 6**).

8. Multiwell (teflon coated) microscope slides with preferably ten windows of at least 6 mm diameter.
9. Fluorescence microscope fitted for epi-fluorescent light with 50 or 100 W mercury high pressure lamp and excitation and barrier filters for FITC and rhodamine/CY-3 and 63× (or 40×, only if the microscope tube has a magnification factor of 1.2) oil or water immersion plan(apo)chromat or fluorite objective (for screening of slides) and 100× oil or water immersion plan(apo)chromat or fluorite objective (for detailed observation of fluorescent cells).
10. Immersion oil (PCB free).
11. Filter paper.
12. Opaque box to contain wet filter paper and IF slides when incubated (wet chamber).
13. Washing trays and holders for microscope slides.

8.3. Methods

The methods described below outline (a) the application of plant extract to test slides, (b) drying and fixation of extract, (c) indirect immunofluorescence (IIF) procedure (incubation with antiserum and conjugate), (d) reading the IIF test, (e) interpretation of the IF reading, and (f) determination of the contamination level.

8.3.1. Preparation of Test Slides

For bacterial cells of pure cultures cell suspensions indicated under **Subheading 8.2, item 7** can be used, following procedure in **Subheading 8.3.1.1**. For plant extract pellets (*see* **Note 4**) that contain low starch or plant debris amounts procedure in **Subheading 8.3.1.1** is also (preferably) followed (here the 1:100 dilution of plant extract is only used to detect false negative reactions in case there are excess antigens, i.e. bacterial cells, present). Plant extracts containing a lot of starch or debris in the sediment should be diluted and procedure in **Subheading 8.3.1.2** be followed. Procedures in **Subheadings 8.3.1.1** and **8.3.1.2** should also be followed for control slides containing naturally infected or spiked plant extract (for preparation of control material *see* **Note 5**).

8.3.1.1. For Pellets of Plant Extract with Relatively Little Starch/Debris Sediment

1. Pipette a measured standard volume (15 μl is appropriate for 6-mm window diameter – scale up volume for larger windows) of a 1/100 dilution of the resuspended plant extract pellet onto the first window of the IF slide.

2. Subsequently pipette a similar volume of undiluted pellet (1/1) onto the remaining windows on the row. The second row can be used as duplicate or for a second sample as presented in **Fig. 8.1a**.

8.3.1.2. For Pellets Rich in Starch or Other Plant Debris Sediment

1. Prepare two decimal dilutions (1/10, 1/100) of the resuspended pellet in pellet buffer. Pipette a measured standard volume (15 µl is appropriate for 6-mm window diameter – scale up volume for larger windows) of the resuspended pellet and each dilution on a row of windows.
2. The second row can be used as duplicate or for a second sample as presented in **Fig. 8.1b**.

8.3.2. Drying and Fixation of Test Slides

1. Dry the droplets at ambient temperature or by warming to temperatures of 40–45°C.
2. Fix the bacterial cells to the slide either by heating (15 min at 50–60°C), flaming (by passing the slides three times slowly through the top of a flame), with 95% ethanol (*see* **Note 6**) or according to specific instructions from the suppliers of the antibodies (*see* **Notes 1**, **2**, **6**, and **9**).

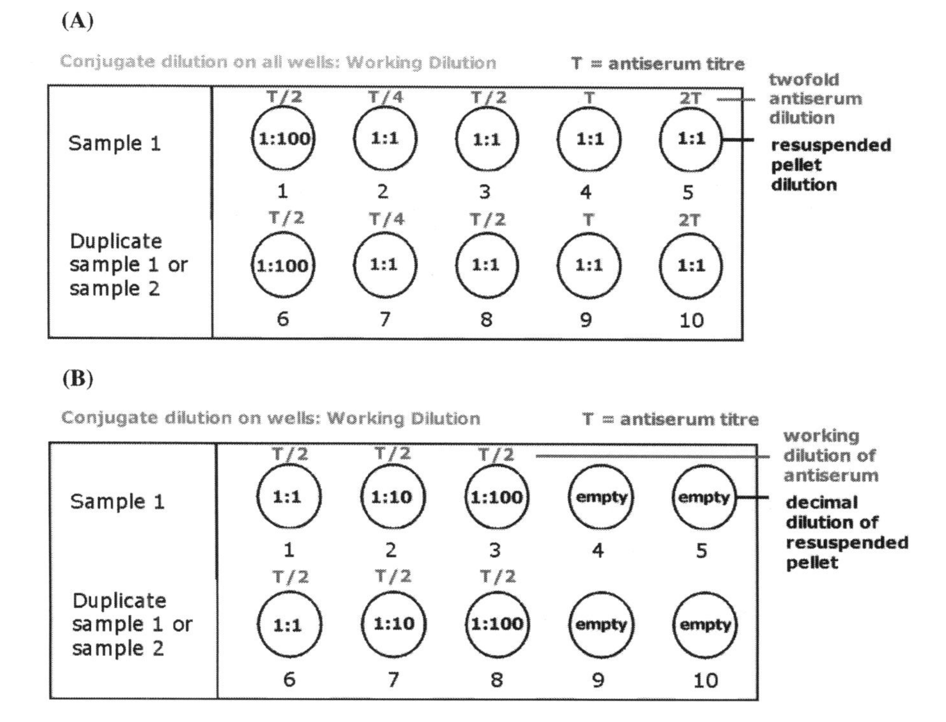

Fig. 8.1. (A) Preparation of IF test slide according to **Subheading 8.3.1.1**. (B) Preparation of IF test slide according to **Subheading 8.3.1.2**.

8.3.3. Indirect ImmunoFluorescence Procedure

The procedure describes the preparation of antiserum and conjugate dilutions when it is not prescribed by the antiserum (*see* **Note 1**) and/or conjugate manufacturer (*see* **Subheading 8.3.3.1**), the incubation and washing procedure (*see* **Subheading 8.3.3.2**) in a moist chamber at ambient temperature and the application of glycerol buffer or anti-fading chemical and coverslips on incubated slides.

8.3.3.1. Preparation of Antiserum Dilutions

Preparation of antiserum dilutions is different for the plant extract preparation procedures mentioned in **Subheadings 8.3.1.1** and **8.3.1.2** as follows (*see* **Notes 2** and 7).

1. When following test slide preparation in **Subheading 8.3.1.1**: Prepare a set of twofold dilutions. The first well should have 1/2 of the titre (T/2) = working dilution, (WD), the others should have 1/4 of the titre (T/4), 1/2 of the titre (T/2), the titre (T) and twice the titre (2T).
2. When following test slide preparation in **Subheading 8.3.1.2**: Prepare the working dilution (WD) of the antibody in IF buffer. The working dilution (1/2 of the titre – T/2) affects the specificity.
3. Prepare conjugate dilution (dilution used to determine the titre of the antiserum or the one according to the instructions of the commercial supplier, usually 1:100).

8.3.3.2. Incubation with Antiserum and Conjugate, Washing Procedure and Preparation for Microscopic Observation

The following procedure should be carried out in the absence of specific instructions from the suppliers of the antibodies, otherwise follow those recommendations:

1. Place the slides on moist tissue paper. Cover each test window completely with the antibody dilution(s). The volume of antibody applied on each window must be at least the volume of the extract applied.
2. Incubate the slides on moist paper under a cover (so-called moist chamber) for 30 (± 10) min at ambient temperature (18–25°C).
3. Shake the droplets off each slide and rinse carefully with IF buffer. Wash by submerging for 5 min in IF buffer-Tween and subsequently in IF buffer. Avoid causing aerosols or droplet transfer that could result in cross-contamination. Carefully remove excess moisture by blotting gently.
4. Arrange the slides on moist paper. Cover the test windows with the dilution of FITC conjugate used to determine the titre. The volume of conjugate applied on the windows must be identical to the volume of antibody applied.
5. Incubate the slides on moist paper under a cover (so-called moist chamber) for 30 (± 10) min at ambient temperature (18–25°C).

6. Shake the droplets of conjugate off the slide. Rinse and wash as before (as mentioned under **Subheading 8.3**). Carefully remove excess moisture.
7. Pipette 5–10 μl of 0.1 M phosphate-buffered glycerol or a commercially antifading mountant on each window and apply a coverslip, avoiding air bubbles (*see* **Note 8**).

8.3.4. Reading the IF Test

1. Examine test slides under an epifluorescence microscope with filters suitable for excitation of FITC, under oil or water immersion at a magnification of 500–1,000. Scan windows across two diameters at right angles and around the perimeter. For samples showing no or low number of cells observe at least 40 microscope fields. Check the positive control slide first. Cells must be bright fluorescent and completely stained at the determined antibody titre or working dilution. The IF test (*see* **Subheading 8.3.3**) must be repeated if the staining is aberrant.
2. Observe for bright fluorescing cells with the characteristic morphology of *R. solanacearum* (or other pathogenic bacterium tested) in the test windows of the test slides (for examples see website http://forum.europa.eu.int/Public/irc/sanco/Home/main) and **Fig. 8.2**. The fluorescence intensity must be equivalent to the positive control strain at the same antibody dilution. Cells with incomplete staining or with weak fluorescence must be disregarded. If any contamination is suspected the test must be repeated. This may be the case when all slides in a batch show positive cells due to the contamination of buffer or if positive cells are found (outside of the slide windows) on the slide coating.
3. There are several problems inherent to the specificity of the immunofluorescence test. Background populations of

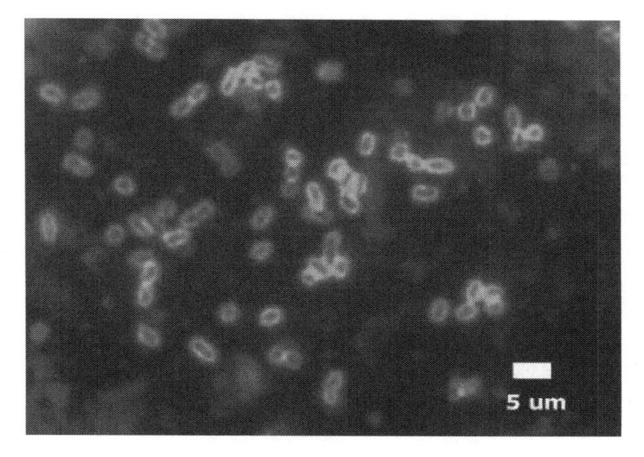

Fig. 8.2. Typical cell morphology of *Ralstonia solanacearum* cells in naturally infected potato tissue extract at a dilution of 1:3,200 of a polyclonal antiserum and using a ×100 planapochromatic oil immersion objective.

fluorescing cells with atypical morphology and cross reacting saprophytic bacteria with size and morphology similar to *R. solanacearum* (or to other pathogen) are likely to occur in plant extract pellets. Consider only fluorescing cells with typical size and morphology at the titre or working dilution of the antibodies as in **Subheading 8.3.3.1**.

8.3.5. Interpretation of the IF Reading

1. If bright fluorescing cells with characteristic morphology are found, estimate the average number of typical cells per microscope field and calculate the number of typical cells per ml of resuspended pellet (*see* **Subheading 8.3.6**). The IF reading is positive for samples with at least 5×10^3 typical cells per ml of resuspended pellet. The sample is considered potentially contaminated and further testing is required.
2. The IF reading is negative for samples with less than 5×10^3 cells/ml of resuspended pellet and the sample is considered negative. Further testing is not required.

8.3.6. Determination of Contamination Level in IF

1. Count the mean number of typical fluorescent cells per field of view (c)
2. Calculate the number of typical fluorescent cells per microscope slide window (C)
3. $C = c \times S/s$ where S = surface area of window of multi well slide and s = surface area of objective field as seen through the ocular
4. Surface area of the window of a multi well slide can be calculated with the formula: $S = \pi D^2/4$ (D = diameter of one well)
5. Surface area of the objective field as seen through the ocular can be calculated with the formula $s = \pi i^2/4G^2K^2$, where i = field coefficient (varies from 8 to 24 depending upon ocular type); K = tube coefficient (1 or 1.25); G = magnification of objective (100×, 40×, etc.)
6. Calculate the number of typical fluorescent cells per ml of re-suspended pellet (N)
7. $N = C \times 1{,}000/y \times F$, where y = volume of re-suspended pellet on each window and F = dilution factor of re-suspended pellet

8.4. Notes

1. Use a validated source of antibodies to *R. solanacearum* (see web site http://forum.europa.eu.int/Public/irc/sanco/Home/main).

2. It is recommended that the titre is determined for each new batch of antibodies. The titre is defined as the highest dilution at which optimum reaction occurs when testing a suspension containing 10^5–10^6 cells/ml of the homologous strain of *R. solanacearum* and using an appropriate fluorescein isothiocyanate (FITC) conjugate according to the manufacturer's recommendations. Validated polyclonal antisera for *R. solanacearum* all had an IF titre of at least 1:2,000. During testing, the antibodies should be used at a working dilution(s) close to or at the titre.
3. Exact composition of extract (pellet) buffers for pathogens other than *Ralstonia solanacearum* should be checked or determined based on literature or own experience.
4. The test should be performed on freshly prepared sample extracts. If necessary, it can be successfully performed on extracts stored at –68 to –86°C under glycerol. Glycerol can be removed from the sample by addition of 1 ml pellet buffer, re-centrifugation for 15 min at 7,000 × *g* and re-suspension in an equal volume of pellet buffer. This is often not necessary, especially if samples are fixed to the slides by flaming.
5. Prepare separate positive control slides of the homologous strain or any other reference strain of *R. solanacearum*, suspended in potato extract, as specified below, and optionally in buffer. Naturally infected tissue (maintained by lyophilisation or freezing at –16 to –24°C) should be used where possible as a similar control on the same slide. As negative controls, aliquots of sample extract which previously tested negative for *R. solanacearum* can be used. Standard reference strains are recommended for use as positive controls or, during optimisation of tests, to avoid misinterpretations due to cross-reactions. Reference strains are commercially available from e.g.: (a) National Collection of Plant Pathogenic Bacteria (NCPPB), Central Science Laboratory, York, UK, (b) Culture Collection of the Plant Protection Service (PD), Wageningen, the Netherlands, and (c) Collection Française de Bactéries Phytopathogènes (CFBP), INRA Station Phytobactériologie, Angers, France.
   - Produce a 48-h culture of a virulent strain of *R. solanacearum* race3/biovar2 (e.g. strain NCPPB 4156 = PD 2762 = CFBP 3857) on basal SMSA or Nutrient Agar medium and suspend in 10 mM phosphate buffer to obtain a cell density of approximately 2×10^8 cfu/ml. This is usually obtained by a faintly turbid suspension equivalent to an optical density of 0.15 at 600 nm. Remove the heel end cores of 200 tubers taken from a white skin variety production known to be free from *R. solanacearum*.

- Process the heel ends as usual and resuspend the pellet in 10 ml pellet (extraction) buffer.
- Prepare ten sterile 1.5-ml microvials with 900 μl of the resuspended pellet. Transfer 100 μl of the suspension of *R. solanacearum* to the first microvial Vortex. Establish decimal levels of contamination by further diluting in the next five microvials. The six contaminated microvials will be used as positive controls. The four non-contaminated microvials will be used as negative controls. Label the microvials accordingly.
- Prepare aliquots of 100 μl in sterile 1.5-ml microvials thus obtaining nine repeats of each control sample. Store at −16 to −24°C until use.
- The presence and quantification of *R. solanacearum* in the control samples should first be confirmed by IF. Perform assays on positive and negative control samples with each series of test samples. *R. solanacearum* must be detected in at least the 10^6 and 10^4 cells/ml of the positive controls and not in any of the negative controls.
- *Commercially available standardised control material.*

The following standard control material is available from the NCPPB culture collection:

  - Freeze dried pellet of potato extract from 200 healthy potato tubers as negative control for all tests.
  - Freeze dried pellet of potato extract from 200 healthy potato tubers containing 10^3–10^4 and 10^4–10^6 cells *R. solanacearum* biovar 2 (strain NCPPB 4156 = PD 2762 = CFBP 3857) as positive controls for serological and PCR tests. Since cell viability is affected during freeze-drying, these are not suitable as standard controls for isolation or bioassay tests.
  - Formalin-fixed suspensions of *R. solanacearum* biovar 2 (strain NCPPB 4156 = PD 2762 = CFBP 3857) at 10^6 cells/ml as positive controls for serological tests.

6. We do not have a good experience with using of 95% ethanol. Cells did not stain as well as with flaming.
7. Calculate and always prepare a little more antiserum dilution than strictly needed for the wells to be covered.
8. Air bubbles can be removed by rolling a toothpick gently over the surface of the cover glass.
9. If necessary, fixed slides may then be stored frozen in a desiccated box for as little time as necessary (up to a maximum of 3 months) prior to further testing.

Acknowledgements

The authors want to thank the contributors of the Annex II as published in Anonymous 2006 to use its information.

References

1. Ploem, J.S. (1973) Immunofluorescence microscopy. In: *Immuno-pathology of the skin: labelled antibody studies*, (Beutner, E.H., ed.), Dowden, Hutchinson and Ross Inc. Stroudsburg, PA, pp. 248–270.
2. Herman, B. (ed.) (1998) *Fluorescent microscopy. Microscopy Handbooks no.40.* BIOS Scientific Publishers Ltd, Oxford, UK.
3. Miller, H.J. (1983) A method for the detection of latent ringrot in potatoes by immunofluorescence microscopy. *Potato Research* 27, 33–42.
4. Janse, J.D. and Ruissen, M.A. (1988) Characterization and classification of *Erwinia chrysanthemi* strains from several hosts in the Netherlands. *Phytopathology* 78, 800–808.
5. Ngwira, N. and Samson, R. (1990) *Erwinia chrsysanthemi*: description of two new biovars (bv 8 and 9) isolated from kalanchoe and maize host plants. *Agronomie* 10, 341–345.
6. Samson, R., Legendre, J.B., Christen, R., Saux, F.-L.S.M., Achouak, W., and Gardan, L. (2005) Transfer of *Pectobacterium chrysanthemi* (Burkholder et al. 1953) Brenner et al. 1973 and *Brenneria paradisiaca* to the genus *Dickeya* gen. nov. as *Dickeya chrysanthemi* comb. nov. and *Dickeya paradisiaca* comb. nov. and delineation of four novel species, *Dickeya dadantii* sp. nov., *Dickeya dianthicola* sp. nov., *Dickeya dieffenbachiae* sp. nov. and *Dickeya zeae* sp. nov. *International Journal of Systematic and Evolutionary Microbiology* 55, 1415–1427.
7. Janse, J.D. (1988) A detection method for *Pseudomonas solanacearum* in symptomless potato tubers and some data on its sensitivity and specificity. *EPPO Bulletin/Bulletin OEPP* 18, 343–351.
8. Elphinstone, J.G., Stead, D.E., Caffier, C., Janse, J.D., López, M.M., Mazzucchi, U., Müller, P., Persson, P., Rauscher, E., Schiessendoppler, E., Sousa Santos, M., Stefani, E., van Vaerenbergh, J.et-al., (2000) Standardization of methods for detection of *Ralstonia solanacearum. Bulletin OEPP/EPPO Bulletin* 30, 391–395.
9. Anon. (1998) EU (1998) Council Directive 98/57/EC of 20 July 1998 on the control of *Ralstonia solanacearum*. Annex II-test scheme for the diagnosis, detection and identification of *Ralstonia solanacearum. Official Journal of the European Communities*, L235, 8–39.
10. Anon. (2006) Commission Directive 2006/63/EC of 14 July 2006 amending Annexes II to VII to Council Directive 98/57/ EC on the control of *Ralstonia solanacearum* (Smith) Yabuuchi et al. *Official Journal of the European Commission* L206, 36–106
11. Van der Wolf, J.M.,Vriend, G.C., Kastelein, P., Nijhuis, E.H., Van Bekkum, P.J., and Van Vuurde, J.W.L. (2000) Immunofluorescence colony-staining (IFC) for detection and quantification of *Ralstonia (Pseudomonas) solanacearum* biovar 2 (race 3) in soil and verification of positive results by PCR and dilution plating. *European Journal of Plant Pathology* 106, 123–133.
12. De Boer, S.H. and Weiczorek, A. (1984) Production of monoclonal antibodies to *Corynebacterium sepedonicum. Phytopathology* 74, 457–465.
13. Griep, R.A., van Twisk, C., van Beckhoven, J.R.C.M., van der Wolf, J.M., and Schots, A. (1998) Development of specific recombinant monoclonal antibodies against the lipopolysaccharide of *Ralstonia solanacearum* race 3. *Phytopathology* 88, 795–803.
14. Anon. (2006) Commission Directive 2006/56/EC of 12 June 2006 amending Annexes to Council Directive 93/85/EC on the control of potato ring rot. *Official Journal of the European Commission* L182, 1–43.
15. EPPO/OEPP (2004) Diagnostic protocols for regulated pests 7/20. *Erwinia amylovora. EPPO Bulletin/Bulletin OEPP* 34, 159–172.

Chapter 9

Detection of Four Major Bacterial Potato Pathogens

Ellen M. Kerr, Greig Cahill and Karen Fraser

Summary

Ring rot, brown rot and blackleg represent major bacterial pathogens of potato. The methods described below are aimed at basic identification of ring rot, brown rot or blackleg in a tuber sample.

Key words: Bacteria, Potato, Tuber, Detection, Ring rot, Brown rot, Blackleg, Selective plating, DNA extraction, DNA purification, Magnetic particle processor, PCR, Real-time PCR, TaqMan®.

9.1. Introduction

Major potato pathogens include the bacterial agents: *Clavibacter michiganensis* subsp. *sepedonicus* (*Cms*) which causes ring rot; *Ralstonia solanacearum* (*Rs*) which causes brown rot; and both *Pectobacterium atrosepticum* (*Pca*) and some *Dickeya* ssp. (*Dcsp*; previously *Erwinia chrysanthemi*), which cause blackleg and soft rots of tubers. Dissemination of these diseases is most likely *via* the introduction of infected tubers, a risk increased by the pathogens ability to latently infect potatoes, potentially evading detection until spread throughout the production system. Rapid and reliable means of detecting bacteria in the tuber are thus essential.

While *Rs* is a quarantine pest worldwide, *Cms*, *Pca* and *Dcsp* are quarantine pests in some countries but have regulated non-quarantine pest status in others *(1–3)*. The significant threat *Rs* or *Cms* poses to potato production in the EU has resulted in the European Union introducing legislation for their control in the form of Directives, which specify detection

R. Burns (ed.), *Methods in Molecular Biology, Plant Pathology, vol. 508*

DOI: 10.1007/978-1-59745-062-1_9

and identification procedures *(4, 5)*. Although there is currently no legislation regarding the control of *Pca* and *Dcsp* in the EU, these diseases can affect tuber quality and crop performance in the field substantially and with the increasing prevalence of the species *D. dianthicola* in Europe, detection and diagnosis of these soft rot-causing diseases is becoming increasingly important *(6)*.

Many techniques have been described for detection of these bacteria *(4, 5, 7–13)*, however, this chapter gives a single detection method for each of these four bacteria from tuber sample (**Fig. 9.1**). All four bacteria are sampled and extracted from the potato tuber simultaneously after which *Rs*, *Pca* and *Dcsp* are treated differently to *Cms*. Culturing on selective media suits the isolation of *Rs*, *Pca* and *Dcsp* as colonies are visible within 48 h. Once *Rs*, *Pca* or *Dcsp* colonies have been isolated, their identity can be rapidly confirmed by conventional PCR and gel electrophoresis. Although *Cms* can be isolated on semi-selective media, its slow-growing nature delays diagnosis and makes it susceptible to overgrowth by saprophytes, which is why a more rapid purification method using a magnetic particle processor is ideal. This purification step coupled with real-time PCR, using a TaqMan® assay, results in a rapid and reliable method of detection. For statutory testing of *Cms* in the EU validated tests are specifically employed, including: immunofluorescence staining assay (IF) for

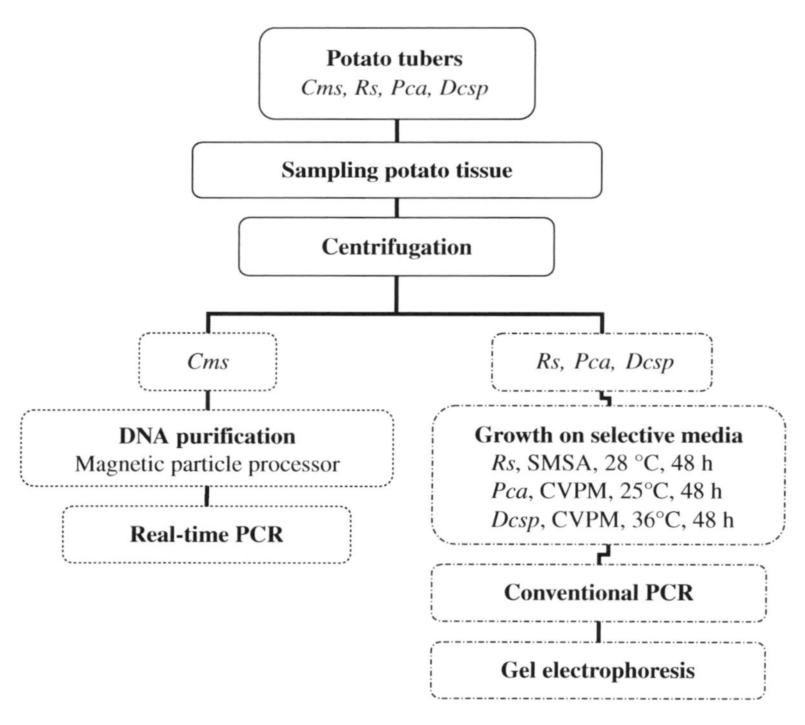

Fig. 9.1. Overview of methods used for detection of *Cms*, *Rs*, *Pca*, and *Dcsp*.

initial screening and eggplant bioassay for confirmation, indication of pathogenicity and subsequent isolation of the pathogen if required. Real-time PCR is used as an additional confirmatory test. Although this chapter provides a guide to the detection of these four major bacterial pathogens of potato, if detection and diagnosis of *Rs* or *Cms* is being performed in a statutory capacity the respective Directives or Standards must be observed *(2–5)*.

9.2. Materials

9.2.1. Bacterial Cells Extraction and Concentration

1. Extraction Buffer: 50 mM phosphate buffer, pH 7.0. Contains 4.26 g of Na_2HPO_4 (anhydrous) and 2.72 g of KH_2PO_4 in 1.0 L of distilled water. Dissolve ingredients, check pH and if buffer is outside pH 6.8–7.2 range, adjust accordingly using NaOH or HCl. Sterilise by autoclaving at 121°C for 15 min. Prior to use, add 0.5 g Lubrol flakes (Nonaethylene glycol monododecyl ether, Sigma), 1.0 g tetrasodium pyrophosphate and 50 g Polyvinylpyrrolidone-40000 (PVP-40) and dissolve.
2. Pellet Buffer: 10 mM phosphate buffer, pH 7.2. Contains 2.7 g $Na_2HPO_4 \cdot 12H_2O$ and 0.4 g $NaH_2PO_4 \cdot 2H_2O$ in 1.0 L distilled water. Dissolve ingredients, check pH and if buffer is outside pH 7.0–7.4 range, adjust accordingly using NaOH or HCl. Sterilise by autoclaving at 121°C for 15 min.
3. Ringer's Solution: two tablets (Oxoid) per 1.0 L of distilled water. Dissolve tablets and sterilise by autoclaving at 121°C for 15 min.
4. MagneSil® KF, Genomic System (Promega). Additional reagents; isopropanol and 95–100% ethanol are required and not included in the kit. Tip combs and 5-tube strips are also required (Thermo Fisher).

9.2.2. Selective and General Plating Media

1. Nutrient agar (Difco) medium (*(5)*; *see* **Note 1**).
2. SMSA medium (*(14)*; *see* **Notes 1** and **2**).
3. CVPM medium (*(15)*; *see* **Notes 1–3**).

9.2.3. DNA Amplification and Analysis

1. Positive reference material (*see* **Note 4**).

9.2.3.1. Conventional PCR of Brown Rot and Blackleg

1. Molecular grade H_2O (Sigma).
2. Forward and reverse primers (*see* **Table 9.1**).
3. *Rs* only: 10× PCR buffer containing 15 mM $MgCl_2$; 20 mM dNTP mix and 5 U/μL AmpliTaq DNA polymerase (Applied Biosystems).

Table 9.1
Primers and probe for conventional and real-time PCR

Target bacterium	Oligonucleotide primers and probe		
		Final conc.	Sequence
Cms (16)	Cms 50-2F	300 nM	5′-CGGAGCGCGATAGAAGAGGA-3′
	Cms 133R	300 nM	5′-GGCAGAGCATCGCTCAGTACC-3′
	TaqMan® probe	100 nM	5′-FAM-AAGGAAGTCGTCGGATGAAGATGCG-TAMRA-3′
Pca (17)	ECA1f	100 nM	5′-CGGCATCATAAAAACACG-3′
	ECA2r	100 nM	5′-GCACACTTCATCCAGCGA-3′
Dcsp (18)	ADE 1	1 μM	5′-GATCAGAAAGCCCGCAGCCAGAT-3′
	ADE 2	1 μM	5′-CTGTGGCCGATCAGGATGGTTTTGTCGTGC-3′
Rs (19)	OLI-1	1 μM	5′-GGGGGTAGCTTGCTACCTGCC-3′
	Y-2	1 μM	5′-CCCACTGCTGCCTCCCGTAGGAGT-3′

4. *Pca/Dcsp* only: JumpStart™ REDTaq® ReadyMix™ reaction Mix for PCR (Sigma).

9.2.3.2. Gel Electrophoresis of Amplified DNA of Brown Rot and Blackleg

1. Agarose.
2. Ethidium bromide (stock solution 10 mg/mL).
3. Loading buffer (*Rs* only; supplied with Promega 100-bp DNA Ladder).
4. 100-bp DNA Ladder.

9.2.3.3. Real-Time PCR Analysis of Ring Rot

1. Molecular grade H_2O (Sigma).
2. Forward and reverse primers and TaqMan® probe (*see* **Table 9.1**).
3. JumpStart™ Taq ReadyMix™ for Quantitative PCR (Sigma). Add 50 μL reference dye (100× solution)/200 reactions of ReadyMix (final reaction volume 25 μL).

9.3. Methods

The methods described below outline (*see* **Subheading 9.3.1**) concentration of bacterial cells (*Cms*, *Pca*, *Dcsp*, and *Rs*) and purification of bacterial DNA (*Cms* only), (*see* **Subheading 9.3.2**) purification of bacteria using selective growth media (*Pca*, *Dcsp*, and *Rs*), (*see* **Subheading 9.3.3**) DNA amplification and analysis by either conventional PCR and subsequent detection using gel electrophoresis (*Pca*, *Dcsp*, and *Rs*) or detection using real-time PCR (*Cms* only).

9.3.1. Bacterial DNA Extraction and Concentration

Bacteria are extracted from the heel-end of the potato (*see* **Note 5**). Potato tuber heel-end cores are shaken in the same extraction buffer regardless of the pathogen of interest. Low-speed centrifugation is used to remove soil and plant debris after which bacterial cells are pelleted using a high-speed centrifugation step. The pelleted samples are then resuspended into one of two buffers: Pellet Buffer (*Cms* and *Rs*) or Ringer's solution (*Pca* and *Dcsp*). At this point, it is possible to PCR these samples without further pre-treatment, however, culturing on a selective medium (*see* **Subheading 9.3.2**) or extracting DNA using a magnetic particle processor (*see* **Subheading 9.3.1.3**) improves detection.

9.3.1.1. Sampling Potato Tissue

1. Use a clean and disinfected potato peeler (*see* **Note 6**) to peel away skin from the heel-end of the potato.
2. Remove a conical sample approximately 5 mm wide from where the stolon was attached to expose 5-mm vascular tissue.
3. Collect the heel-end cores (up to 200 cores in 40 mL Extraction Buffer) in disposable, sealable 100-mL beakers (greiner bio-one). The heel-end cores should preferably be processed immediately. If this is not possible, store the cores in the container without adding Extraction Buffer at 4°C for no longer than 72 h (or alternatively store at room temperature for no longer than 24 h). After adding the Extraction Buffer, shake the cores for 16–24 h on an orbital shaker (100 rpm, 4°C).

9.3.1.2. Bacterial Concentration by Centrifugation

1. Take the potato cores in Extraction Buffer and decant the supernatant into a 50-mL centrifuge tube (or smaller if using a smaller initial volume).
2. Spin the supernatant if excessively cloudy and clarify using a low speed (at no more than 180 g, 4°C, 10 min) to remove plant and soil debris. Pour off the supernatant into a new 50-mL centrifuge tube.
3. Spin at a higher speed (10,000 × *g*, 4°C, 10 min) to pellet the bacterial cells. Gently pour off and discard the supernatant without disturbing the pellet.
4. Re-suspend the bacterial pellet in 1.5 mL Pellet Buffer (*see* **Note 7**).
5. Store at 4°C for up to 48 h. For longer term storage, add glycerol (to final concentration 20%, v/v), vortex and store at –20°C (weeks) or at –70°C (months).

9.3.1.3. DNA Purification Using a Magnetic Particle Processor

The difficulty and time taken in isolating ring rot from a tuber sample, coupled with the need to remove any PCR inhibiting substances, has led to the use of genomic DNA purification using MagneSil® Paramagnetic Particles. The MagneSil® KF Genomic System, used in conjunction with a magnetic particle processor (BioSprint 15, Qiagen), facilitates a quick, easy and semi-automated clean up method.

1. Immediately prior to processing samples should be heated in a hot block for 10 min at 100°C. At this stage samples are in Pellet buffer.
2. Set up 5-tube strips as per Manufacturer's instructions. Directions are provided in the Technical Bulletin (http://www.promega.com/tbs/tb322/tb322.pdf) supplied with the kit. When assembling the 5-tube strip volumes used (starting with left-most tube, numbered 1), which deviate slightly from those suggested in the Technical bulletin, are as follows (*see* **Note 8**):

   Tube 1 = 500 μL sample, 100 μL MagneSil® KF magnetic beads, 500 μL Lysis buffer KF

   Tube 2 = 1 mL salt wash

   Tube 3 = 1 mL alcohol wash

   Tube 4 = 1 mL alcohol wash

   Tube 5 = 60 μL Nuclease-free water
3. Run the "PromegaGenomic" program (this program must be installed onto the instrument and instructions on downloading the relevant software can be found in the Technical Bulletin).
4. After the program has finished, transfer the purified DNA from tube 5 to a 1.5 mL Eppendorf tube, avoiding any magnetic particles that may be at the bottom of the well.
5. Store at –20°C.

9.3.2. Selective and General Plating for Brown Rot and Blackleg

Isolating candidate bacteria from tuber material can be problematic owing to the numerous saprophytic micro-organisms present within a soil fraction that may quickly outgrow the bacteria of interest *(20)*. Plating on a selective medium preferentially increases the brown rot or blackleg populations in the sample extracts thus improving sensitivity of detection. Suspect colonies may then be re-isolated on the selective media (*Rs*) or a general media (*Pca*, *Dcsp*) prior to PCR. Blackleg-causing bacteria *Pca* and *Dcsp* are both plated on CVPM but growth at specific incubation temperatures, 25°C for *Pca* and 37°C for *Dcsp*, confirms the presence of the organism. As such, two sets of plates should be prepared, one set to be incubated at 25°C and the other at 37°C.

1. Prepare a dilution series from 10^0 to 10^{-3}. This is to ensure background saprophytes are diluted out.
2. Spread 100 μL per plate (*see* **Table 9.2** for media type required; *see* **Note 9**) of each sample and series of dilution. A single streak plate of a positive control from a suspension of 10^6 colony forming units (cfu)/mL of a type strain should be included for *Rs*, *Pca*, and *Dcsp* (*see* **Note 4**). For *Pca* and *Dcsp*, an additional dilution series of a positive control from a suspension of approximately 10^4–10^1 cfu/mL should be prepared.
3. *Rs*: Plates are incubated at 28°C and checked after 48 h for typical colonies (milky white, flat, irregular and fluidal initially and then a pink to dark red centre may develop after 72 h). Any suspect colonies should be re-isolated back onto SMSA to check again for typical colony growth as well as obtaining a pure culture prior to PCR analysis. Colonies will have developed by 6 days if they are present, providing the positive control show the media is suitable for the growth of the pathogen by testing prior to use and there is no suspicion of inhibition or antagonism by other bacteria.

   Pca/Dcsp: duplicate plates are prepared. Incubate one set at 25°C (*Pca*) and the other at 36°C (*Dcsp*). By 48 h, characteristic pits (deep cavities in CVPM owing to pectolytic activity of the bacterium) will have formed in the media if the bacterium is present. Re-isolation of a pure culture on nutrient agar results in raised creamy circular colonies without pitting.

Table 9.2
Media type, use and incubation conditions used for brown rot and blackleg.

Target organism	Media type	Media use	Incubation temperature/time
Pca, *Dcsp* (15)	CVPM	Selective	Pca 25°C, 48 h
	Nutrient agar	General	*Dcsp* 36°C, 48 h
Rs (14)	SMSA	Selective	28°C, 48 h and every day for 6 days thereafter

Selective media is for isolation of bacterium from tuber and re-isolation of suspect colonies; general media is used for re-isolation of suspect colonies grown on selective media

9.3.3. DNA Amplification and Analysis

9.3.3.1. Conventional PCR of Brown Rot and Blackleg

Negative controls should include a sample extract that has previously tested negative for the bacterium, extraction Buffers and PCR reaction mix controls. Generation of the expected amplicon from these would suggest contamination. Positive controls which include a spiked sample and a positive of pure culture are also advisable.

Sample Preparation

1. Prepare a suspension of approximately 10^6 cells/mL by resuspending a single colony in molecular grade water. The suspension should appear slightly cloudy.
2. Transfer 100 μL of the suspension into a screw-top 1.5 mL vial and incubate in a hot block for 5 min at 100°C.
3. Boiled cell samples should be frozen immediately and stored at –20°C until required.

Master Mix Preparation

The PCR master mix should be prepared in a contamination-free environment. Preferably the master mix should be prepared in a DNA template-free lab and then the sample added in a separate lab.

1. Make up the master mix (**Table 9.3**; *see* **Notes 10** and **11**).
2. Add 1 μL (*Pca*, *Dcsp*) or 2 μL (*Rs*) boiled cell sample to the PCR reaction mix (final volume 25 μL). The sample should be amplified both undiluted and diluted (1:10 in molecular grade H_2O; *see* **Note 12**). For reference, a known negative (molecular grade H_2O) and positive control (*Pca* and *Dcsp* positives for blackleg, *Rs* for brown rot) should also be included.
3. Seal the tube and perform amplifications under the following conditions **(Table 9.4)** in a thermal cycler.
4. Tubes can be held at 4°C.

Gel Electrophoresis

After PCR amplification, agarose gel electrophoresis is used to resolve the PCR amplicon of interest.

1. Prepare an agarose gel in 1× Tris–borate EDTA (TBE) buffer *(21)*. For *Rs* 2% (w/v) and for *Pca* or *Dcsp* 1.5% (w/v) agarose gels are required. Add ethidium bromide (final concentration 0.5 μg/mL) prior to pouring.
2. For *Rs*, load 12 μL of amplified DNA extraction mixture with 3 μL loading buffer and for *Pca* or *Dcsp* load 15 μL of amplified DNA extraction mixture (*see* **Note 10**). Gels are run at 5 V/cm in 1× TBE buffer. Include a 100-bp ladder.
3. Examine gel under short wave UV transillumination ($\lambda = 302$ nm) for the amplicon of interest (*Rs* = 288 bp; *Pca* = 690 bp; *Dcsp* = 420 bp.). Follow manufacturer's safety guidelines when using UV transilluminator.

Table 9.3
Master mix reagents and final concentrations where appropriate for: (A) *Rs* (B) *Pca* and *Dcsp*

(A) Reagent	*Rs*	
	Vol. per reaction (μL)	Final conc.
Molecular grade H_2O	17.65	–
10× PCR buffer (contains 15 mM $MgCl_2$)	2.5	1×
dNTP mix (20 mM)	0.25	200 μM
Primer OLI-1 (25 μM)	1.0	1 μM
Primer Y-2 (25 μM)	1.0	1 μM
AmpliTaq DNA polymerase (5 U/μL)	0.1	0.5 U
Sample (cell suspension)	2.0	–

(B) Reagent	Pca		Dcsp	
	Vol. per reaction (μL)	Final conc.	Vol. per reaction (μL)	Final conc.
Molecular grade H_2O	9.5	–	9.5	–
REDTaq®ReadyMix™ (2×)	12.5	1×	12.5	1×
Primer ECA1f (2.5 μM)	1.0	0.1 μM (*see* **Note 11**)	–	–
Primer ECA2r (2.5 μM)	1.0	0.1 μM (*see* **Note 11**)	–	–
Primer ADE1 (25 μM)	–	–	1.0	1 μM
Primer ADE2 (25 μM)	–	–	1.0	1 μM
Sample (cell suspension)	1.0	–	1.0	–

Table 9.4
Conventional PCR conditions

Target organism	Step 1	Step 2	Step 3
Rs:	96°C, 2 min	94°C, 20 s; 68°C, 20 s; 72°C, 30 s (35 cycles)	72°C, 10 min
Pca	95°C, 5 min	94°C, 30 s; 62°C, 45 s; 72°C, 45 s (40 cycles)	72°C, 8 min
Dcsp	–	94°C, 1 min; 72°C, 2 min (25 cycles)	72°C, 8 min

These conditions may need to be modified depending on the thermal cycler used

Table 9.5
Master mix reagents and final concentrations where appropriate for *Cms*.

Reagent (for *Cms*)	Volume per PCR (μL)	Final concentration
Molecular grade H_2O	4.5	
Taq ReadyMix™ (2×)	12.5	1×
25 mM $MgCl_2$	2.0	3.5 mM (including ReadyMix)
Primer Cms 50-2F (7.5 pmol/μL)	1.0	300 nM
Primer Cms 50-133RF (7.5 pmol/μL)	1.0	300 nM
Probe Cms 50-53T Probe (2.5 pmol/μL)	1.0	100 nM
Sample (cell suspension)	3.0	–

Taq ReadyMix™ (2×) also contains 1.5 mM $MgCl_2$

9.3.3.2 Real-Time PCR of Ring Rot

Real-time PCR master mixes should be prepared using the same precautions as conventional PCR master mixes (see subheading "Master Mix Preparation").

Setting up Reaction Mixture

1. Make up the master mix as follows (**Table 9.5**), allowing for sufficient quantities to run each sample in triplicate.
2. Add 3 μL sample (from **Subheading 9.3.1.3**) to 22 μL TaqMan® reaction mix. For reference, a known negative (molecular grade H_2O) and positive control should be included (*see* **Note 4**).
3. Seal the plate and perform amplifications under standard TaqMan® conditions in a Applied Biosystems 7900 HT Fast Real-Time PCR System (or equivalent): 50°C for 2 min; 95°C for 10 min; 95°C for 15 s, 60°C for 1 min (40 cycles of each).

9.4. Notes

1. Media and stock solutions of antibiotics should be stored at 4°C in the dark for no longer than 1 month. Plates should be dried briefly before use to remove surface condensation.
2. If large numbers of samples are being processed, 6-L batches of SMSA and CVPM media can be prepared at a time, making

the volumes of antibiotic quantities easier to handle (6 L of medium makes about 220 CVPM plates or 300 SMSA plates). Each new batch of media should be tested by plating with a reference strain resulting in the formation of typical colonies in standard conditions (*Rs* 28°C, 48 h; *Pca* 25°C, 48 h; *Dcsp* 36°C, 48 h).

3. The supply of sodium polypectate can be problematic *(9, 22)*. New batches of sodium polypectate may fulfil the supplier's criteria but may not to work for media production. It is essential to test all new batches to ensure compatibility. One current sodium polypectate product is GENU® pectin type X-914-02, from CP Kelco ApS, DK-4623 Lille Skensved, Denmark.
4. Positive controls are recommended for reference when testing of any of the pathogens. A licence is required for holding quarantine pathogens in the UK but it is possible to obtain reference material *(4, 5)*. Heat killed cells will suffice for Real-time PCR and PCR analysed products, however, live cultures will be required for brown rot and blackleg plate work. Presently we use blackleg strains from our own collection. Recommended type strains can be obtained from culture collections such as the National Collection of Plant Pathogenic Bacteria (NCPPB; http://www.ncppb.com/); these include *Pca* NCPPB 549, *Dcsp* NCPPB 402, *Cms* NCPPB 4053 and *Rs* NCPPB 4156.
5. The method outlined here details the preparation of bacteria from core-end samples. Erwinia soft rots infect potato tubers *via* a different route to *Cms* and *Rs* and so higher numbers are detectable using a mechanical peeler method *(9)* although it is still possible to detect *Pca* or *Dcsp* using this method.
6. A disposable scalpel can be used if there is a small number of cores per sample but for 200 tuber samples it is easier to use a peeler provided the peeler is cleaned and disinfected between each sample (disinfect by soaking in 10% hypochlorite solute for 10 min, then thoroughly rinse and flame). In statutory testing, 200 tubers is the maximum per sample, as testing larger numbers have associated problems *(4, 5)*; however, fewer than 200 tubers in the sample is acceptable.
7. The sample is split at this point to 500 μL for *Rs*, 500 μL for *Cms* and 500 μL left for reference. Prior to storage at –70°C, sterile glycerol can be added to a final concentration of 20% and mixed well. Avoid repeatedly freeze-thawing the sample. To test for *Pca* or *Dcsp*, **steps 3** and **4** can be repeated at any stage and the sample re-suspended in Ringer solution rather than in Pellet Buffer.

8. Sample volume can also be reduced (e.g. to 100 µL) if you do not have 500 µL available. The final volume of nuclease-free water that the purified DNA sample is collected in can also be reduced (200 µL recommended in the Promega Technical Bulletin). To ensure that the entire purified sample is deposited in the nuclease-free sample from the magnetic beads, it is not advisable to reduce the final volume much below 60 µL.
9. CVPM can be used both as single- and double-layer plates *(9, 15, 22)*. Colonies take 4–5 days to develop on single layer plates but can be visualised in 2 days using the double-layer method; however, high bacterial populations can result in the media liquefying. Double-layer works well for initial plating and further isolation but although pitting clearly indicates where colonies have formed, colonies should be purified on nutrient agar prior to PCR.
10. The final reaction volume is 25 µL in each case but the master mix for amplifying *Rs* (A) uses separate PCR buffer, dNTP mixture and Taq polymerase whereas the master mix for *Pca* and *Dcsp*(B) utilises JumpStart™ REDTaq® ReadyMix™ which contains these constituents already. JumpStart™ REDTaq® ReadyMix™ also contains a loading buffer and thus additional loading buffer is not required at **step 2** in the subheading "Gel Electrophoresis".
11. Final concentrations of primers ECA1f and ECA2r were reduced from 0.5 µM to 0.1 µM as this was found to reduce non-target amplification.
12. Too much bacteria can inhibit the PCR reaction. Amplification of a 1:10 diluted sample in addition to a neat sample reduces the likeliness of a failed amplification owing to excess template. The inclusion of an internal control can also mitigate false negatives but are not discussed in this chapter as internal controls are not available for all of the organisms.

Acknowledgments

We would like to thank colleagues in Diagnostics and Molecular Biology for their help and support, especially Vince Mulholland for reviewing the manuscript. This work was supported by the Flexible Fund project (FF/05/02/01): "Identification and reduction of risks posed by potato ring rot disease to the Scottish potato industry," financed by the Scottish Executive Environment & Rural Affairs Department, under Plant Health Licence No. PH/18/2007.

References

1. Anon. (2006) *Clavibacter michiganensis* subsp. *sepedonicus* (Potato ring rot): CAB International (CABI) Crop Protection Compendium. http://www.cabicompendium.org/cpc/home.asp.
2. Anon. (2003) Requirement for importation of potatoes into a NAPPO member country. North American Plant Protection Organization (NAPPO) Regional Standard for Phytosanitary Measures (RSPM No. 3) http://www.nappo.org/Standards/REVIEW/RSPM3Rev-e.pdf.
3. Anon. (2004) Potato. Commodity-specific phytosanitary measures (PM 8). EPPO Bull 34; 463–477. http://archives.eppo.org/EPPOStandards/PM8_COMMODITY/pm8-01(1)-e.pdf.
4. Anon. (2006) Commission Directive 2006/56/EC of 12 June 2006 amending the Annexes to Council Directive 93/85/EEC on the control of potato ring rot. Official J Eur Union L182 (1); 43. http://eur-lex.europa.eu/LexUriServ/site/en/oj/2006/l_182/l_18220060704en00010043.pdf.
5. Anon. (2006) Commission Directive 2006/63/CE of 14 July 2006 amending Annexes II to VII to Council Directive 98/57/EC on the control of *Ralstonia solanacearum* (Smith) Yabuuchi et al. Official J Eur Union L206 (36); 106. http://eur-lex.europa.eu/LexUriServ/site/en/oj/2006/l_206/l_20620060727en00360106.pdf.
6. Elphinstone, J.G. and Toth, I.K. (2007) *Erwinia chrysanthemi* (*Dickeya* spp.). The Facts. Oxford: British Potato Council.
7. Anon. (2006) *Clavibacter michiganensis* subsp. *sepedonicus*. EPPO Bull. 36(1);99–109. http://www.blackwell-synergy.com/doi/pdf/10.1111/j.1365-2338.2006.00919.x.
8. Anon. (2004) *Ralstonia solanacearum*. EPPO Bull. 34; 173–178. http://www.blackwell-synergy.com/doi/pdf/10.1111/j.1365-2338.2004.00715.x.
9. Pérombelon, M.C.M. and van der Wolf, J.M. (2002) Methods for the detection and quantification of *Erwinia carotovora* subsp. *atroseptica* (*Pectobacterium carotovorum* subsp. *atrosepticum*) on potatoes: a laboratory manual. Scottish Crop Research institute Occasional Publication No. 10. revised version 2002. http://www.scri.sari.ac.uk/TIPP/documents/ErwinMa.pdf.
10. Smid, E.J., Jansen, A.H.J. and Gorris, L.G.M. (1995) Detection of *Erwinia carotovora* subsp. *atroseptica* and *Erwinia chrysanthemi* in potato tubers using polymerase chain reaction. *Plant Pathol* 44; 1058–1069.
11. Toth, I.K., Hyman, L.J. and Wood, J.R. (1999) A one step PCR-based method for the detection of economically important soft rot *Erwinia* species on micro-propagated potato plants. *J Appl Microbiol* 87 (1); 158–166.
12. Bach, H.J., Jessen, I., Schloter, M., and Munch, J.C. (2003) A TaqMan-PCR protocol for quantification and differentiation of the phytopathogenic *Clavibacter michiganensis* subspecies. *J Microbiol Methods* 52; 85–91.
13. Weller, S.A., Elphinstone, J.G., Smith, N.C., Boonham, N. and Stead, D.E. (1999) Detection of *Ralstonia solanacearum* strains with a quantitative multiplex, Real-Time, flourogenic PCR (TaqMan) assay. *Appl Environ Microbiol* 66; 2853–2858.
14. Elphinstone, J.G., Hennessy, J., Wilson, J.K. and Stead, D.E. (1996) Sensitivity of detection of *Ralstonia solanacearum* in potato tuber extracts. *EPPO Bull* 26; 663–678.
15. Bdliya, B.S. (1995) Studies on the detection and identification of soft rot causing *Erwinia* species (*Erwinia carotovora* ssp *atroseptica* (van Hall) Dye, *Erwinia carotovora* ssp *carotovora* (Jones) Bergey et al. and *Erwinia chrysanthemi* Burkholder et al.) on potato tubers. Ph.D. Thesis, University of Göttingen, Germany, 148 pp.
16. Schaad, N.W., Berthier-Schaad, Y., Sechler, A. and Knorr, D. (1999) Detection of *Clavibacter michiganensis* subsp. *sepedonicus* in potato tubers by BIO-PCR and an automated real-time fluorescence detection system. *Plant Dis* 83 (12); 1095–1100.
17. De Boer, S.H. and Ward, L.J. (1995) PCR detection of *Erwinia carotovora* subsp. *atroseptica* associated with potato tissue. *Phytopathology* 85 (8); 854–858.
18. Nassar, A., Darrasse, A., Lemattre, M., Kotoujansky, A., Dervin, C., Vedel, R. and Bertheau, Y. (1996) Characterization of *Erwinia chrysanthemi* by Pectinolytic Isozyme Polymorphism and Restriction Fragment Length Polymorphism Analysis of PCR-Amplified Fragments of *pel* Genes. *Appl Environ Microbiol* 62 (7); 2228–2235.
19. Seal, S.E., Jackson, L.A., Young, J.P.W. and Daniels, M.J. (1993) Detection of *Pseudomonas solanacearum*, *Pseudomonas syzygii*, *Pseudomonas pickettii* and Blood Disease Bacterium by partial 16S rRNA sequencing: construction of Oligonucleotide primers for sensitive detection by polymerase chain reaction. *J Gen Microbiol* 139; 1587–1594.

20. Hyman, L.J., Birch, P.R.J., Dellagi, A., Avrova, A.O. and Toth, I.K. (2000) A competitive PCR-based method for the detection and quantification of *Erwinia carotovora* subsp. *atroseptica* on potato tubers. *Lett Appl Microbiol* 30 (4); 330–335.
21. Sambrook, J. and Russell, D. (2001) *Molecular Cloning: A Laboratory Manual*, 3rd edition. Cold Spring Harbor, New York: Cold Spring Harbor Laboratory Press.
22. Ahmed, M.E.E. (2001) Detection and effects of latent contamination of potato tubers by soft rot bacteria, and investigations on the effect of hydrogen peroxide on lipopolysaccharides of *Erwinia carotovora* in relation to acquired resistance3 against biocides. Ph.D. Thesis, Institute of Plant Pathology and Plant Protection of the Georg-August-University, Göttingen, Germany. http://webdoc.sub.gwdg.de/diss/2001/ahmed/ahmed.pdf.

Chapter 10

Erwinia amylovora: Modern Methods for Detection and Differentiation

Antonet M. Svircev, Won-Sik Kim, Susan M. Lehman, and Alan J. Castle

Summary

Erwinia amylovora is the causative agent of fire blight, a very destructive disease of numerous members of the rosaceae. The primary route of infection for host species, including commercially grown apple and pear, is the newly opened blossom. Susceptibility of flowers to infection for only a few days creates narrow window for infection. Not surprisingly, the risk of disease is related to *E. amylovora* population size. As a result, methods that supply quick, accurate and sensitive quantification of the pathogen population are important tools for determining the need for and the efficacy of disease control intervention. Plating samples and assessing colony-forming units constitutes an accurate and sensitive but slow method. End-point PCR is quick and sensitive but is not particularly amenable to quantification. We describe a real-time PCR procedure that provides all requirements. This method is based on chromosomal genes rather than on the pEa29 plasmid and so can be used to measure isolates that have been cured of the plasmid. The method has been used very successfully in directly quantify whole *E. amylovora* cells, in a variety of tissues from the orchard environment.

Key words: Fire blight, Erwinia pyrifoliae, dualplex, real time PCR, Malus, Pyrus, Rubus

10.1. Introduction

Erwinia amylovora (Burrill) Winslow et al. infects members of the rosaceous family, particularly the economically important *Malus*, *Pyrus* and *Rubus* spp. Differential identification of *E. amylovora* and closely related phytopathogenic bacteria can involve extensive combinations and permutations of morphological, biochemical, serological, and DNA-based tests *(1–10)*. DNA-based methods such as DNA: DNA hybridization and rDNA intergenic sequence analyses have become a critical component of the systematic classification of bacterial species.

R. Burns (ed.), *Methods in Molecular Biology, Plant Pathology, vol. 508*

DOI: 10.1007/978-1-59745-062-1_10

There is a wealth of information from which to create highly specific, sensitive, robust, rapid, and quantitative methods for more efficient differentiation of phytopathogens in the laboratory.

We outline DNA-based methods for the isolation and differential identification of *E. amylovora* from plant tissue. This chapter provides methodologies based on qualitative (*see* **Subheading 10.3.3**) and quantitative (*see* **Subheading 10.3.4**) PCR-based identification of *E. amylovora*. Particular emphasis is placed on molecular differentiation between *E. amylovora* and *Erwinia pyrifoliae (11)*. *E. pyrifoliae* is a distinct but very closely related species that shares many morphological, biochemical, and genetic characteristics with *E. amylovora*. Sample preparation is critical for the success of these methods and so we also describe methods for the preparation of suitable PCR templates directly from a variety of different plant tissues. Traditionally, identification of this pathogen often relies heavily on a multi-step process that involves enrichment, isolation of presumptive colonies, followed by a number of tests that include PCR. Isolation of *E. amylovora* from asymptomatic tissue becomes cumbersome and time consuming due to a battery of recommended tests.

The emphasis in this chapter is on the direct use of PCR-based technologies **(Fig. 10.1)** to identify and quantify the pathogen while avoiding the multi-step approach outlined in the right-hand section of **Fig. 10.1**. Our protocols may be used on both asymptomatic and symptomatic tissue infected with *E. amylovora*. The real time PCR protocol is used on a routine basis in our laboratory

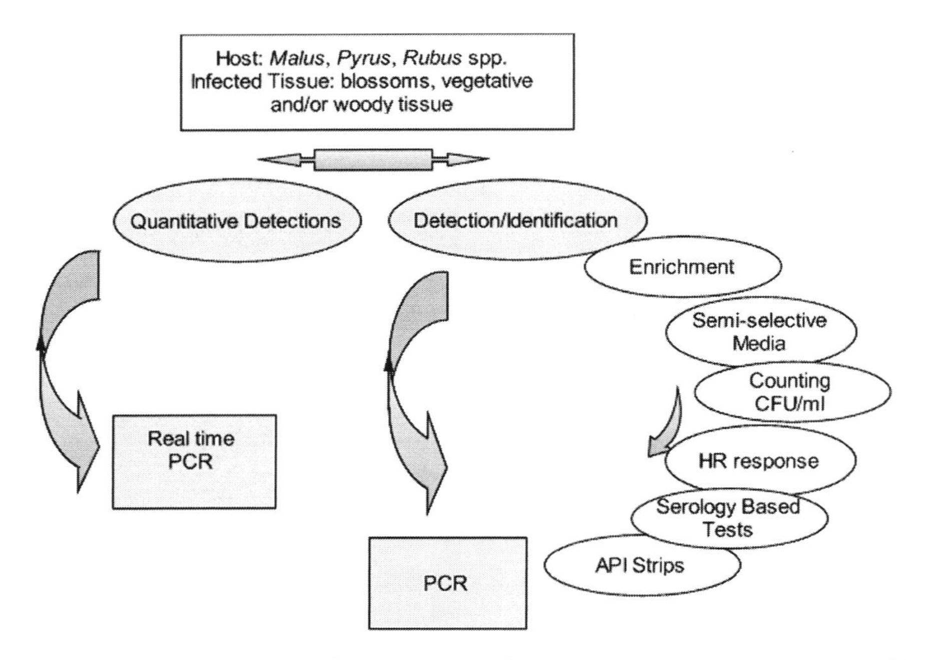

Fig. 10.1. Schematic representation of the PCR-based methods for the quantitative and traditional based PCR identification of *E. amylovora*. The protocols described on the extreme right hand side (clear *semi-circles*) represent the multitude of techniques that are often used in literature.

with symptomatic and asymptomatic tissues. Once the pathogen is identified as *E. amylovora*, it can be isolated in pure culture (*see* **Subheading 10.1**) using any number of semi-selective media described in Schaad et al. *(8)*. Using the pathogenicity tests that employ host tissue and/or green fruit sections *(8)*, Koch's postulates still need to be followed to confirm that, in fact, you have isolated *E. amylovora*.

10.2. Materials

10.2.1. Isolation from Tissue

1. Sterile distilled water.
2. 70% (v/v) ethanol.
3. 5.25% (w/w) sodium hypochlorite.
4. Semi-selective media for *E. amylovora* (*see* **Note 1**).

10.2.2. Sample Preparation

1. Sterile water or Direct Pathogen Extraction Buffer (DiPEB) (*see* **Note 2**).
2. Sample tubes (usually microcentrifuge tubes; snap-cap culture tubes are also convenient for larger or composite samples).
3. Water bath-style sonicator.
4. Sterile tissue dissection implements, such as scalpels.
5. 70% (v/v) ethanol.

10.2.3. Qualitative Molecular Detection and Differentiation (Conventional PCR)

1. PCR-grade water.
2. Oligonucleotide primers.
3. Deoxyribonucleotides (dNTPs).
4. PCR-grade $MgCl_2$.
5. *Taq* polymerase and manufacturer-supplied buffer.
6. Standard thin-walled PCR tubes.
7. Thermal cycler.
8. Agarose and gel electrophoresis equipment.
9. Ethidium bromide, 0.5 μg/mL in distilled water.
10. UV gel visualization equipment.

10.2.4. Quantitative Molecular Detection and Differentiation (Real-Time PCR)

1. Real-time PCR machine.
2. Oligonucleotide primers.
3. TaqMan® probes and low TE buffer (5 mM Tris–HCl, pH 8.0; 0.1 mM EDTA).
4. dNTPs.
5. PCR-grade $MgCl_2$.

6. *Taq* polymerase and manufacturer-supplied buffer.
7. PCR-grade water.
8. PCR tubes and caps suitable for real-time PCR applications.

10.3. Methods

The methods below describe: (a) the isolation of culturable *E. amylovora* from field samples, primarily woody or vegetative tissues and blossoms and, (b) the preparation of samples for the molecular detection of culturable and non-culturable *E. amylovora* from plant tissue, (c) the qualitative molecular detection and differentiation of *E. amylovora* using conventional PCR, and (d) the quantitative molecular detection and differentiation of *E. amylovora* using real-time PCR.

10.3.1. Isolation on Semi-Selective Media

10.3.1.1. Surface Disinfection of Vegetative and Woody Tissue

1. Immerse 10 s in 70% ethanol.
2. Treat 10–15 min in 1/10 dilution of 5.25% (w/w) sodium hypochlorite.
3. Wash 3× sequentially in sterile distilled water.
4. Using a sterile scalpel, cut tangentially through the tissue. Ideal tissue to sample is the zone immediately adjacent to the symptomatic tissue. Place on top of semi-selective medium.

Leaves

1. Immerse 10 s in 70% ethanol.
2. Treat 3–5 min in 1/10 dilution of 5.25% (w/w) sodium hypochlorite.
3. Cut through the leaf and place the cut surface onto the semi-selective medium.

10.3.2. Sample Preparation

A great deal of research has been devoted to the search for rapid, sensitive, high-throughput methods of preparing samples suitable for direct analysis by various PCR techniques. A selection of these methods is described below, including: (a) preparation from laboratory cultures and (b) preparation from symptomatic and asymptomatic plant tissue. These methods are designed for direct detection without further DNA extraction steps (*see* **Note 3**).

10.3.2.1. From Laboratory Cultures

If the goal is to test an organism that has been isolated in pure culture (i.e. following the procedures outlined in **Subheading 10.3.1**), it is more than sufficient to simply pick a single colony from a nutrient agar plate and to resuspend the cells in 0.5 mL of sterile distilled water or a PCR compatible buffer (phosphate buffer or DiPEB), in a microcentrifuge tube (*see* **Note 4**). The initial denaturation step of the amplification cycle will adequately lyse the cells, making the target genome accessible to the PCR primers and polymerases.

10.3.2.2. From Plant Material

The following methods are optimized for general monitoring purposes using asymptomatic or suspect plant samples. Tissues showing clear symptoms can be processed by the same methods. If the concentration of the *E. amylovora* target is likely to be low, the sample may be concentrated by centrifugation (5,000 × *g* for 5 min) and resuspend in a smaller volume. Resuspending in fresh water or buffer can also help to eliminate soluble PCR inhibitors. It is generally preferable to use freshly prepared samples for PCR applications; however processed samples can be stored at –20°C for later use.

General (Stems, Bark, and Leaves)

1. Surface-sterilize tissue sample by immersion for 10 s in 70% ethanol. Allow tissue to dry before proceeding.
2. With sterile implements, cut tissue into small pieces, being sure to transect xylem and phloem vessels. Return tissue to sterile tube.
3. Add 1 mL sterile, distilled water or DiPEB and vortex briefly.
4. Let it sit at room temperature for 15–20 min.
5. If the processed sample is not to be used immediately, remove plant tissue, or remove the supernatant to clean, sterile microcentrifuge tube. This prevents further diffusion of potentially PCR-inhibiting plant compounds.

Blossoms (Surface Populations)

1. Remove the petals and petioles from each blossom and collect the remaining blossom tissues (hypanthium, anthers, and stigma) in a sterile tube of suitable size. Blossoms can be processed individually or as composite samples of multiple blossoms.
2. Add 1 mL of DiPEB per blossom.
3. Sonicate the sample for 3–5 min.
4. Vortex, and remove plant tissue using sterile forceps, toothpicks, or any other convenient method. Alternatively, if multiple pieces of plant tissue are present, remove 1 mL of supernatant to a clean 1.5-mL microcentrifuge tube.
5. Collect the cells at 5,000 × *g* for 10 min.
6. Resuspend the cells in a final volume of 20–30 μL of DiPEB.

Anther Sacs

1. Collect 0.1 g of apple anther sacs.
2. Add 800 μL of DiPEB and vortex briefly.
3. Sonicate the sample for 3–5 min.
4. Vortex briefly.
5. Take the supernatant as much as possible without debris.
6. Transfer to a clean 1.5-mL microcentrifuge tube.
7. Centrifuge at 5,000 × *g* for 10 min.
8. Resuspend the cells in a final volume of 20–30 μL of DiPEB.

Isolated Pollen

1. Collect 0.1 g of pollen.
2. Add 400 μL of DiPEB and vortex briefly.
3. Sonicate the sample for 3–5 min.
4. Vortex briefly.
5. Leave at room temperature for 5 min to separate the supernatant from pollens by gravity.
6. Carefully take the supernatant without debris.
7. Transfer to a clean 1.5-mL microcentrifuge tube.
8. Centrifuge at 5,000 × *g* for 10 min.
9. Resuspend the cells in a final volume of 20–30 μL of DiPEB.

10.3.3. Qualitative Molecular Detection with Conventional PCR

Multiple primer sets and PCR protocols have been proposed for the specific detection of *E. amylovora* (*see* **Note 5**). The following sections describe the (a) primers, (b) reaction conditions, and (c) reaction product visualization needed to detect *E. amylovora* and to differentiate it from *E. pyrifoliae*. These protocols are adapted from Bereswill et al. *(2)*, Kim et al. *(12)*, and Kim et al. *(13)*.

10.3.3.1. Oligonucleotide Primers

E. amylovora (12)
AMS1: 5′-GGC AGA AGT TGT GAG CA-3′
AMS2: 5′-AAA CAG GTG CGC CGA ATA-3′
Amplicon size: 1.2 kbp

These primers are based on the chromosomal gene *amsB*, which is part of the operon that synthesizes a host-specific pathogenicity factor *(2)*. These primers are not known to have any cross-reaction with closely related species *(12)*, and the size of the amplification product is convenient for differentiation from the very closely related *E. pyrifoliae* when used with the primers listed below (*see* **Note 6**). The sensitivity of this assay is about 100 cfu per reaction (unpublished data).

E. pyrifoliae (13)
EP16A: 5′-AGA TGC GGA AGT GCT TCG-3′
EPIG2c: 5′-ACC GTT AAG GTG GAA TC-3′
Amplicon size: 0.7 kbp

These primers are based on the 16S–23S ITS region. There is enough sequence divergence in this region that it can be used to differentiate between *E. amylovora* and *E. pyrifoliae (11)*. The sensitivity of this assay is particularly high, about 20 cfu per reaction *(13)*, likely because the target region is present in multiple copies per genome.

Oligonucleotide primers do not need to be HPLC purified after synthesis; desalting is sufficient. Master stocks and working stocks of primers should be suspended in PCR-grade water and stored at –20°C. Primers are generally quite stable under these conditions.

10.3.3.2. Preparation of the PCR

When testing a given sample for both *E. amylovora* and *E. pyrifoliae* it is advisable to conduct the amplification reaction for each organism in a separate reaction. Combining two different primer sets in one reaction, may result in a reduction in sensitivity and amplification efficiency. However, because the *E. pyrifoliae*-specific amplicon and the *E. amylovora*-specific amplicon are of different sizes, the individual reaction products can be combined and run in a single well during electrophoresis, if desired.

1. Prepare enough master mix for the number of samples to be processed, plus for 1–2 extra samples. For best results the master mix should be freshly prepared as needed, and stored on ice during use. If the polymerase manufacturer supplies buffer containing 15 mM $MgCl_2$, that buffer can be used and the independent addition of $MgCl_2$ can be eliminated. Also, if the sample being tested is suspended in DiPEB buffer, Tween-20 is not needed in the PCR master mix. In both of these cases, the volume of water in the master mix must be adjusted accordingly.
2. Aliquot 45 µL of the master mix into each PCR tube and add 5 µL of the prepared sample for each 50 µL reaction.

Table 10.1
Master mix preparation for conventional PCR

Reagent	Concentration of working stock	Volume per reaction (µL)	Final concentration
PCR-grade water	–	30.3	–
Polymerase Buffer	10×	5	1×
dNTPs (dATP, dCTP, dGTP, dTTP)	10 mM each	1	0.2 mM each
$MgCl_2$[a]	25 mM	3	1.5 mM
Tween-20[a]	–	0.5	1%
Primer A	10 pmol/µL	2.5	25 pmol
Primer B	10 pmol/µL	2.5	25 pmol
Taq polymerase	5 U/µL	0.2	1 U
Total volume		45	

[a]See step 1 in **Subheading 10.3.3.2**

Table 10.2
Thermal profile for *E. amylovora*-specific and/or *E. pyrifoliae*-specific PCR

1 cycle	Denaturation: 95°C, 1 min
35 cycles	Denaturation: 95°C, 15 s
	Annealing: 52°C, 15 s
	Extension: 72°C, 30 s

Total time: approximately 2 h

3. Centrifuge PCR tubes briefly (10–20 s) to ensure that the sample and reagents are properly mixed.
4. Run amplification reaction as described in **Table** 10.2.

10.3.3.3. Analysis of PCR Products Using Agarose Gel Electrophoresis

The amplification reaction described above can be visualized by running as much of each reaction product as possible on a 1% agarose gel for approximately 1.5 h at 100 V, staining with a 0.5 μg/mL ethidium bromide solution, and photographing the gel under a UV light, according to standard methods *(14)*. **Figure 10.2** represents agarose gel electrophoresis of PCR products. Each lane contains 15 μL of the 50 μL reaction volume. Lanes contain: *L* 1-kbp ladder (MBI Fermentas); (1) 7.5×10^6 cfu per reaction of *E. amylovora* Ea6-4 *(15)*; (2) 4.5×10^6 cfu per reaction of *E. pyrifoliae* 1/96 *(16)*; (3) 1:1 mixture of reaction products from lanes 1 and 2. The 1.2 kbp amplification product from *E. amylovora* is easily distinguishable from the 0.7 kbp *E. pyrifoliae* amplification product (*see* **Note 6**). Note the slight shadow bands >0.7 kbp produced by the *E. pyrifoliae* primers. Since such shadow bands are faint compared to the specific product they should not be a concern. Kim et al. *(13)* report that these can be eliminated by raising the annealing temperature to 57°C, but that this will reduce sensitivity; no shadow bands were reported when cells were isolated from plant tissue.

10.3.4. Quantitative Molecular Detection with Real-Time PCR

The following section describes the differential detection of *E. amylovora* using real-time PCR with TaqMan® chemistry. This type of PCR detection system is based on the addition of a short oligonucleotide probe to which a fluorophore molecule and a quencher molecule are attached. As the advancing polymerase hydrolyzes the probe, the fluorophore is released from the influence of the quencher and it will fluoresce. As more copies are made, the total fluorescence in the tube increases. The most straightforward application of this technique for *E. amylovora* detection is to conduct singleplex reactions with the *E. amylovora* primers and

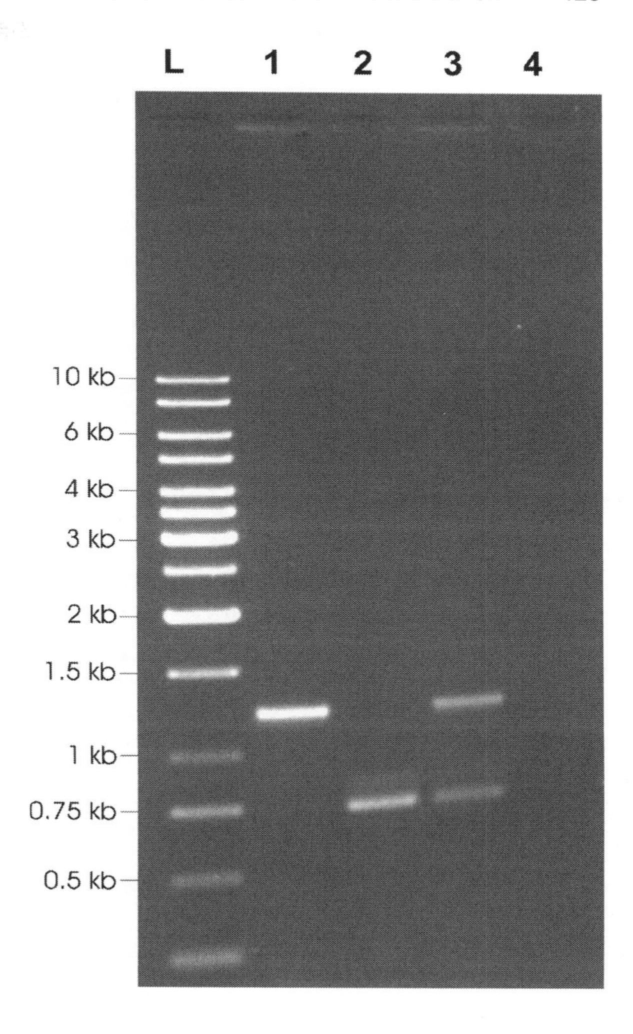

Fig. 10.2. Agarose gel electrophoresis of PCR products. Lanes contain: L 1-kbp ladder (MBI Fermentas); (1) 7.5 × 10^6 cfu per reaction of *E. amylovora* Ea6-4; (2) 4.5 × 10^6 cfu per reaction of *E. pyrifoliae* 1/96; (3) Dualplex from *E. amylovora* and *E. pyrifoliae* mixture. The 1.2 kbp amplification product from *E. amylovora* is easily distinguishable from the 0.7 kbp *E. pyrifoliae* amplification product.

probe and to draw a qualitative conclusion based on the presence or absence of a fluorescence signal above the threshold value.

There are two main advantages to TaqMan® real-time PCR. Additional target-specific probe provides better specificity than SYBR Green based real-time PCR which requires dissociation curve analysis, and an extra confirmation cycle to distinguish original amplification from unspecific amplification. If all primers and probes have the same annealing temperature and exhibit no cross-reaction, it is possible to simultaneously quantify multiple organisms in a single reaction by using different fluorophores for each species-specific probe. The initial quantity of each target can be calculated based on the number of cycles needed for that fluorescence signal to reach a threshold level where it begins to increase exponentially (the C_t value). This section describes (a) the primers and probes needed to differentiate between *E. amylovora* and *E. pyrifoliae* *(17)*, (b) preparation and running of the real-time amplification reaction, (c) construction of a standard curve for quantification, and (d) interpretation of results. The real-time PCR machine should be calibrated for use with FAM and CY5-labelled probes, according to the manufacturer's instructions.

10.3.4.1. Oligonucleotide Primers and Taqman® Probes

Master stocks and working stocks of primers and probes should be stored at –20°C. Master stocks of primers can be suspended in PCR-grade water at 100 pmol/μL and stored for an extended period of time. Working stocks of each primer pair at can be suspended in PCR-grade water as a 1:1 mixture of the forward (F) and reverse (R) primers (2.5 pmol/μL each), and stored for several months. Probes (P) are light-sensitive and all stocks should be stored in amber microcentrifuge tubes. Master stocks of probes can be suspended in low Tris–EDTA buffer (TE; 10 mM Tris, 1 mM EDTA pH 7.5) at 100 pmol/μL. Working stocks at 2.5 pmol/μL should also be prepared in low TE buffer. When properly stored, probe stocks are stable for several months, but a variety of factors can affect that stability. Infrequently used probe stocks should be tested periodically for changes in fluorescence patterns.

E. amylovora (17):

Ea-lscF: 5′-CGC TAA CAG CAG ATC GCA-3′

Ea-lscR: 5′-AAA TAC GCG CAC GAC CAT-3′

Ea-lscP: Cy5-5′-CTG ATA ATC CGC AAT TCC AGG ATG-3′ IAbRQ

Amplicon size: 105 bp

E. pyrifoliae (17):

Ep-hrpwF: 5′-CGC TAA CCC GAC TGT GCT-3′

Ep-hrpwR: 5′-TGA AGG TTT GCC CTT TGC-3′

Ep-hrpwP: FAM-5′-ATG ACA CCA TCA TCG TAA AGG CGG-3′ BHQ-1

Amplicon size: 77 bp

Both of these sets of primers and TaqMan® probes were derived from the chromosomal DNA of *E. amylovora* and *E. pyrifoliae* (*see* **Note 7**). The sensitivity of these reactions is about 20 cfu per reaction, corresponding to 8×10^2 cfu/mL of bacterial suspension or plant tissue extract.

10.3.4.2. Preparation of the PCR

1. Prepare enough master mix for the number of samples to be processed, plus 1–2 extra samples. For best results the master mix should be freshly prepared as needed, and stored on ice during use.
2. Aliquot 23 μL of the master mix into each PCR tube and add 2 μL of the prepared sample for each 25 μL reaction.
3. Centrifuge PCR tubes briefly (5–10 s) to ensure that the sample and reagents are properly mixed. It is important that bubbles in the reaction mixture be minimized as they can interfere with detection of the fluorescence signal.
4. Run the real-time PCR amplification according to the thermal profile in **Table 10.4.**

Table 10.3
Master mix preparation for dualplex detection and differentiation of *E. amylovora* and *E. pyrifoliae*, or for singleplex detection of either

Reagent	Concentration of working stock	Final concentration	Volume per reaction	
			Singleplex reaction (μL)	Dualplex reaction
Water	–	–	13.2	10.2 μL
Primer mix	2.5 pmol/μL each	200 nM each	2.0	2.0 μL each mix
Probe	2.5 pmol/μL	100 nM	1.0	1.0 μL each
Polymerase buffer	10×	1×	2.5	2.5 μL
Taq polymerase	5.0 U/μL	1.5 U	0.3	0.3 μL
dNTPs (dATP, dCTP, dGTP, dTTP)	10.0 mM	0.8 mM	2.0	2.0 μL
$MgCl_2$	25.0 mM	2 mM	2.0	2.0 μL
Total			23.0	23.0 μL

Table 10.4
Thermal profile for singleplex or dualplex differentiation of *E. amylovora* and/or *E. pyrifoliae* by real-time PCR

1 cycle	Denaturation: 95°C, 3 min
40 cycles	Denaturation: 95°C, 10 s
	Extension: 60°C, 16 s (2 endpoint data collections)

Total time: approximately 70 min

10.3.4.3. Analysis of Results

The determination that a sample is positive for the target organism depends on whether the C_t value for a given reaction is smaller than a pre-determined threshold value. That threshold value should be determined based on the lower limit of a standard curve created based on the manufacturer's instructions for the real-time PCR machine and software in question (*see* **Fig. 10.3**). A simple tenfold dilution series of a known quantity of the target organism is sufficient for most purposes, and a twofold dilution series may help refine the lower range of the curve.

Once a standard curve is defined, the technique can be used simply to determine the presence or absence of the target, or the initial quantity of the target can be calculated using the C_t value

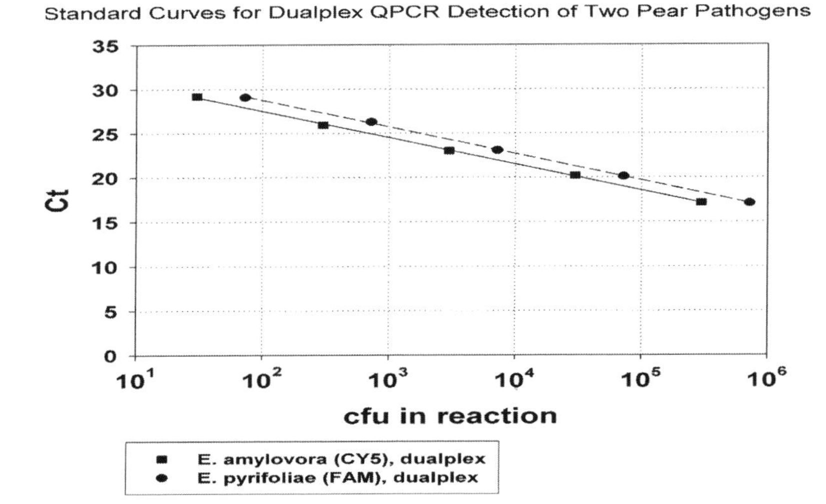

Fig. 10.3. Interpreting real-time PCR detection and differentiation of *E. amylovora*. Sample standard curves for *E. amylovora* (*squares, solid line*) and *E. pyrifoliae* (*circles, dashed line*) in a dualplex reaction, based on a series of tenfold dilutions only. According to these curves, both *Erwinia* spp. would be considered to be present if the C_t value of a sample reaction were less than 30 cycles. Each point is based on the average of at least two replicate reactions, in order to account for any variation in reaction preparation. Standard curves for each data set was determined by linear regression on the logarithm of initial quantity such that $C_t = (a)\log(\text{initial quantity}) + b$.

from each sample and the regression equation of the standard curve. Interpreting real-time PCR detection and differentiation of *E. amylovora* is shown in (**Fig. 10.3**).

10.4. Notes

1. The isolation of *E. amylovora* on differential and semi-selective media is well described by Schaad et al. *(8)*. Any one of the described semi-selective media may be used for the isolation of *E. amylovora*. In our laboratory, the semi-selective Modified Miller and Scroth medium (MMS) *(18)* is routinely used for isolation purposes. The medium is green in colour and *Erwinia* spp. form bright orange colonies on the surface.
2. Direct Plant Extract Buffer (DiPEB) (Cat. No. ACC 00690, Agdia, Elkhard, IN, USA).
3. In most cases the described sample preparation techniques will allow direct PCR detection, even if the suspension medium used is simply water *(13, 19, 20)*. If a substantial problem with PCR inhibitors is encountered, there are some reasonably fast options for preparatory DNA extraction. The European and Mediterranean Plant Protection Organization standard

method involves suspension of macerated tissue in an extraction buffer, and precipitation of DNA from the supernatant *(1)*. More recently, Stöger et al. *(9)* describe modifications to the REDExtract-N-Amp™ Plant PCR Kit protocols that make it suitable for rapid, high-throughput DNA extraction from apple and pear tissues. They achieved very high sensitivity in subsequent PCR analyses with pEA29-based primers (*see* **Note 6**). The manufacturer specifies that this kit does not work well with apple and pear leaf tissue.

4. Bacterial suspensions can also be collected from MMS *(18)* plates, and the suspension used directly for PCR. Colonies collected from some types of selective or semi-selective media may potentially contain PCR inhibitors, making it important that a positive control strain grown on the same media be included in initial tests.
5. Another primer pair based on the *amsB gene* has been described *(2)*. These primers strictly require a lower annealing temperature and longer cycle times than the ones described in **Subheading 10.3.3.1**, making differentiation from *E. pyrifoliae* less convenient since two separate reaction cycles would be necessary to accommodate the different thermal profiles. However, these primers are highly specific and could be used if confirmation of *E. pyrifoliae* is not a concern. Alternative *E. pyrifoliae*-specific primers have been published by Kim et al. *(13)*. These primers are slightly less sensitive than those given in **Subheading 10.3.3.1**, and are less convenient for differentiating between *E. pyrifoliae* and *E. amylovora* because the amplicon size is the same as the AMS1 and AMS2c primers given in **Subheading 10.3.3.1**.
6. Several other reasonable sets of PCR primers for the specific detection of *E. amylovora* have been published. However, most suffer from specific drawbacks that make them less suitable for this purpose than the primers recommended in **Subheading 10.3.2.2**. Bereswill et al. *(3)*, McManus and Jones *(21)*, and Llop et al. *(5)* describe primers based on the pEa29 plasmid, which is absent in several laboratory strains and is now also known to be absent in some natural *E. amylovora* isolates *(22)*. The exact genomic target of the primers described by Taylor et al. *(10)* is not known. The primers described by Maes et al. *(23)*, which are based on 23S rDNA, produced an amplification product from an isolate not thought to be *E. amylovora* *(7)*. Often there is not enough divergence in the 16S rRNA genes, themselves, for primers based on those sequences to definitively distinguish between closely related species, even with further analysis of the amplification product (such as restriction analysis). Therefore the 16S–23S intergenic transcribed sequence is a much better target *(4, 11)*.

7. Salm and Geider *(24)* have also published protocols and sequences for real-time PCR detection of *E. amylovora* using TaqMan® chemistry, based on sequences from the pEA29 plasmid. Sequences may not detect all isolates of *E. amylovora* *(21)*.
   E. amylovora (24):
   P29TF: 5′-CAC TGA GGT GCC GTT G-3′
   P29TR: 5′-CGC CAG GAT AGT CGC ATA-3′
   P29TM: FAM-5′-TAC CTC CGC AGC CGT CAT GG-3′
   Thermal Profile: 95°C for 15 min; 40 cycles of 94°C for 15 s and 52°C for 1 min.
   Amplicon size: 112 bp

Acknowledgments

This work was supported by the Pest Management Centre of Agriculture and Agri-Food Canada, Pesticide Risk Reduction and Minor Use Programs and the Natural Sciences and Engineering Research Council of Canada (NSERC) for post-doctoral and pre-doctoral scholarship funding to Drs. W.-S. Kim and S.M. Lehman.

References

1. Anon. (2004) EPPO Standards Diagnostic Protocols for Regulated Pests: *Erwinia amylovora*. *Bulletin* **34**, 155–171.
2. Bereswill, S., Bugert, P.B.I., and Geider, K. (1995) Identification of the fire blight pathogen, *Erwinia amylovora*, by PCR assays with chromosomal DNA. *Appl. Environ. Microbiol.* 61, 2636–2642.
3. Bereswill, S., Pahl, A., Belleman, P., Zeller, W., and Geider, K. (1992) Sensitive and species-specific detection of *Erwinia amylovora* by polymerase chain reaction analysis. *Appl. Environ. Microbiol.* 58, 3522–3526.
4. Jeng, R.S., Svircev, A.M., Myers, A.L., Beliaeva, L., Hunter, D.M., and Hubbes, M. (2001) The use of 16S and 16S–23SrDNA to easily detect and differentiate common Gram-negative orchard epiphytes. *J. Microbiol. Methods* 44, 69–77.
5. Llop, P., Bonaterra, A., Penalver, J., and Lopez, M.M. (2000) Development of a highly sensitive nested-PCR procedure using a single closed tube for detection of *Erwinia amylovora* in asymptomatic plant material. *Appl. Environ. Microbiol.* 66, 2071–2078.
6. Llop, P., Caruso, P., Cubero, J., Morente, C., and Lopez, M.M. (1999) A simple extraction procedure for efficient routine detection of pathogenic bacteria in plant material by polymerase chain reaction. *J. Microbiol. Methods* 37, 23–31.
7. Rosello, M., Penalver, J., Llop, P., Gorris, M.T., Chartier, R., Garcia, F., Monton, C., Cambra, M., and Lopez, M.M. (2006) Identification of an *Erwinia* sp. different from *Erwinia amylovora* and responsible for necrosis on pear blossoms. *Can. J. Plant Pathol.* 28, 30–41.
8. Schaad, N.W., Jones, J.B., and Chun, W. (2001) Laboratory Guide for the Identification of Plant Pathogenic Bacteria, The American Phytopathological Society, St. Paul, MN.
9. Stoger, A.S., Schaffer, J., and Ruppitsch, W. (2006) A rapid and sensitive method for direct detection of *Erwinia amylovora* in symptomatic and asymptomatic plant tissues by polymerase chain reaction. *J. Phytopathol.* 154, 469–473.

10. Taylor, R.K., Guilford, P.J., Clark, R.G., Hale, C.N., and Forster, R.L.S. (2001) Detection of *Erwinia amylovora* in plant material using novel polymerase chain reaction (PCR) primers. *N Z J. Crop Hort. Sci.* 29, 35–43.
11. Kim, W.S., Gardan, L., Rhim, S.L., and Geider, K. (1999) *Erwinia pyrifoliae* sp. *nov.*, a novel pathogen that affects Asian pear trees (*Pyrus pyrifolia* Nakai). *Int. J. Syst. Bacteriol.* 49, 899–905.
12. Kim, W.-S., Hildebrand, M., Jock, S., and Geider, K. (2001) Molecular comparison of pathogenic bacteria from pear trees in Japan and the fire blight pathogen *Erwinia amylovora*. *Microbiology* 147, 2951–2959.
13. Kim, W.-S., Jock, S., Paulin, J.P., Rhim, S.L., and Geider, K. (2001) Molecular detection and differentiation of *Erwinia pyrifoliae* and host range analysis of the Asian pear pathogen. *Plant Dis.* 85, 1183–1188.
14. Sambrook, J., and Russell, D.W. (2001) Molecular Cloning: A Laboratory Manual, Cold Spring Harbour Laboratory Press, Cold Spring Harbour, NY.
15. Gill, J.J., Svircev, A.M., Smith, R.J., and Castle, A.J. (2003) Bacteriophages of *Erwinia amylovora*. *Appl. Environ. Microbiol.* 69, 2133–2138.
16. Rhim, S.L., Volksch, B., Gardan, L., Paulin, J.P., Langlotz, C., Kim, W.-S., and Geider, K. (1999) *Erwinia pyrifoliae*, an *Erwinia* species different from *Erwinia amylovora*, causes a necrotic disease of Asian pear trees. *Plant Pathol.* 48, 514–520.
17. Lehman, S.M., Kim, W.S., Castle, A.J., and Svircev, A.M. (2008) Duplex real-time polymerase chain reaction reveals competition between *Erwinia amylovora* and *E. pyrifoliae* on pear blossoms. *Phytopathology* 98, 673–679.
18. Brulez, W., and Zeller, W. (1981) Seasonal changes in epiphytic *Erwinia amylovora* on ornamentals in relation to weather conditions and the course of infection. *Acta Hort.* 117, 37–43.
19. Bogs, J., Richter, K., Kim, W.S., Jock, S., and Geider, K. (2004) Alternative methods to describe virulence of *Erwinia amylovora* and host–plant resistance against fireblight. *Plant Pathol.* 53, 80–89. 0032–0862.
20. Jock, S., Rodoni, B., Gillings, M., Kim, W.-S., Copes, C., Merriman, P., and Geider, K. (2000) Screening of ornamental plants from the Botanic Gardens of Melbourne and Adelaide for the occurrence of *Erwinia amylovora*. *Aust. Plant Pathol.* 29, 120–128.
21. McManus, P.S., and Jones, A.L. (1995) Detection of *Erwinia amylovora* by nested PCR and PCR-dot-blot and reverse-blot hybridizations. *Phytopathology* 85, 618–623.
22. Llop, P., Donat, V., Rodriguez, M., Cabrefiga, J., Ruz, L., Palomo, J.L., Montesinos, E., and Lopez, M.M. (2006) An indigenous virulent strain of *Erwinia amylovora* lacking the ubiquitous plasmid pEA29. *Phytopathology* 96, 900–907.
23. Maes, M., Garbeva, P., and Crepel, C. (1996) Identification and sensitive endophytic detection of fire blight pathogen *Erwinia amylovora* with 23S ribosomal DNA sequences and the polynerase chain reaction. *Plant Pathol.* 45, 1138–1149.
24. Salm, H., and Geider, K. (2004) Real-time PCR for detection and quantification of *Erwinia amylovora*, the causal agent of fireblight. *Plant Pathol.* 53, 602–610.

Chapter 11

The Use of Fluorescent In Situ Hybridization in Plant Fungal Identification and Genotyping

Mohamed Hijri

Summary

FISH is a widely used technique in many laboratories not only for cytogenetic studies, but also in other biological fields. It requires a combination of skills in molecular biology, cytogenetics, immunocytochemistry, microscopy and cellular imaging analysis.

Key words: Fluorescence, DNA, Microscopy, Probes, Genome.

11.1. Introduction

Fluorescence in situ hybridization (FISH) has become an essential technique in molecular cytogenetics, which involves the combination of molecular biology and cytogenetics. FISH can be used to detect and localize the presence or absence of specific DNA sequences on genomes. It uses fluorescent probes or haptene-labeled probes that bind only to those parts of the chromosome with which they show a high degree of sequence similarity. Fluorescence microscopy or confocal microscopy can be used to find out where the fluorescent probe is bound to the chromosome; DNA probes are labeled with different colored fluorescent tags to visualize one or more specific regions of the genome. FISH can either be performed as a direct approach to mitotic or meiotic chromosomes and interphase nuclei, or as an indirect approach, which involves the use of comparative genomic hybridization analysis to asses the entire genome for chromosomal rearrangement.

R. Burns (ed.), *Methods in Molecular Biology, Plant Pathology, vol. 508*

DOI: 10.1007/978-1-59745-062-1_11

FISH is a widely used technique in many laboratories on not only cytogenetic studies, but also in other biological fields. It requires a combination of skills in molecular biology, cytogenetics, immunocytochemistry, microscopy and cellular imaging analysis. In this chapter, I have aimed to provide efficient and effective details of protocols for FISH currently used mostly for fungal cytogenetics. The methods are given with concentrations, times and alternatives which I know work successfully for many fungi. In some filamentous fungi, karyotypic data obtained using conventional light microscopy are not consistent with the electrophoretic karyotypes produced by pulsed field gel electrophoresis. Methods described here can therefore be used for determining chromosome numbers and their morphology, which constitutes a powerful tool for identification of some plant fungal pathogens *(1, 2)*.

11.2. Materials

11.2.1. Slide Pretreatment

1. Proteinase K stock solution ready to use: 20 mg/ml (Fermentas, Germany). The recommended working concentration for Proteinase K is 1–5 µg/ml in 20 mM Tris–HCl, pH 7.5, 2 mM $CaCl_2$; make fresh as required (This step is optional: not recommended for metaphase FISH).
2. 20× standard saline citrate stock solution: 3.0 M NaCl, 0.3 M Na-citrate; set up with distilled water, adjust to pH 7.0, autoclave, and store at room temperature.
3. RNase stock solution (*see* **Note 1**): 10 mg/ml of RNase type A (Roche Diagnosis, Switzerland); set up with sterile distilled water; aliquot and store at –20°C. RNase working solution: dilute to 100 µg/ml in 2× SSC; make fresh as required.
4. Pepsin stock solution 10% (w/v): dissolve 100 mg pepsin (Sigma) in 1 ml of sterile distilled water at 37°C; aliquot and store at –20°C. (Optional: not recommended for metaphase FISH).
5. Pepsin buffer: Add 1 ml of 1 M HCl to 99 ml of distilled water and incubate at 37°C for about 20 min, then add 50 ml of the pepsin stock solution 10% (w/v) and leave the coplin jar at 37°C; make fresh as required. (Optional: not recommended for metaphase FISH).
6. 1× Phosphate Buffered Saline (PBS): 8 g NaCl, 0.2 g KCl, 1.44 g Na_2HPO_4, 0.24 g KH_2PO_4 in 800 ml of distilled water. Adjust the pH to 7.4 with HCl. Add water to 1 L. Sterilize by autoclave and store at room temperature.

7. Formaldehyde buffer: 3% (v/v) paraformaldehyde (37%; Sigma) in 1× PBS; make fresh as required. (Optional: not recommended for metaphase FISH.)
8. Tris–HCl: 20 mM Tris–HCl, 2 mM $CaCl_2$, 50 mM $MgCl_2$, pH 7.5.

11.2.2. Probe Labeling

Several methods are used to label probes: Nick translation, Random Priming and PCR (*see* **Note 2**). Dig PCR probe labeling is described in this section.

1. PCR Dig Probe Synthesis Mix (Roche Diagnosis, Switzerland), 10× conc., 125 μl of a mixture containing dATP, dCTP, dGTP (2 mM each), 1.3 mM dTTP, 0.7 mM DIG-11-dUTP, alkali-labile, pH 7.0.
2. PCR 10× buffer with $MgCl_2$.
3. dNTP mixture (2 mM each).
4. *Taq* DNA Polymerase.
5. DNA template.

11.2.3. Fluorescence In Situ Hybridization

11.2.3.1. Slide Denaturation

1. Denaturation buffer (preferred): 70% (v/v) deionized formamide, 10% (v/v) sterile distilled water, 10% (v/v) 20× SSC, 10% (v/v) phosphate buffer (see below); make fresh as required.
2. OR: Denature solution: 70% (v/v) formamide, 2× SSC (pH 7.0), 0.1 mM EDTA, pH 7.0 (Add 175 ml formamide, 25 ml 20× SSC (pH 7.0), 50 ml 0.5 M EDTA, pH 7.0 and 50 ml purified water to make 250 ml solution and mix thoroughly. Verify that the pH is 7.0–7.5 by measuring the pH at ambient temperature. Between uses, store covered at 4°C. Discard after 7 days).
3. Deionized formamide: Add 5 g of ion exchanger Amberlite MB1 (Sigma) to 100 ml of formamide (Merck); stir for 2 h (room temperature) and filter twice through Whatman no. 1 filter paper. Aliquot and store at –20°C.
4. Phosphate buffer: prepare 0.5 M Na_2HPO_4 and 0.5 M NaH_2PO_4, mix these two solutions (1:1) to get pH 7.0, then aliquot and store at –20°C.

11.2.3.2. Hybridization Mixture

1. Deionized Formamide (100%).
2. 20× standard saline citrate stock solution: 3.0 M NaCl, 0.3 M Na-citrate (SSC).
3. 50% dextran sulfate in water. Dissolve by heating at 70°C for several hours. Aliquot and store at –20°C.
4. 10% (w/v) sodium dodecyl sulfate (SDS) in water, filter (0.22 μm) sterilize and store at room temperature.

5. Salmon or Herring sperm DNA 1 μg/μl (Sigma). Autoclave and store at –20°C.
6. Purified labeled probe (5 ng/μl).

11.2.3.3. Post-hybridization and Detection Washing

1. Washing solution 1: 50% (v/v) Deionized formamide (Merck), 10% (v/v) 20× SSC, 40% (v/v) distilled water; make fresh as required.
2. Washing solution 2: 2× SSC and 0.05% Tween 20 (low stringency) or 0.1× SSC and 0.05% Tween 20 (high stringency).
3. Washing solution 3: 2× SSC.
4. Blocking solution: 1–3% (w/v) BSA in PBS, 0.05% (v/v) Triton X-100 (make up fresh).
5. Detection solution (different detection reagents can be combined to detect different labels simultaneously in multiple FISH experiments).
   (a) 20 μg/ml anti-DIG-fluorescein conjugate antibody (Roche, Switzerland), in PBS, 0.5% BSA pH 7.4 for Digoxigenin probes.
   (b) 20 μg/ml Streptavidin Texas-Red conjugate), in PBS, 0.5% BSA pH 7.4 for biotin probes.
6. Antifade solution: 100 mg *p*-phenylenediamine dihydrochloride (Sigma)) in 10 ml PBS. Adjust to pH 8.0 with 0.5 M carbonate–bicarbonate buffer (0.42 g $NaHCO_3$ in 10 ml dH_2O, adjust pH to 9.0 with NaOH). Add to 90 ml with glycerol. Filter with 0.22 μm membrane to remove undissolved particles. If necessary, add 0.5–1 mg/ml DAPI). Ready-to-use antifade solutions are available commercially, such as FluoroGuard Antifade Reagent (Bio-Rad cat. No. 170-3140).
7. Counterstaining solution: 1 μg/ml propidium iodide (Molecular Probes) or 2 μg/ml DAPI (4′,6-diamidino-2-phenylindol; Sigma) or TOTO3 (Molecular Probes). The counterstaining solution can be combined with antifade solution.
8. Epifluorescent microscope equipped with digital camera or confocal laser scanning microscope (preferred).

11.3. Methods

The methods described below outline (a) cytological preparation, (b) preparation of DNA for use as a probe, (c) Fluorescent in situ hybridization (d) and fluorescence microscopy.

11.3.1. Cytological Preparation

FISH is usually carried out on a chromosome or nuclear spread on a microscope slide. This method yields well separated chromosomes or nuclei with good morphology. However, cytological techniques to study mitotic chromosomes have proved difficult when applied to fungi. Blocking mitosis in metaphase, when chromosomes are condensed and easy to identify according to their morphological features, is not successful with many fungi due to the coenocytic organization of hyphae for Zygomycota and Glomeromycota. Karyotypes of filamentous ascomycetes have been determined cytologically. Conventional cytology by light microscopy has been the main methodology for this purpose, and observations were mostly made on meiosis in asci. Contrary to plants and animals, mitotic observations are very scarce in fungi because metaphase chromosomes are rarely seen during somatic divisions and chromosome sizes are much smaller than meiotic bivalents.

11.3.1.1. Germ Tube Burst Method

Specimens of mitotic chromosomes were prepared by Germ tube burst method (GTBM) (more detail can be found in Taga et al. *(1)*, references therein). The conditions specific for the plant pathogenic fungus *Nectria haematococca* are as follows.

1. Harvest in water fresh macroconidia produced on carnation leaf agar, then wash twice with potato dextrose broth (PDB) by centrifugation, and finally suspend in PDB at a concentration of approximately 5×10^5 ml.
2. Place on microscope slide covered with a thin layer of poly-l-lysine, 100 μl conidial suspension and incubate under humid conditions at 25°C for 4–5 h.
3. Dip the slides in water to wash off the PDB, and wipe away the excess water with filter paper, leaving the germlings wet.
4. Immerse the slices in fixative (methanol: acetic acid 17: 3 (v/v)) for 30 min and flame dry the slides. Mark the area on the opposite side of the slide using a tungsten-point lab pen to etch the glass.
5. Treat the specimens with 100 μg/ml RNase A in 2× SSC for 1 h at 37°C and dehydrate through an ethanol series (70–80–99%), 5 min each.
6. Store the dried specimens at room temperature until they are used.

11.3.1.2. Interphase Nuclear Spread

This protocol is recommended for coenocytic fungi forming multinucleate spores (*see* **Fig. 11.1**) *(3, 4)*.

1. Crush a single or few spores on super frost/plus slides and then dry overnight.

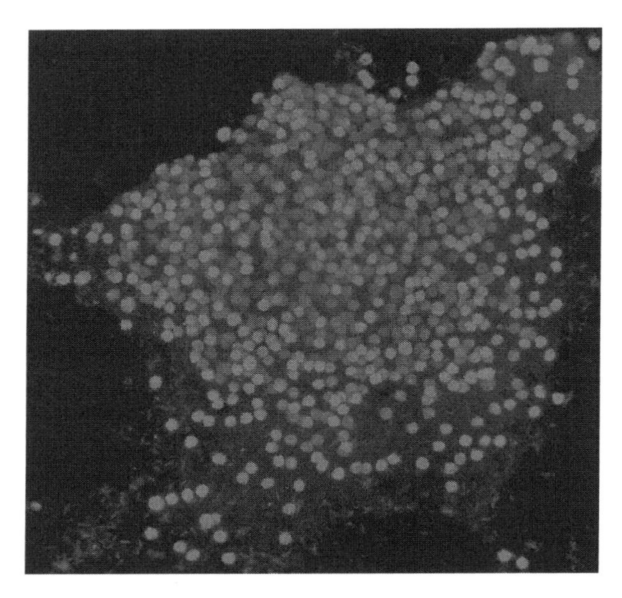

Fig. 11.1. Nuclear spread with interphasic nuclei of the arbuscular mycorrhizal fungus *Scutellospora castanea* (Glomeromycota) shown by fluorescent propidium iodide.

2. Mark the area on the opposite side of the slide using a tungsten-point lab pen to etch the glass.
3. Incubate the slides with 100 µl of DNase-free RNase (100 µg/ml) in 2× SSC for 1 h at 37°C. Rinse in 2× SSC for 5 min at room temperature.
4. Incubate the slides in a proteinase K solution (1 µg/ml proteinase K, 20 mM Tris–HCl, pH 7.5, 2 mM $CaCl_2$, at 37°C) for 10 min. Rinse in a 20 mM Tris–HCl, pH 7.5, 2 mM $CaCl_2$, 50 mM $MgCl_2$ at room temperature for 5 min, then dehydrate in ethanol series (70–80–99%), 5 min each.
5. Store the dried specimens at room temperature until use.

11.3.2. DNA Probe Preparation

DNA or RNA can be used to obtain the probes for FISH. PCR products and DNA oligonucleotides are the preferable probes at present. Cloned DNA sequences and total genomic DNA are also common. PCR is a fast, easy and efficient method for preparing probes using haptene-dUTP labeled nucleotides such as Digoxigenin-11-dUTP, Biotin-16 (or 11)-dUTP, fluorescein-dUTP or rhodamine-dUTP, etc. (*see* **Note 2**). Commercial kits, for example PCR DIG Probe Synthesis Kit, are available from Roche Diagnosis. More details for probe labeling can be found in Schwarzacher and Heslop-Harris on 2000 *(5)*.

11.3.2.1. PCR Labeling of Cloned DNA Sequences

To perform several parallel reactions, prepare a master mix containing water, buffer, $MgCl_2$, primers and *Taq* DNA Polymerase in a single tube, which can then be aliquoted into individual tubes. dNTPs, Dig-dNTPs (dATP, dCTP, dGTP (2 mM each), 1.3 mM dTTP, 0.7 mM DIG-11-dUTP, alkali-labile, pH 7.0) and template DNA solutions are then added. This method of setting reactions minimizes the possibility of pipetting errors and saves time by reducing the number of reagent transfers.

1. Gently vortex and briefly centrifuge all solutions after thawing.
2. Add, in a thin-walled PCR tube, on ice:
3. Gently vortex the sample and briefly centrifuge to collect all drops from walls of the tube.
4. Overlay the sample with half volume of mineral oil. This step may be omitted if the thermal cycler is equipped with a heated lid.
5. Place samples in a thermocycler and start PCR. Amplification parameters depend greatly on the template, primers and amplification apparatus used:
   (a) The initial denaturation should be performed over an interval of 1–3 min at 95°C if the GC content is 50% or less.
   (b) Denaturation step for 0.5–2 min at 94–95°C.
   (c) Usually the optimal primer annealing temperature is 5°C lower than the melting temperature of primer-template DNA duplex. Incubate for 0.5–2 min.
   (d) Perform the extending step at 68–72°C. Recommended extending time is 1 min for the synthesis of PCR frag-

Reagent	**Quantity, for 50 µl of PCR**	**Final concentration**
Sterile deionized water	Variable	
10× *Taq* buffer	5 µl	1×
Dig-dNTP mix (dATP, dCTP, dGTP (2 mM each), 1.3 mM dTTP, 0.7 mM Dig-11-dUTP)	5 µl	0.2 mM dATP, dCTP, dGTP of each, 0.13 mM dTTP 0.07 mM Dig-11-dUTP
Or 2 mM dNTP mix	5 µl	0.2 mM of each
Forward primer	Variable	0.1–1 µM
Reverse primer	Variable	0.1–1 µM
Taq DNA Polymerase	Variable	1.25 u/50 µl
25 mM $MgCl_2$	Variable	1–4 mM
Template DNA	Variable	10 pg to 1 µg

ments up to 2 kb. When larger DNA fragments are amplified, the extending time is usually increased by 1 min for each 1 kb.

(e) The number of PCR cycles depends on the amount of template DNA in the reaction mix and on the expected yield of the PCR product. For less than 10 copies of template DNA, 40 cycles should be performed. If the initial quantity of template DNA is higher, 25–35 cycles are usually sufficient.

(f) Perform a final extention at 72°C for 5–15 min to fill in the protruding ends of newly synthesized PCR products.

11.3.2.2. Checking Digoxigenin Labeled Probe

Two methods can be used to test the labeled probe: gel electrophoresis and dot-blot hybridization (*see* **Fig. 11.2**) (for more details on dot-blot hybridization see Hijri and Sanders *(6)*, Sambrook et al. *(7)*).

1. Prepare agarose gel (concentration depends on the DNA size, usually 1% is efficient for DNA of 0.5–2 kb).
2. Mix 5 μl of PCR product of the labeling reaction and control reaction and 1 μl of loading buffer (50% glycerol, 0.1% Bromophenol blue, 1× TAE).
3. 3. Run the electrophoresis and stain the gel using ethidium brimide (alternatively, this can be added to the loading buffer).
4. Compare the banding pattern between labeled DNA and non-labeled DNA. Dig-labeled DNA will migrate less then the non-labeled DNA (*see* **Fig. 11.3**).
5. Purify labeled DNA probe using regular PCR purification kits or use lithium chloride for purification as described by Sambrook et al. *(7)*.

11.3.3. Procedures for Fluorescence In Situ Hybridization

Procedures for *in situ hybridization* of labeled DNA probes to chromosome spread or interphasic nuclei are described in **Subheadings 11.3.3.1—11.3.3.6**. This includes hybridization, detection and visualization.

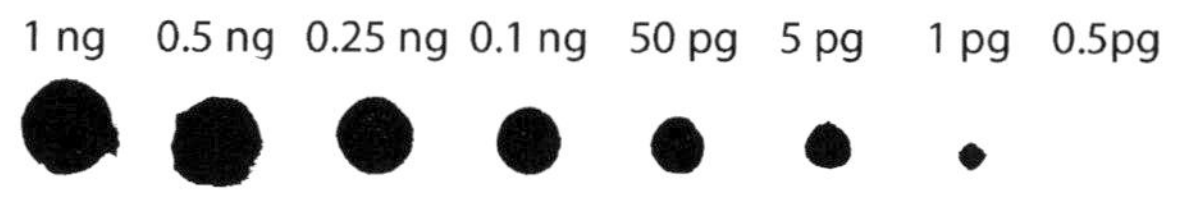

1 ng 0.5 ng 0.25 ng 0.1 ng 50 pg 5 pg 1 pg 0.5pg

Fig. 11.2. Dot blot molecular hybridization used to test probe labeling. PCR product of the DNA probe was deposited into the membrane in serial dilution (numbers correspond to the amount of DNA deposited per spot) and hybridized with the labeled probe. This shows that the probe was successfully labeled.

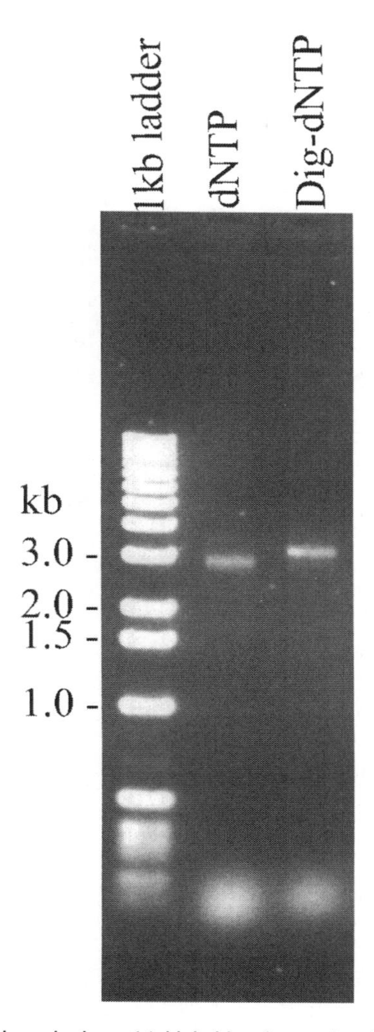

Fig. 11.3. Gel electrophoresis. *Lane 1* 1-kb ladder size marker. *Lane 2* PCR amplification of *Bip* gene of the arbuscular mycorrhizal fungus *Glomus intraradices* (Glomeromycota) cloned fragment into plasmid vector. *Lane 3* PCR amplification of *Bip* gene labeled with digoxigenin showing a migration shift in comparison to non-labeled fragment on lane 2.

11.3.3.1. Pretreatment of Slides

1. (Optional) If the slide is not dry enough, dehydrate the slide by immersing the slide into 100% ethanol for 1 min. Air dry.
2. Pretreat the slides with 200 μl diluted RNase solution (100 μg/ml) in the marked area of the slide, cover with a plastic coverslip (*see* **Note 3**) and incubate for about 30–60 min at 37°C in a humid chamber (*see* **Note 4** and **Fig. 11.4**).
3. Remove the coverslip carefully and wash the slides with 2 × SSC for 5 min.

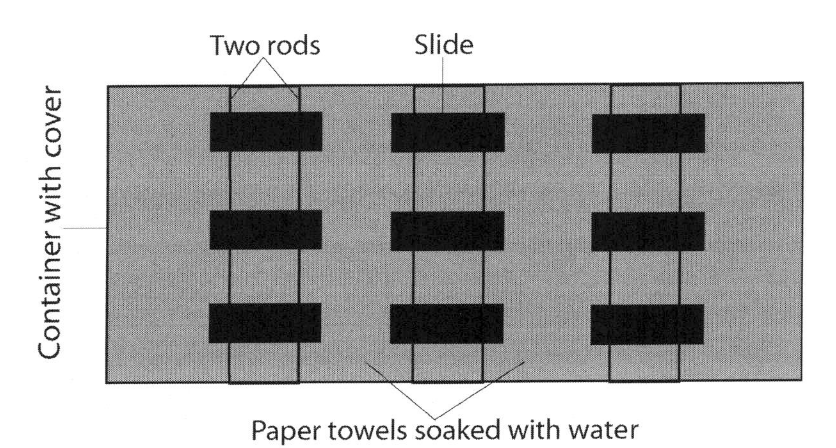

Fig. 11.4. Scheme of typical humid chamber made with a container chamber with cover and used for in situ hybridization.

4. Pepsin treatment: (Optional: recommended for interphasic nuclei) (*see* **Note 5**).
   (a) Add 200 μl diluted (with 0.01 M HCl) pepsin to each slide and cover with a plastic coverslip.
   (b) Incubate the slides at 37°C for 5–10 min.
   (c) Remove the coverslip and wash in 1× PBS twice for 5 min at RT with gentle shaking.
5. Place the slides in PBS/50 mM $MgCl_2$ for 5 min, then in paraformaldehyde fixative in PBS/50 mM $MgCl_2$/1% formaldehyde for 10 min in the fume hood.
6. Wash in PBS twice for 5 min (with agitation).
7. Dehydrate the slide with 70, 90, and 100% ethanol for 1–2 min each and allow to air dry. Slides can be stored desiccated at 4°C for up to 1 month before use.

11.3.3.2. Preparation of Hybridization Mixture and Probe Denaturation

Prepare hybridization mixture for a volume of 100 μl (typical hybridization mixture volume per slide is 20–30 μl per slide) as follows:

1. 50 μl of formamide (final concentration 50%).
2. 10 μl of 20× SSC (final concentration 2×).
3. 20 μl of 50% dextran sulfate (final concentration 10%).
4. 10 μl of 1 μg/μl salmon or herring sperm DNA (final concentration 0.1 μg/μl).
5. 5. 1 μl of 10% SDS (final concentration 0.1%).
6. Purified labeled DNA probe (final concentration 5 μg/μl).

7. Complete with 100 µl with sterile distilled water, and mix by pipetting up and down or vortex gently.
8. Denature the hybridization mixture by boiling in water for 5 min, then place immediately on ice.
9. Proceed with denaturation and hybridization. Avoid bubble formation. Seal the coverslip with rubber cement and denature the hybridization mix at 90°C for 5–10 min. Place the probe immediately on ice. Transfer the denatured probe to 37°C for 15 min – 2 h for prehybridization.

11.3.3.3. Denaturation and Hybridization

1. Add 25 µl of the denatured hybridization mixture to each marked area on the slides; cover with a small vinyl coverslip. Avoid bubble formation by carefully placing the coverslip.
2. Seal the coverslip with rubber cement to avoid the evaporation of the hybridization mixture.
3. Heat the slides to denature the target, preferably on an in situ hybridization thermal-cycler or heated block, at 90°C for 5–10 min *(8)*.
4. Allow the slides to cool down to 37°C.
5. Keep the slides in a humid chamber at 37–42°C overnight for hybridization.

11.3.3.4. Washing Slides

Prepare post-hybridization washing solutions 1, 2, and 3 (200 ml each typically for 2 times washing). Use staining jar of 100 ml volume for washing the slides. Slides must be totally covered with solutions.

1. Place the wash tanks containing washing solution 1 in a 45°C water bath for at least 30 min prior to use.
2. Remove rubber cement using forceps. Coverslips can then be removed either by floating off in 2× SSC or gently tipping them off into the glass disposal bin (never pull them off).
3. Wash the slides with washing solution 1 at 45°C for 5–10 min. This step can be repeated up to tree times.
4. Wash the slides with washing solution 2 at RT for 5 min.
5. Wash the slides with washing solution 3 for 5 min at RT (*see* **Note 6**).

11.3.3.5. Signal Detection

Make sure the slides are not dried any point during the detection and washing steps.

1. Place slides in blocking solution and inoculate for 5–15 min at 37°C.
2. Add 100 µl of detection solution (20 µg/ml anti-Dig-FITC conjugate antibody or 20 µg/ml Streptavidin Texas-Red conjugate (*see* **Note 7** and **Fig. 11.5**)) onto slides and cover with

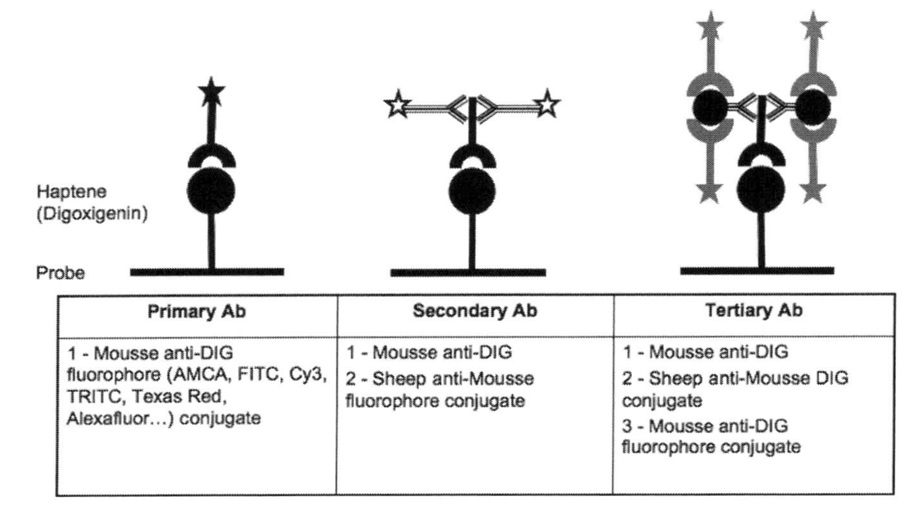

Primary Ab	Secondary Ab	Tertiary Ab
1 - Mousse anti-DIG fluorophore (AMCA, FITC, Cy3, TRITC, Texas Red, Alexafluor...) conjugate	1 - Mousse anti-DIG 2 - Sheep anti-Mousse fluorophore conjugate	1 - Mousse anti-DIG 2 - Sheep anti-Mousse DIG conjugate 3 - Mousse anti-DIG fluorophore conjugate

Fig. 11.5. Antibody combinatorial schemes show that digoxigenin-labeled DNA (Dig is taken here as an example but any other haptene can be used) can be detected with a variety of labeled antibodies (using different fluorophores) and that the signal can be further amplified using secondary and tertiary antibodies.

plastic coverslip. Keep the slides in a moist chamber in dark for 30–60 min at 37°C.

3. Float off the plastic coverslip in blocking solution and inoculate for 5 min at 37°C, then proceed with the counterstaining.
4. This step is optional before counterstaining. Dehydrate the slides with 70, 90, and 100% ethanol for 1–2 min each and allow to air dry.

11.3.3.6. Visualizing the Hybridization

1. Apply 100 μl of mixed Counterstaining solution and antifade solution on the marker area of slides and cover with a large thin glass coverslip, avoiding bubble formation (*see* **Note 8**).
2. Store in the dark if not used; otherwise, examine the slide at once under fluorescence microscope or confocal microscope (*see* **Fig. 11.6** and **Note 9**).

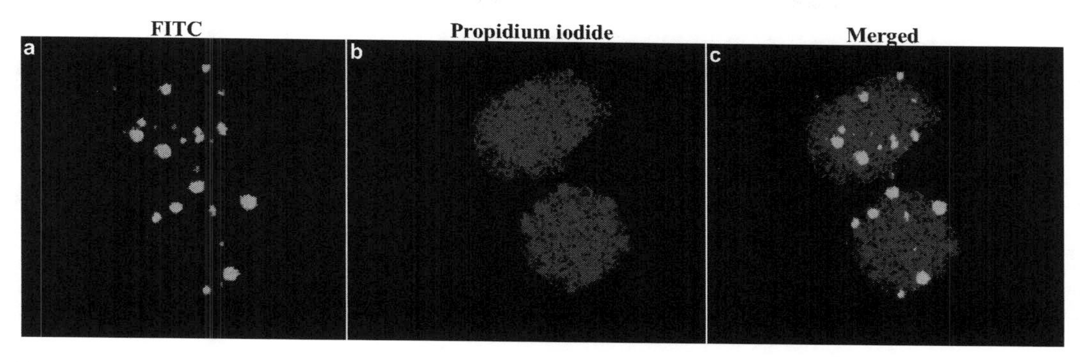

Fig. 11.6. Fluorescence in situ hybridization of $(TTAGGG)_{50}$ telomere probe (*9*) in interphasic nuclei of the arbuscular mycorrhizal fungus *Glomus intraradices* (Glomeromycota) using DIG-labeled product of PCR amplification as probe. (**a**) Fluorescein isothiocyanate (FITC) signals are visualized as green spots; (**b**) Nuclei were counterstained with propidium iodide (PI); (**c**) Merged image of FITC hybridization signals and PI counterstaining. Scale bar represents 1 μm.

11.4. Notes

1. The RNase must be DNase free. DNase should be inactivated by boiling the RNase stock in water for 10–15 min. Alternatively, purchase DNase-free RNase.
2. PCR has become the method of choice for labeling probes using different haptene-dUTP (biotin-dUTP, digoxigenin-dUTP, DNP-dUTP, Cy3-dUTP or DEAC-dUTP). First, PCR uses very little template DNA, but yields large amounts of well-labeled probe. However, the PCR labeling procedure requires optimization, which can be costly in time and consumables. Alternatively, other methods such as nick translation, random priming, end-labeling in vitro translation, can be performed. Many labeling kits are available commercially.
3. Plastic coverslips can be prepared by cutting 20 × 30 mm pieces from transparent plastic autoclavable waste disposal bags or microwave cook-in bags available in the market.
4. Use any covered container that will hold paper towels soaked with water and two rods to hold hybridization slides horizontally in the humidified chamber (*see* **Fig. 11.5**).
5. Pepsin is used to remove the cytoplasm covering and help the probe to access its target. Over-digestion can also cause problems (loss of cells from the slide), so only use when absolutely necessary.
6. Adjust stringency as required. FISH is typically carried out in a range of 70–90%, described as low and high stringency respectively.

7. Antibody combinatorial schemes allow, for example, multiple color detection and signal amplification for two or more probes labeled with different haptene-dUTPs. When detecting the haptenes, one can choose only one fluor-labeled antibody for each probe (for example: anti-Dig-FITC conjugate antibody or Streptavidin Texas-Red conjugate respectively for Dig and biotin detection) or one can use primary, secondary and even tertiary antibodies to amplify the hybridization signals. When combining antibodies in multicolor detection schemes, it is important to pay special attention to potential unwanted antibody-antibody interactions or cross-reactions or fluorescent cross-talk. It is also useful to purchase antibodies previously affinity-purified for interactions with other animal species used. **Figure 11.5** shows antibody combinations allowing signal amplification of digoxigenin-labeled DNA that can be detected with a variety of labeled antibodies (using different fluors) in FISH. This signal can be further amplified using secondary and tertiary antibodies.
8. Counterstaining step can be done separately before mounting slides with an antifade solution. Slides should then be incubated in counterstaining solution (DAPI or PI) for 5–10 min in the dark, washed briefly in detection buffer and then mounted with antifade solution.
9. The confocal laser scanning microscope (CLSM) provides for digitized optical sectioning of prepared samples and can be operated in fluorescence (confocal), reflectance (confocal), and transmittance (non-confocal) modes. This instrument has the capacity to collect data from samples using up to three extremely sensitive photomultiplier tubes and digital laser scanning. A CLSM will always give a better image than can be obtained with a standard epifluorescence microscope. The improvement is spectacular in the case of large whole-mount specimens. Indeed, it is often impossible to distinguish any interior details with conventional microscopy, yet a perfectly clear image of an optical section can be obtained using confocal imaging.

Acknowledgments

I thank all members of my group at the Plant Biology Research Institute, University of Montreal, for fruitful discussions. I also thank Erin Zimmerman for the constructive comments.

References

1. Taga, M., Murata, M., Saito, and H. (1998) Comparison of different karyotyping methods in filamentous ascomycetes–a case study of *Nectria haematococca*. *Mycological Research* 102, 1355–1364.
2. Tsuchiya, D., Matsumoto, A., Covert, S.F., Bronson, C.R., and Taga, M. (2002) Physical mapping of plasmid and cosmid clones in filamentous fungi by fiber-FISH. *Fungal Genetics and Biology* 37, 22–28.
3. Kuhn G., Hijri M., and Sanders I.R. (2001). Evidence for the evolution of multiple genomes in arbuscular mycorrhizal fungi. *Nature* 414, 745–748.
4. Trouvelot, S., van Tuinen, D., Hijri, M., and Gianinazzi-Pearson, V. (1999) Visualisation of ribosomal DNA loci in spore interphasic nuclei of glomalean fungi by fluorescence *in situ* hybridization. *Mycorrhiza* 8: 203–206.
5. Schwarzacher, T., and Heslop-Harrison, P. (2000) Practical *in situ* hybridization. , Bios Scientific Oxford: Bios Scientific Ltd Press.
6. Hijri M., and Sanders IR. (2004) The arbuscular mycorrhizal fungus *Glomus intraradices* is haploid and has a small genome size in the lower limit of eukaryotes. *Fungal Genetics and Biology* 41(3), 253–261.
7. Sambrook, J., Fritsch, E.F., and Maniatis, T., (2001) Molecular cloning: a laboratory manual 3rd edition, Vols. 1, 2 and 3. NY: Cold Spring Harbor Laboratory Press.
8. Heslop-Harrison, J.S., Schwarzacher, T., Anamthawat-Jonsson, T., Leith, A.R., Shi, M., and Leith, I.J. (1991) *In situ* hybridization with automated chromosome denaturation. *Technique* 3, 109–115.
9. Hijri, M., Niculita, H., and Sanders, I.R. (2007) Molecular characterization of chromosome termini of the multi-genomic arbuscular mycorrhizal fungus *Glomus intraradices* (Glomeromycota). *Fungal Genetics and Biology* 41, 1380–1386.

Chapter 12

Use of Molecular Methods for the Detection of Fungal Spores

Elaine Ward

Summary

Traditional methods for the isolation and identification of fungal spores can be time-consuming and laborious. DNA-based methods for fungal detection can be used to detect the spores of plant-pathogenic fungi. Air borne spores can be collected and identified by PCR allowing identification of the species.

Key words: Spores, DNA, PCR, Pathogen, Fungi, Identification.

12.1. Introduction

Traditional methods for detecting and enumerating fungal spores often rely on microscopic and cultural techniques that are time-consuming and laborious and require skilled and highly specialized expertise. Also, microscopy is often unreliable for identification of the small, non-descript spores that are produced by many species, whilst cultural techniques are unsuitable for detection of fungi that are slow growing, or non-culturable *in vitro* and the choice of the medium may select the species recovered. DNA-based methods offer greater potential for sensitive and specific detection and quantification of fungal spores and are increasingly becoming a valuable tool in diagnosis and research. DNA-based techniques, particularly the polymerase chain reaction (PCR) allow minute quantities of a pathogen to be detected, e.g. a single fungal spore *(1, 2)*. These methods have opened up new opportunities to study and understand the biology of plant pathogenic fungi, pathogen population structure and dynamics,

R. Burns (ed.), *Methods in Molecular Biology, Plant Pathology, vol. 508*

DOI: 10.1007/978-1-59745-062-1_12

host–pathogen interactions, gene flow in pathogen populations and inoculum movement.

Many different protocols are available for extracting DNA, and choosing the most appropriate method for any particular application depends on the type and quantity of the material used. In some situations it is possible to obtain fungal DNA of sufficient quality and quantity by simply boiling spores, fungal mycelium or infected plant material in Tris–HCl buffer (10 mM, pH 8) for 10 min, centrifuging for 2 min at high speed, and directly using a small aliquot of the remaining liquid in a subsequent assay *(3, 4)*. More often however, the DNA must be extracted from the sample and purified prior to use in molecular protocols. Most DNA extraction methods start by disrupting the spores mechanically and using detergents such as sodium dodecyl sulphate (SDS) to help release cell contents. Various chemical treatments are then applied to remove other cell components and purify the DNA. Phenol or phenol:chloroform extractions are often employed to remove proteins. Ethanol and isopropanol are frequently used to precipitate and concentrate the DNA, and RNaseA can be used to remove RNA from the sample. The protocols below are based on modifications of the procedure of Lee and Taylor *(1)*, and we also routinely use the method of Fraaije et al. *(5)*. However, a variety of other methods have also been used for extraction of DNA from fungal spores *(6–8)*.The Methods used often need to be modified for particular plant species or environments (e.g. soil) where co-extraction of inhibitory chemicals can cause problems in the subsequent detection of molecular protocols. This is sometimes achieved by the use of commercial kits that contain resins to purify the DNA *(9, 10)*.

Monitoring of airborne inoculum can be used to study the spread of plant disease and aid in assessment of an epidemic development. This has traditionally involved recognising differences in spore morphology by microscopy, or by culture-based tests. However, recently molecular techniques have begun to be used in the detection of airborne inoculum. A wide variety of methods are available for collecting fungal spores prior to analysis, including cyclone samplers, impaction samplers (e.g. "Hirst-type" and rotating arm samplers), liquid impingers and filter-based samplers *(11, 12)*. To date most reports of detection of airborne microorganisms by PCR have been in the area of medical microbiology, however there have been a few reports of detection of airborne spores of fungal plant pathogens by PCR, including those of *Botrytis cinerea*, *Fusarium circinatum*, *Leptosphaeria maculans*, *Mycosphaerella graminicola*, *Pyrenopeziza brassicae* and *Sclerotinia sclerotiorum (5, 13–15)*.

Most molecular diagnostic assays involve the use of PCR, a method for rapidly synthesising (amplifying) millions of copies of specific DNA sequences. Various publications describe the principles of PCR,that cover practical details, and review its applications in plant pathology *(11, 16–22)*. Variations of the technique include nested PCR, which allows more sensitive detection, real-time PCR, which allows quantification of pathogen inoculum, and multiplex PCR, which allows detection of several pathogens simultaneously.

12.2. Materials (*See* Note 1)

1. Sterile distilled water.
2. 0.1% Nonidet P-40.
3. Ballotini beads 8.5 grade, 400–455 μm diameter (Jencons-PLS, Leighton Buzzard, UK) – acid washed (*see* **Note 2**).
4. FastPrep® machine (Savant Instruments, Holbrook, New York, USA) or Ball mill (e.g. Glen Creston, Stanmore England M280) (*see* **Notes 3** and **4**).
5. 2× Lee and Taylor lysis buffer (made freshly approximately every 2 weeks): 100 mM Tris–HCl pH 7.4 (37°C), 100 mM EDTA, 6% SDS, 2% β-mercaptoethanol.
6. Vortexer.
7. Water bath.
8. Phenol:chloroform (1:1).
9. Microcentrifuge, e.g. Eppendorf 5415D.
10. Disposable polypropylene microcentrifuge tubes: 1.5 mL conical; 2 mL screw-capped.
11. Glycogen (20 μg/mL) Roche diagnostics Ltd, Lewes, UK) or GlycoBlue (15 mg/mL, Ambion, Austin, USA).
12. 6 M ammonium acetate.
13. Isopropanol.
14. TE – 10 mM Tris–HCl pH 7.5 (25°C), 0.1 mM EDTA.
15. RNaseA – 10 mg/mL in TE (heat treated – stand in a boiling water bath for 15 min) (*see* **Note 6**).
16. Phenol solution equilibrated with 10 mM Tris–HCl pH 8, 1 mM EDTA (Sigma Aldrich, Gillingham, UK).
17. Ethanol (70%).
18. For spore trapping: Miniature cyclone air sampler or Burkard 7-day recording volumetric spore trap (both from Burkard Manufacturing Co., Ltd, Rickmansworth, UK; http://www.burkard.co.uk/instmts.htm). The 7-day recording

spore trap also requires Melinex tape, hexane, Vaseline and paraffin wax.

19. dNTPs – a mixture of dATP, dCTP, dGTP and dTTP (10 mM each, MBI Fermentas, York, UK), stored at –20°C.
20. Thermostable DNA polymerase (e.g. *Taq*, and reaction buffer supplied by manufacturer). We typically use *Taq* from MBI Fermentas (York, UK), which is supplied with a 10× reaction buffer of 100 mM Tris–HCl pH 8.8 (25°C), 500 mM KCl and 0.8% Nonidet P40. Separate stocks of BSA (20 mg/mL) and $MgCl_2$ (25 mM) are also supplied for adding to the reaction. All of these reagents are stored at –20°C.
21. Oligonucleotide primers (custom-made, Sigma-Genosys, Haverhill, UK) re-suspended to a concentration of 100 μM using sterile distilled water and stored at –20°C.
22. PCR tubes (thin walled).
23. Thermal cycler, e.g. Applied Biosystems GeneAmp 9700.
24. Agarose, e.g. Molecular biology grade agarose from Helena Biosciences (Sunderland, UK).
25. TBE buffer – 50 mM Tris, 50 mM boric acid, 1 mM EDTA. Dilute when needed from a 10× stock.
26. Ethidium bromide – 0.5 mg/mL stock.
27. Gel loading mixture – 40% (w/v) sucrose, 0.1 M EDTA, 0.15 mg/mL bromophenol blue.
28. Horizontal electrophoresis equipment (e.g. Biorad Wide Mini Sub Cell).
29. U.V. transilluminator and camera suitable for photographing agarose gels, e.g. Syngene Gene Genius Bioimaging system.

12.3. Methods

The methods given below describe general procedures for obtaining DNA from spores of fungi and other eukaryotic microbes, such as the oomycetes and the plasmodiophorids, and its subsequent testing in PCR assays. Modifications that may be needed to extract DNA from different types and sources of material are also described. The volumes and number of tubes used per sample may need to be varied, depending on the type of sample and the number of spores being processed.

12.3.1. Extraction of DNA from Spores(e.g. of Sclerotinia sclerotiorum) in Liquid Suspension

This method has been used to prepare DNA from spores, including those of *Sclerotinia sclerotiorum*, *Leptosphaeria maculans*, *Pyrenopeziza brassicae*, *Penicillium roqueforti* and *Polymyxa* spp. (*2, 13, 14, 23, 24*).

1. Collect or re-suspend the spores in sterile distilled water or 0.1% Nonidet P-40 and transfer to 2 mL screw-capped microcentrifuge tubes in 400-μL aliquots.
2. Add 0.4 g acid-washed Ballotini beads and agitate in a Fast-Prep® machine at 6 m/s for 2 periods of 40 s, with 2 min on ice between.
3. Add 400 μL 2× Lee and Taylor lysis buffer, vortex and incubate at 65°C for 1 h.
4. Add 800 μL phenol:chloroform (1:1) to each tube and vortex briefly.
5. Centrifuge 15 min at top speed in the microfuge (16,000 rcf/13,200 rpm), then carefully transfer as much as possible of the top aqueous layer to a clean tube. Do not disturb the debris at the interface.
6. Add 1 μL (20 μg) glycogen (or 2 μL (30 mg) GlycoBlue), 40 μL 6 M ammonium acetate, 600 μL isopropanol, invert gently to mix and spin for 2 min. For preparation of DNA from small numbers of spores, the addition of glycogen as a carrier for the DNA during isopropanol precipitation is essential; without it, the DNA does not reliably precipitate and is easily lost. Also, incubation of the mixture at –20°C for 10–60 min before centrifugation may improve recovery of DNA but can result in a less pure sample. For preparations from large numbers of spores these measures are not necessary.
7. Remove supernatant.
8. Add 50 μL RNaseA in TE to the pellet and incubate at 37°C for 15 min.
9. Pool into 200 μL samples or add 150 μL TE, then add 200 μL phenol, vortex and centrifuge at top speed for 6 min. Carefully transfer the top aqueous layer to a fresh tube.
10. Add 10 μL 6 M ammonium acetate, 600 μL isopropanol and incubate at –20°C for 10 min.
11. Centrifuge at top speed for 10 min, discard the supernatant.
12. Add 800 μL cold 70% ethanol, centrifuge for 2 min and discard the supernatant.
13. Centrifuge for 10 s and remove the remaining liquid with a Micropipette.
14. Allow the pellet to dry for 20 min in a fume hood and then re-suspend it in 50–200 μL TE or sterile distilled water.

12.3.2. Extraction of DNA from Airborne Spores(e.g. of Botrytis cinerea) Collected on Miniature Cyclone Air Samplers

In the miniature cyclone air sampler, which is a relatively new design, spores are collected from the air into 1.5-mL microcentrifuge tubes *(25)*. This sampler is ideal for downstream molecular analysis because it collects a dry sample directly into a vessel suitable for use with PCR. Therefore there is no need to remove the sample from a substrate as is necessary with more conventional spore samplers (*see* **Subheading 12.3.3**). Fungal spores that have been analysed by this method include *Penicillium roqueforti (2)* and *Botrytis cinerea (14)* (*see* **Note 9**).

1. Operate the miniature cyclone at a sampling rate of 20 L/min. The sampling time will depend on the expected concentration of spores in the air.
2. Re-suspend the material collected into the 1.5-mL microcentrifuge tube in 0.8 mL of Nonidet P-40 by vortexing for 2 min. Transfer 400 μL to each of 2 screw-capped microfuge tubes.
3. Follow the steps in **Subheading 12.3.1** from **step 2**.

12.3.3. Extraction of DNA from Airborne Spores (e.g. of Sclerotinia sclerotiorum) Collected on Burkard Volumetric Spore Samplers

The Burkard 7-day recording volumetric spore trap (Burkard Manufacturing Co., Rickmansworth, UK) is a Hirst-type spore sampler, which collects airborne particles on a wax-coated Melinex tape attached to a slow rotating drum. Trapped particles can be recorded over a 7-day period and the trap is operated with an air throughput of 14.4 m^3 per day *(26–28)*. The method described below has been used for analysing spores of *Penicillium roqueforti (24)*, *S. sclerotiorum (14)*, *Leptosphaeria maculans* and *Pyrenopeziza brassicae (13)* and *Mycosphaerella graminicola (5)*. Rotating arm-type spore traps *(29, 30)*, which collect spores on a coated tape, can also be used to collect material for DNA analysis using the same method *(24)* (*see* **Note 10**).

1. Attach clear polyester film tape (Melinex, Burkard Manufacturing Co., Rickmansworth, UK) to the Burkard spore trap then coat the tape with a thin film of a 5:1 mixture of paraffin wax and petroleum jelly (Vaseline). This is done by dissolving the paraffin wax/petroleum jelly mixture in hexane and applying the solution thinly to the tape using a fine brush or a thin plastic spatula. The hexane evaporates leaving a thin layer of wax on the tape.
2. After every 7 days, remove the exposed Melinex tape from the trap and cut into sections representing known 12-h periods (9.5 × 24 mm), then cut these daily tape sections in half along their centrelines in the direction of rotation.
3. Place one of these halves in a screw-capped 2-mL plastic tube for DNA extraction. The other half can be permanently mounted onto a slide using Gelvatol containing Trypan blue stain *(27)*

for microscopic examination, if needed. Alternatively, the other half of the tape may also be used for DNA extraction.

4. Add 400 μL 0.1% Nonidet P-40 and 0.4 g Ballotini beads to the tape samples to be used for DNA extraction.
5. Agitate for two periods of 40 s, at 6 m/s in a FastPrep machine, with 2 min on ice in between.
6. Transfer the disrupted spore suspension to a fresh tube.
7. Follow the steps in **Subheading 12.3.1** from **step 3**.

12.3.4. Detection of Spore DNA Using PCR-Based Methods

The most commonly used target for fungal detection is the DNA encoding the ribosomal RNA genes, but other targets include genes for beta-tubulin, elongation factor 1 alpha, calmodulin, chitin synthase, actin, ras protein, mating type determination and cytochrome *b (10, 31–34)*.The design of new primers will be facilitated as more sequences become available of genes other than those encoding ribosomal RNA, for example as a result of initiatives such as the AFTOL (Assembling the Fungal Tree of Life) project (http://www.aftol.org/, *(35)*). There is an on-line searchable database of primer sets useful for the detection of plant pathogenic fungi: http://www.spaadbase.com)*36)*.

Targets for diagnostic assays may also be developed by screening random regions of the genome to find sequences that are unique to a particular taxon. This can be done by using techniques such as RAPD (random amplified polymorphic DNA) analysis to generate randomly amplified fragments from the genome. The patterns of bands generated from related fungi are compared, and any potentially diagnostic fragments are sequenced and used to design specific SCAR (sequence characterised amplified region) primers *(37, 38)*.

The following PCR protocol uses consensus (Universal) primers to amplify the internal transcribed spacer (ITS) regions of fungal ribosomal RNA genes *(39)*. Under the conditions described here, it can also be used to detect other eukaryotic microbes of interest to plant pathologists, namely the oomycetes (e.g. *Phytophthora* and *Pythium* spp.) and the plasmodiophorids (e.g. *Plasmodiophora*, *Spongospora* and *Polymyxa* spp.), but may also detect plant DNA. Primers ITS4 and ITS5 *(39)* amplify a region from the 3′ end of the 18S rRNA gene to the 5′ end of the 28S rRNA gene that includes the 5.8S rRNA gene and the two ITS regions. This PCR is useful for providing a positive control to ensure that amplifiable DNA has been extracted from the sample. In combination with RFLP analysis, it can also be used to aid identification of organisms *(3, 23)*. These and other consensus rRNA gene primers are also useful in studies of fungal community DNA (e.g. from plant roots, soil, airborne samples) and for identifying "unknown" microorganisms in samples without any any prior knowledge of what they might be. The amplified rRNA genes are cloned into a plasmid vector and sequenced. Computer programs such as BLAST *(40)* and FASTA *(41)* are then used to identify the organisms in the sample based on the closest matches with sequences of known organisms in the Genbank/EMBL databases *(42–45)*.

1. Thaw the reagents and DNA preparations and keep on ice until required. Prepare a reaction mixture sufficient for the number of samples to be tested. Each 100 μL of this "master mix" contains: 75 μL sterile distilled water, 10 μL 10× PCR buffer, 6 μL 25 mM $MgCl_2$, 2 μL dNTP mix, 1 μL each primer, 0.5 μL 20 mg/mL BSA and 0.2 μL *Taq* polymerase (5 U/μL). Vortex and centrifuge the mixture briefly then distribute 25 μL aliquots into suitable PCR tubes and place on ice.
2. Add DNA (10–100 ng or 1–2 μL of a suitable dilution) to the tubes, using the pipette tip to mix the tube contents briefly.
3. Place the tubes in the PCR machine and run the appropriate cycling programme: 30 cycles of 30 s at 94°C, 1 min at 42°C and 2 min at 72°C, followed by 10 min at 72°C.
4. When the cycling is complete, analyse 5 μL of each sample by agarose gel electrophoresis (*see* **Note 11**).
5. To prepare the gel, add 1% agarose to 1× TBE buffer and heat to dissolve (e.g. using a microwave oven).
6. Add 100 μL of 0.5 mg/mL ethidium bromide per 100 mL agarose, mix and allow to cool to approximately 60°C.
7. Pour into a suitable gel tray with comb and allow to set for approximately 20 min.
8. Remove the comb and place the gel in a horizontal electrophoresis tank containing 1× TBE, so that the wells are just covered with buffer.
9. Add 3 μL gel-loading mixture to 5 μL of the PCR samples, transfer to the wells of the agarose gel, and electrophorese.
10. Visualize using U.V. transilluminator and photograph.

12.4. Notes

1. Gloves should be worn throughout these procedures and particular care should be taken when handling phenol and ethidium bromide (consult safety data sheets). Both of these are suspected mutagens and phenol is corrosive. Steps involving

phenol, chloroform and β-mercaptoethanol should be performed in a fume hood.

2. It is preferable to acid-wash the Ballotini beads before use to exclude possible contamination. A 25 mm depth of beads is placed into a 250-mL beaker and covered with 50% hydrochloric acid. The beaker is covered with cling film and the beads are stirred, using a magnetic stirrer, overnight. The acid is then discarded and the beads are rinsed thoroughly with water until the pH is approximately 7. The beads are transferred to 20-mL glass bottles, then autoclaved and dried in an oven.
3. Using the FastPrep® machine, agitation conditions of two or three periods of 40 s at 5 or 6 m/s, with 2 min on ice between each 40-s period, have been found to be efficient for disruption of most spores tested, including those of *Penicillium roqueforti*, *Leptosphaeria maculans*, *Pyrenopeziza brassicae*, *Sclerotinia sclerotiorum* and *Mycosphaerella graminicola (5, 13, 14, 24)*, but optimisation may be needed when testing a new species or spore type for the first time.
4. If a FastPrep® machine, is not available, spore disruption can be achieved in a Ball mill (e.g. Glen Creston, Stanmore England M280). The process takes longer, e.g. 5 min has been used for *Polymyxa graminis* resting spores *(23)*, and 8 min for *Penicillium roqueforti (2)*. Other equipment that can be used for spore disruption includes the Mikro-Dismembrator II (B. Braun, Meslungen, Germany) *(10)*.
5. Some spores that are thin-walled, e.g. zoospores of *Polymyxa graminis* or *Olpidium brassicae*, do not need to be disrupted using Ballotini beads prior to DNA extraction, and vortexing in lysis buffer alone is enough to release the cell contents effectively *(23, 46)*. In these cases, transfer suspensions of spores in sterile water to 2-mL microfuge tubes in 400 μL aliquots and follow the steps in **Subheading 12.3.1** from **step 3**.
6. The RNaseA treatment and subsequent phenol extraction and isopropanol precipitation steps (**steps 8–11** inclusive in **Subheading 12.3.1**) may be omitted depending on the type of sample being processed.
7. There are a variety of published methods that could be used to prepare DNA from fungal spores, but when adapting a protocol to use with bead-beating methods, avoid using high concentrations of detergent that might cause excessive frothing.
8. A variety of manufacturers also make DNA purification kits that are suitable for preparation of DNA from fungal spores, e.g. UltraClean soil DNA isolation kit (MoBio Laboratories,

Solana Beach, USA) *(10)*. We have found SureClean (Bioline Ltd., London, Cat. No. BIO37042) to be particularly effective at purifying DNA, prepared by other methods, that has proved recalcitrant to PCR.

9. For airborne spores, the most appropriate type of air sampler to use depends on the application. The miniature cyclone sampler is well suited to indoor use and the collection of the sample directly into 1.5-mL microcentrifuge tubes is ideal for use with DNA extraction protocols. However, the collection efficiency characteristics of this device has not been studied in detail. The Hirst-type traps are more suitable for use in plant pathology field studies, particularly where spores need to be monitored over daily, weekly or monthly periods. However, the sample processing (which involves extraction from coated tapes) is more complicated than with the miniature cyclone. Liquid impinger-type spore traps, e.g. the SKC Biosampler® (SKC Inc. Eighty-Four, PA, USA) have also been used to collect spores for DNA analysis *(12, 47, 48)*. These spore traps have a high sampling efficiency, which is useful where small numbers of microorganisms are being monitored, but they are difficult to use in the field and processing of the material for PCR (which requires concentration of the sample) which is more complex. Filter samplers may also be appropriate for collecting airborne spores for DNA analysis; they have been used to collect airborne bacteria for this purpose *(44)*.
10. When extracting DNA from spore traps, it is advisable to use appropriate negative controls for the DNA extraction process to demonstrate that the collection and extraction equipment (e.g. Melinex tapes, Vaseline:paraffin wax coating) and reagents are all free from microbiological contamination.
11. Agarose gel electrophoresis is described in more detail elsewhere *(20)*.

Acknowledgements

I thank Penny Hirsch, Brian Kerry and Alastair McCartney for constructive comments on the manuscript. Rothamsted Research receives grant-aided support from the Biotechnology and Biological Sciences Research Council.

References

1. Lee, S. B. and Taylor, J. W. (1990) Isolation of DNA from fungal mycelia and single spores. In PCR Protocols. A Guide to Methods and Applications. (Innis, M. A., Gelfand, D. H., Sninsky, J. J. and White, T. J., eds.), Academic Press, San Diego, CA, pp. 282–287.
2. Williams, R., Ward, E. and McCartney, H. A. (2001) Methods for integrated air sampling and DNA analysis for the detection of airborne fungal spores. *Appl. Environ. Microbiol.* 67, 2453–2459.
3. Ward, E. and Akrofi, A. Y. (1994) Identification of fungi in the *Gaeumannomyces–Phialophora* complex by RFLPs of PCR-amplified ribosomal DNAs. *Mycol. Res.* 98, 219–224.
4. Morales, F. J., Ward, E., Castano, M., Arroyave, J. A. and Adams, M. J. (1999) Emergence of rice stripe necrosis virus in South America. *Eur. J. Plant Pathol.* 105, 643–650.
5. Fraaije, B. A., Cools, H. J., Fountaine, J., Lovell, D. J., Motteram, J., West, J. S. and Lucas, J. A. (2005) Role of ascospores in further spread of QoI-resistant cytochrome b alleles (G143A) in field populations of *Mycosphaerella graminicola. Phytopathology* 95, 933–941.
6. Zolan, M. E. and Pukkila, P. J. (1986) Inheritance of DNA methylation in *Coprinus cinereus. Mol. Cell. Biol.* 6, 195–200.
7. Raeder, U. and Broda, P. (1985) Rapid preparation of DNA from filamentous fungi. *Lett. Appl. Microbiol.* 1, 17–20.
8. Volossiouk, T., Robb, E. J. and Nazar, R. N. (1995) Direct DNA extraction for PCR-mediated assays of soil organisms. *Appl. Environ. Microbiol.* 61, 3972–3976.
9. Sreenivasaprasad, S., Sharada, K., Brown, A. E. and Mills, P. R. (1996) PCR-based detection of *Colletotrichum acutatum* on strawberry. *Plant Pathol.* 45, 650–655.
10. Mauchline, T., Kerry, B. R. and Hirsch, P. R. (2002) Quantification in soil and the rhizosphere of the nematophagous fungus *Verticillium chlamydosporium* by competitive PCR and comparison with selective plating. *Appl. Environ. Microbiol.* 68, 1846–1853.
11. McCartney, H. A., Foster, S. J., Fraaije, B. A. and Ward, E. (2003) Molecular diagnostics for fungal plant pathogens. *Pest Manag. Sci.* 59, 129–142.
12. Peccia, J. and Hernandez, M. (2006) Incorporating polymerase chain reaction-based identification, population characterization, and quantification of microorganisms into aerosol science: a review. *Atmos. Environ.* 40, 3941–3961.
13. Calderon, C., Ward, E., Freeman, J., Foster, S. J. and McCartney, H. A. (2002) Detection of airborne inoculum of *Leptosphaeria maculans* and *Pyrenopeziza brassicae* in oilseed rape crops by polymerase chain reaction (PCR) assays. *Plant Pathol.* 51, 303–310.
14. Freeman, J., Ward, E., Calderon, C. and McCartney, H. A. (2002) A polymerase chain reaction (PCR) assay for the detection of inoculum of *Sclerotinia sclerotiorum. Eur. J. Plant Pathol.* 108, 877–886.
15. Schweigkofler, W., O'Donnell, K. and Garbelotto, M. (2004) Detection and quantification of airborne conidia of *Fusarium circinatum*, the causal agent of pine pitch canker from two California sites by using a real-time PCR approach combined with a simple spore trapping method. *Appl. Environ. Microbiol.* 70, 3512–3520.
16. Ward, E. (1994) Use of the polymerase chain reaction for identifying plant pathogens. In Ecology of Plant Pathogens. (Blakeman, J. P. and Williamson, B., eds.), CAB International, Wallingford, UK, pp. 143–160.
17. Edel, V. (1998) Polymerase chain reaction in mycology: an overview. In Applications of PCR in Mycology. (Bridge, P. D., Arora, D. K., Reddy, C. A. and Elander, R. P., eds.), CAB International, Wallingford, UK, pp. 1–20.
18. Martin, R. R., James, D. and Lévesque, C. A. (2000) Impacts of molecular diagnostic technologies on plant disease management. *Ann. Rev. Phytopathol.* 38, 207–239.
19. Ward, E., Foster, S. J., Fraaije, B. A. and McCartney, H. A. (2004) Plant pathogen diagnostics: immunological and nucleic-acid based approaches. *Ann. Appl. Biol.* 145, 1–16.
20. Sambrook, J., Fritsch, E. F. and Maniatis, T. (1989) Molecular Cloning. A Laboratory Manual, Cold Spring Harbor Laboratories, Cold Spring Harbor, New York.
21. Innis, M. A., Gelfand, D. H., Sninsky, J. J. and White, T. J., Eds. (1990) PCR Protocols. A Guide to Methods and Applications. Academic Press, San Diego, CA.
22. Dieffenbach, C. W. and Dveksler, G. S., Eds. (1995) PCR Primer. A Laboratory Manual, Cold Spring Harbor Laboratory, Cold Spring Harbor, New York.

23. Ward, E., Adams, M. J., Mutasa, E. S., Collier, C. R. and Asher, M. J. C. (1994) Characterisation of *Polymyxa* species by restriction analysis of PCR-amplified ribosomal DNA. *Plant Pathol.* 43, 872–877.

24. Calderon, C., Ward, E., Freeman, J. and McCartney, H. A. (2002) Detection of airborne fungal spores sampled by rotating-arm and Hirst-type spore traps using polymerase chain reaction assays. *J. Aerosol Sci.* 33, 283–296.

25. Emberlin, J. and Baboonian, C. (1995) Proceedings of the *16th European Congress of Allerology and Clinical Immunology*, Bologna, Italy, pp. 39–43.

26. Lacey, J. and Venette, J. (1995) Outdoor sampling techniques. In Bioaerosols Handbook (Cox, C. S. and Wathes, C. M., eds.), CRC Lewis, London, pp. 407–471.

27.British Aerobiology Federation (1995) *Airborne pollen and spores: a guide to trapping and counting*, British Aerobiology Federation, Worcester, UK.

28. Lacey, M. and West, J. (2006) The Air Spora – A Manual for Catching and Identifying Airborne Biological Particles, Springer, Dordrecht, The Netherlands.

29. McCartney, H. A., Lacey, M. E. and Rawlinson, C. J. (1986) Dispersal of *Pyrenopeziza brassicae* spores from an oil-seed rape crop. *J. Agric. Sci. Camb* 107, 299–305.

30. McCartney, H. A., Fitt, B. D. L. and Schmechel, D. (1997) Sampling bioaerosols in plant pathology. *J. Aerosol Sci.* 28, 349–364.

31. Fraaije, B. A., Lovell, D. J., Rohel, E. A. and Hollomon, D. W. (1999) Rapid detection and diagnosis of *Septoria tritici* epidemics in wheat using a polymerase chain reaction/ Pico green assay. *J. Appl. Microbiol.* 86, 701–708.

32. Carbone, I. and Kohn, L. M. (1999) A method for designing primer sets for speciation studies in filamentous ascomycetes. *Mycologia* 91, 553–556.

33. Foster, S. J., Singh, G., Fitt, B. D. L. and Ashby, A. M. (1999) Development of PCR based diagnostic techniques for the two mating types of *Pyrenopeziza brassicae* (light leaf spot) on winter oilseed rape (*Brassica napus* ssp. *oleifera*). *Physiol. Mol. Plant Pathol.* 55, 111–119.

34. Fountaine, J. M., Shaw, M. W., Napier, B., Ward, E. and Fraaije, B. A. (2007) Application of real-time and multiplex PCR assays to study leaf blotch epidemics in barley. *Phytopathology* 97, 297–303.

35. James, T. Y., Kauff, F., Schoch, C. L., Matheny, P. B., Hofstetter, V., Cox, C. J., Celio, G., Gueidan, C., Fraker, E., Miadlikowska, J., Lumbsch, H. T., Rauhut, A., Reeb, V., Arnold, A. E., Amtoft, A., Stajich, J. E., Hosaka, K., Sung, G.-H., Johnson, D., O'Rourke, B., Crockett, M., Binder, M., Curtis, J. M., Slot, J. C., Wang, Z., Wilson, A. W., Schuessler, A., Longcore, J. E., O'Donnell, K., Mozley-Standridge, S., Porter, D., Letcher, P. M., Powell, M. J., Taylor, J. W., White, M. M., Griffith, G. W., Davies, D. R., Humber, R. A., Morton, J. B., Sugiyama, J., Rossman, A. Y., Rogers, J. D., Pfister, D. H., Hewitt, D., Hansen, K., Hambleton, S., Shoemaker, R. A., Kohlmeyer, J., Volkmann-Kohlmeyer, B., Spotts, R. A., Serdani, M., Crous, P. W., Hughes, K. W., Matsuura, K., Langer, E., Langer, G., Untereiner, W. A., Lucking, R., Buedel, B., Geiser, D. M., Aptroot, A., Diederich, P., Schmitt, I., Schultz, M., Yahr, R., Hibbett, D. S., Lutzoni, F., McLaughlin, D. J., Spatafora, J. W. and Vilgalys, R. (2006) Reconstructing the early evolution of Fungi using a six-gene phylogeny. *Nature* 443, 818–822.

36. Ghignone, S. and Migheli, Q. (2005) The database of PCR primers for phytopathogenic fungi. *Eur. J. Plant Pathol.* 113, 107–109.

37. Nicholson, P., Lees, A., Maurin, N., Parry, D. and Rezanoor, H. N. (1996) Development of a PCR assay to identify and quantify *Microdochium nivale* var. *nivale* and *Microdochium nivale* var. *majus* in wheat. *Physiol. Mol. Plant Pathol.* 48, 257–271.

38. Nicholson, P., Simpson, D. R., Weston, G., Rezanoor, H. N., Lees, A. K., Parry, D. W. and Joyce, D. (1998) Detection and quantification of *Fusarium culmorum* and *Fusarium graminearum* in cereals using PCR assays. *Physiol. Mol. Plant Pathol.* 53, 17–37.

39. White, T. J., Bruns, T., Lee, S. and Taylor, J. (1990) Amplification and direct sequencing of fungal ribosomal RNA genes for phylogenetics. In PCR Protocols. A Guide to Methods and Applications. (Innis, M. A., Gelfand, D. H., Sninsky, J. J. and White, T. J., eds.), Academic Press, San Diego, CA, pp. 315–322.

40. Altschul, S. F., Madden, T. L., Schaffer, A. A., Zhang, J. H., Zhang, Z., Miller, W. and Lipman, D. J. (1997) Gapped BLAST and PSI-BLAST: a new generation of protein database search programs. *Nucleic Acids Res.* 25, 3389–3402.

41. Pearson, W. R. and Lipman, D. J. (1988) Improved tools for biological sequence

comparison. *Proc. Natl Acad. Sci. U S A* 85, 2444–2448.

42. Smit, E., Leeflang, P., Glandorf, B. and van Elsas, J. D. (1999) Analysis of fungal diversity in the wheat rhizosphere by sequencing of cloned PCR-amplified genes encoding 18S rRNA and temperature gradient gel electrophoresis. *Appl. Environ. Microbiol.* 65, 2614–2621.
43. Vandenkoornhuyse, P., Baldauf, S. L., Leyval, C., Straczek, J. and Young, J. P. W. (2002) Extensive fungal diversity in plant roots. *Science* 295, 2051.
44. Hughes, K. A., McCartney, H. A., Lachlan-Cope, T. A. and Pearce, D. A. (2004) A preliminary study of airborne microbial biodiversity over peninsular Antarctica. *Cell. Mol. Biol.* 50, 537–542.
45. Paez-Rubio, T., Viau, E., Romero-Hernandez, S. and Peccia, J. (2005) Source bioaerosol concentration and rRNA gene-based identification of microorganisms aerosolized at a flood irrigation wastewater reuse site. *Appl. Environ. Microbiol.* 71, 804–810.
46. Ward, E., Kanyuka, K., Motteram, J., Kornyukhin, D. and Adams, M. J. (2005) The use of conventional and quantitative real-time PCR assays for *Polymyxa graminis* to examine host plant resistance, inoculum levels and intraspecific variation. *New Phytol.* 165, 875–885.
47. Haugland, R. A., Vesper, S. J. and Wymer, L. J. (1999) Quantitative measurement of *Stachybotrys chartarum* conidia using real time detection of PCR products with the TaqMan fluorogenic probe system. *Mol. Cell. Probes* 13, 329–340.
48. Vesper, S., Dearborn, D. G., Yike, I., Allen, T., Sobolewski, J., Hinkley, S. F., Jarvis, B. B. and Haugland, R. A. (2000) Evaluation of *Stachybotrys chartarum* in the house of an infant with pulmonary hemorrhage: quantitative assessment before, during and after remediation. *J. Urban Health* 77, 68–85.

Chapter 13

Identification of Phytophthora fragariae var. rubi by PCR

Alexandra Schlenzig

Summary

The following chapter describes a PCR method for the identification of the raspberry root rot pathogen *Phytophthora fragariae* var. *rubi*. Furthermore, a nested PCR suitable for the detection of the pathogen in infected raspberry roots and validated against the "Duncan bait test" (EPPO Bull 35:87–91, 2005) is explained. Protocols for different DNA extraction methods are given which can be transferred to other fungal pathogens.

Key words: Phytophthora fragariae var. rubi, Raspberry root rot, PCR detection.

13.1. Introduction

Over the last decade, polymerase chain reaction has become an important tool for plant pathologists. It makes it possible to test for a pathogen even in the absence of obvious symptoms and can support the identification process in cases of doubt.

The preferred targets for fungal pathogens are the ribosomal DNA genes. The reason for that is that they are present in each cell as multiple copies increasing therefore the sensitivity of the PCR test. Furthermore, they contain the very variable internal transcribed spacer (ITS) regions, which are relatively species-specific and can be used for differentiation between related fungi *(1, 2)*. More recently, the beta-tubulin gene *(3)* or the cytochrome oxidase gene *(4)* have become popular targets to discriminate between particular fungal species. There are a great number of these DNA sequences available on public databases facilitating the development of specific primers.

R. Burns (ed.), *Methods in Molecular Biology, Plant Pathology, vol. 508*

DOI: 10.1007/978-1-59745-062-1_13

PCR as identification tool is often performed on fungal cultures whereas PCR as detection method is performed on infected host plants. The latter is more difficult since the host tissue might contain inhibitors interfering with the PCR. Furthermore, the small proportion of target fungal DNA has to be detected against the high amount of plant DNA.

This chapter will describe the use of PCR methods for both purposes using the example of the raspberry root rot pathogen *Phytophthora fragariae* var. *rubi*.

13.2. Materials

13.2.1. DNA Extraction

13.2.1.1. DNA Extraction from Fungal Cultures

1. Extraction buffer (50 ml):
   - 10 ml 1 M Tris–HCl (pH 7.5–8).
   - 2.5 ml 5 M NaCl.
   - 2.5 ml 0.5 M EDTA (pH 8).
   - 0.25 g sodium dodecyl sulphate (SDS).
   - 35 ml distilled water.

   (Tris stabilises the pH, NaCl delivers Na^+ ions to neutralise the negatively charged DNA, SDS is a detergent breaking down cell membranes disconnecting the lipids in them, and EDTA chelates metal ions that might disturb the PCR.)
2. PVPP (Polyvinylpolypyrrolidone).
3. Isopropanol.
4. Ethanol.
5. Rotations shaker.
6. Micro pestle.
7. Sulphuric acid washed sand (Merck).

13.2.1.2. DNA Extraction from Infected Raspberry Roots

1-6. As above.

7. Liquid nitrogen.
8. Multi spin separation columns (Thistle Scientific, Glasgow).

13.2.2. PCR

13.2.2.1. PCR on Fungal Cultures

1. Thermocycler.
2. Horizontal gel and electrophoresis equipment.
3. Agarose.
4. Ethidium bromide.
5. Oligonucleotide primers.
6. Taq DNA Polymerase inclusive Buffer.
7. Deoxynucleotides (dNTPs).
8. PCR tubes.

13.2.2.2. PCR on Infected Raspberry Roots

1-5. As above.

6. PuRe 'Ready to go' PCR beads (Amersham Bioscience).

13.3. Methods

13.3.1. DNA Extraction

There are many different DNA extraction methods described in literature and most of them will work sufficiently in most situations. All of them will be suitable for the DNA extraction from pure fungal cultures which is usually trouble-free. Sometimes even no DNA extraction at all is necessary – the PCR can simply be performed with pure mycelium or spores added to the PCR mix. It is more difficult to detect a pathogen within host tissue since that contains components that might inhibit the PCR, like starch, pectin and polyphenols. Especially woody tissue, roots or soil can cause problems and a DNA purification step is necessary at the end of the extraction. Some extraction methods contain the use of hazardous chemicals like phenol and it is always advisable to try one of the more harmless methods first and come back to these methods only as a last resort.

Independent from the chosen extraction method the breaking up of the cells is essential. As soon as the cells are broken up enzymes will start to degrade the DNA. To prevent this the sample either should be ground in extraction buffer, which contains enzyme inhibiting substances, or the sample should be kept frozen during the grinding process. For mycelium or soft plant tissue grinding directly in microcentrifuge tubes with micro pestles is very effective, using sterile sand as a grinding agent. Roots or woody material usually is ground frozen in a mortar with liquid nitrogen.

13.3.1.1. DNA Extraction from Fungal Cultures (see Note 1)

1. Making of extraction buffer. The buffer can be made in advance and stored at room temperature for at least 6 months.
2. Scrape a little mycelium from a fungal culture and transfer it into a 1.5-ml microcentrifuge tube. Add a pinch of sterile sand and 600 µl of extraction buffer and grind lightly using a micro pestle. Add another 600 µl of extraction buffer.
   Caution: Excessive grinding can destroy the DNA!
3. Vortex and place tubes on a rotator for 10 min.
4. Centrifuge at maximal speed for 8 min.
5. Pipette supernatant (~1 ml) into a new 2-ml tube. Do not pipette up any of the solid material.

6. To precipitate the DNA add an equivalent amount (~1 ml) of 100% isopropanol (from the freezer) to the supernatant and put the tubes into the freezer for about 10 min. Invert 3–4 times during incubation to mix.
7. Centrifuge at maximal speed for 5 min.
8. Discard supernatant carefully. You should see a small DNA pellet down the side of the tube.
9. Refill with 500 µl 70% ethanol to wash the pellet and centrifuge at 16,000 × *g* for 5 min.
10. Discard supernatant carefully. Dry tubes upside down on a tissue at room temperature until no residues of ethanol are left (about 1 h). This is important because any ethanol residues would inhibit the PCR.
11. Resuspend pellet in 100 µl sterile distilled water. Vortex. Preheated water (65°C) improves the process.

13.3.1.2. DNA Extraction from Infected Raspberry Roots

1. If possible pick out unhealthy looking roots from the root stocks: roots with black discoloration, roots with less than normal fine roots (*see* **Note 2**). Wash them thoroughly in a sieve under running water. Pat them dry between layers of clean tissue because freezing water would interfere with the grinding.
2. Grind 2.5–3 g of roots to a fine powder in a mortar using liquid nitrogen. Fill a microspoon full of powder into a new 2-ml microcentrifuge tube, add the equivalent of PVPP (Polyvinylpolypyrrolidone) and 1.2 ml of extraction buffer to it. Do not allow the sample to thaw! Pre-cool mortar, pestle and spatula in liquid nitrogen before use.

3-11. As above.

12. Purify DNA through multi spin columns.

12.1 Fill empty columns to within 1 cm from the top with insoluble PVPP (see above).

12.2 Load columns with 400 µl sterile water, place them in the flat bottomed Eppendorf (supplied with kit) and spin 2,400 × *g* for 5 min. Discard flow through (*see* **Note 3**).

12.3 Place columns in new 1.5-ml Eppendorf. Load the 100 µl of DNA evenly over the PVPP.

12.4 Spin at 1,500 × *g* for 5 min. The flow through is your purified DNA.

12.5 Store in freezer until you need it for the PCR.

13.3.2. PCR Amplification

13.3.2.1. PCR on Fungal Cultures

Perform PCR in 20 µl volumes. Remember to include a positive control (*P. fragariae* var. *rubi* DNA) and a negative control (water).

The primers are based on the ITS region of the rDNA and have been developed at the Scottish Crop Research Institute *(5)*. Their sequence is

DC1 (forward): 5′ act tag ttg ggg gcc tgt ct 3′

DC5 (reverse): 5′ cgc cga ctg gcc aca cag 3′

1. Make up a master mix containing for each sample:
   - 14.47 μl distilled water
   - 2 μl 10× Buffer (as delivered with polymerase)
   - 2 μl 1.25 mM dNTP solution (Sigma)
   - 0.2 μl forward primer solution (20 μM)
   - 0.2 μl reverse primer solution (20 μM)
   - 0.13 μl Taq DNA polymerase (5 U/μl) (Sigma)
2. Distribute master mix between PCR tubes (19 μl per tube) and add 1 μl of template DNA per tube.
3. Amplify DNA under following PCR conditions: Initial denaturation at 94°C for 2 min followed by 30 cycles of 94°C for 30 s, 66°C for 30 s, 72°C for 60 s followed by 72°C for 10 min.
4. Prepare a 1% horizontal agarose gel adding ethidium bromide (10 mg/ml stock solution) so that the final concentration of ethidium bromide in the gel is 0.5 μg/ml. Run 5 μl of the amplification product + 1 μl of loading dye at about 80 V for about 1 h.

The expected band size for the PCR product is 533 bp (*see* **Note 4**).

Caution: Ethidium bromide is toxic, a potent mutagen, and possibly carcinogen. As it can be absorbed through the skin it is important to wear nitrile gloves whenever handling it. Latex gloves are porous to it!

13.3.2.2. PCR on Infected Raspberry Roots

Compared to pure fungal cultures the amount of *Phytophthora* DNA in infected raspberry roots is very low. To increase the sensitivity a so called nested PCR is performed. This means that the amplification product of a first PCR is amplified again in a second PCR with another set of primers. The second primers bind within the first PCR product and produce another, shorter PCR product. This method not only increases the sensitivity of a PCR considerably but also increases the specificity because if the first set of primers has amplified an unspecific product it is very unlikely that this product will be amplified again in the second PCR.

Using a PCR product as a template means that the method is also very sensitive to carry over contamination!

For the first PCR use primers DC6 (forward) 5′ gag gga ctt ttg ggt aat ca 3′ *(5)* and ITS4 (reverse) 5′ tcc tcc gct tat tga tat gc 3′ *(6)*. The PCR is performed in 25 μl volumes using PuRe

'Ready to go' PCR beads (Amersham Bioscience). These beads contain all the ingredients for the PCR so that only primers, template DNA and water have to be added.

Remember to include a positive control (*P. fragariae* var. *rubi* DNA) and a negative control (water).

1. Make up a master mix containing for each sample:
   - 23.5 µl distilled water
   - 0.25 µl forward primer solution (20 µM)
   - 0.25 µl reverse primer solution (20 µM)
2. Distribute master mix between PCR tubes of the kit containing the beads (24 µl per tube) and add 1 µl of template DNA per tube.
3. Amplify DNA under following PCR conditions: Initial denaturation at 94°C for 2 min followed by 30 cycles of 94°C for 30 s, 56°C for 30 s, 72°C for 60 s followed by 72°C for 10 min.
4. Prepare a 1% agarose gel and run 5 µl of the amplification product as above.

This first PCR amplifies Downy Mildews, *Pythiums* and *Phytophthora*. The expected band size is about 1,310 bp.

For the second PCR use primers DC1 and DC5 (see above).

1. Make up a master mix containing for each sample:
   - 24 µl distilled water
   - 0.25 µl forward primer solution (20 µM)
   - 0.25 µl reverse primer solution (20 µM)
2. Use new PCR tubes from the kit. Distribute master mix (24.5 µl per tube) and add 0.5 µl of the PCR product from the first PCR per tube.
3. Amplify DNA under following PCR conditions: Initial denaturation at 94°C for 2 min followed by 30 cycles of 94°C for 30 s, 66°C for 30 s, 72°C for 60 s followed by 72°C for 10 min.
4. Prepare a 1% agarose gel and run 5 µl of the amplification product as above.

The second PCR is specific for *Phytophthora fragariae*. The expected band size is 533 bp. (*see* **Note 5**). The PCR is not able to distinguish between the two pathovars of *P. fragariae*, *rubi* and *fragariae*, but since both pathogens are specific to their host, it does not cause problems.

13.4. Notes

1. This DNA extraction method is an example for of a quick and easy method and works well for extraction from fungal cul-

tures and from raspberry roots. It was also successfully used by the author for the detection of other fungi, for example *Monilinia* from plum fruits or *Phytophthora* spp. from *Rhododendron* leaves. However, for the where there are cases where problems are encountered with some host material or where a higher quality of DNA is required an alternative method is given below. Both protocols can only be examples as many other methods have been published and a number of DNA extraction kits are available from different suppliers.

CTAB extraction method, exemplified for DNA extraction of mycelium

(a) Making of extraction buffer (50 ml):
- 7 ml 5 M NaCl,
- 2.5 ml 1 M Tris–HCl (pH 8),
- 1 ml 0.5 M EDTA,
- 0.5 ml 2-Mercapto-ethanol (100%),
- 10 ml 5% CTAB,
- 29 ml distilled Water.

(b) Scrape a little mycelium from a fungal culture and transfer it into a 1.5-ml microcentrifuge tube. Add a pinch of sterile sand and 600 μl of extraction buffer and grind lightly using a micro pestle.

(c) Keep tubes at 65°C for about 1 h. Shake tubes about every 15 min.

(d) After incubation add 400 μl chloroform/isoamylalcohol 24:1 (ready to buy).

(e) Shake for 15 min on a rotations shaker.

(f) Centrifuge for 10 min at maximal speed.

(g) The content of the tube will be separated in two layers. Transfer upper layer into a fresh tube without disturbing the lower layer.

(h) Add 500 μl extraction buffer and 500 μl chloroform/isoamylalcohol.

(i) Shake for 15 min on a rotations shaker.

(j) Centrifuge for 10 min at maximal speed.

(k) Transfer upper layer into a fresh tube.

Remark: **Steps** h–k provide additional cleaning of the DNA and are not always necessary, especially when extraction is performed from pure cultures.

(l) Add 500 μl ice cold isopropanol. Incubate tubes for 5 min in freezer.

(m) Centrifuge for 5 min at maximal speed. Discard supernatant carefully. You might see a small DNA pellet down the side of the tube.

(n) Add 500 μl ice cold ethanol (70%).

(o) Centrifuge for 5 min at maximal speed. Discard supernatant carefully.

(p) Dry tubes upside down at room temperature on a tissue for about 1 h until no residues of ethanol are left.

(q) Resuspend DNA pellet in 50 μl sterile distilled water. Preheated water (65°C) improves the process.

(r) Store DNA at –20°C until use.

2. The sample size which can be processed in a PCR is fairly small. Especially for field samples only a fraction of the original sample can be tested. There is a risk that the pathogen is missed if the sub-sample taken for PCR is chosen randomly *(6)*. Usually the aim of a field test is to determine whether the pathogen is present or not. If symptomatic roots are noticed in a sample they should be selected for PCR testing rather than following a strict statistic procedure. This will increase the likelihood of finding the fungus.

3. If you do not have a flow through load more water and spin again until you get one. Otherwise the column is not saturated with water and will absorb the DNA sample in the next step.

4. To ensure that a negative result is not caused by a loss of the DNA during the DNA extraction it is advisable to check samples for the presence of amplifyable DNA after extraction. This can be done by running a PCR with general primers, for example ITS4 (5 tcc tcc gct tat tga tat gc 3) and ITS5 (5 gga agt aaa agt cgt aac aag g 3). Those primers will amplify not only fungal DNA but also plant DNA resulting in a double band for infected samples and a single band for healthy plant material.

5. After the second round of the nested PCR the band of the positive control often shows some background smear or some faint multiple banding apart from the main band of 533 bp. This is due to the high concentration of fungal DNA. Diluting the PCR product of the positive control after the first round 1:100 usually helps if it is seen as a problem.

References

1. White, T.J., Burns, T., Lee, S., and Taylor, J. (1990) Amplification and direct sequencing of fungal ribosomal RNA genes for phylogenetics, in: *PCR Protocols: a Guide to Methods and Applications* (Innis, M.A., Gelfand, D.H., Sninsky, J.J. and White, T.J. eds.), Academic Press, London (GB), pp. 315–322.

2. Cooke, D.E.L., Drenth, A., Duncan, J.M., Wagels, G., and Brasier, C.M. (2000) A molecular phylogeny of *Phytophthora* and related oomycetes. *Fungal Genet. Biol.* 30, 17–32.

3. Fraaje, B., Lovell, D.J., Rohel, E.A., and Hollomon, D.W. (1999) Rapid detection and diagnosis of *Septoria tritici* epidemics

in wheat using a polymerase chain reaction/PicoGreen assay. *J. Appl. Microbiol.* 86, 701–708.

4. Martin, F.N., Tooley, P.W., and Blomquist, C. (2004) Molecular detection of *Phytophthora ramorum*, the causal agent of Sudden Oak Death in California, and two additional species commonly recovered from diseased plant material. *Phytopathology* 94, 621–631.
5. Bonants, P., Hagenaar-de Weerdt, M., van Gent-Pelzer, M., Lacourt, I., Cooke, D., and Duncan, J. (1997) Detection and identification of *Phytophthora fragariae* by polymerase chain reaction. *Eur. J. Plant Pathol.* 103, 345–355.
6. Schlenzig A., Cooke D.E.L., and Chard J.M. (2005) Comparison of a baiting method and PCR for the detection of *Phytophthora fragariae* var. *rubi* in certified raspberry stocks. *EPPO Bull.* 35, 87–91.

Chapter 14

Detection of Double-Stranded RNA Elements in the Plant Pathogenic Fungus *Rhizoctonia solani*

Nikki D. Charlton, Stellos M. Tavantzis, and Marc A. Cubeta

Summary

Many species of fungi have been shown to harbor double-stranded RNA (dsRNA) elements. A single fungal isolate of *Rhizoctonia solani* may have as many as five different dsRNA elements within them. The presence of specific dsRNA elements influence pathogenicity in host plants.

Key words: *Rhizoctonia solani*, dsRNA, Anastomosis group, Hypovirulence, Potato.

14.1. Introduction

Many species of fungi have been shown to harbor double-stranded RNA (dsRNA) elements. Apart from a few intensively studied species, the role(s) that dsRNA elements play in the biology and ecology of fungi are not well understood. In many instances, fungi that harbor dsRNA elements often exhibit no or minimal observed phenotypic differences when compared with isolates of the same species lacking them. dsRNA elements were first identified in the plant pathogenic soil fungus *Rhizoctonia solani* by Castanho and Butler in 1978 *(1)* and have been studied for more than 30 years. They are commonly found in field populations of *R. solani* anastomosis group 3 (AG-3), a widely distributed fungal pathogen of potato, and a single isolate may have as many as five different dsRNA elements ranging in size from 1.2 to 25 kb *(2)*. One specific dsRNA found in *R. solani* AG-3 (called M2) has been shown to influence the disease causing activity of the fungus, a phenomenon referred to as hypovirulence *(3)*.

R. Burns (ed.), *Methods in Molecular Biology, Plant Pathology, vol. 508*

DOI: 10.1007/978-1-59745-062-1_14

The M2 dsRNA element has also been hypothesized to play a role in the utilization of the carbon source quinic acid for its saprobic growth *(4)*. The detection of the M2 dsRNA element in *R. solani* AG-3 via gel electrophoresis and reverse-transcription PCR (RT-PCR) will be the major focus of this chapter. Subsequent information on sequencing and analysis of sequence data is also presented. Fungal virologists commonly deploy these methods to investigate dsRNA elements in filamentous fungi. The techniques presented in this chapter are also applicable to scientists interested in the study of dsRNA elements in other fungi.

14.2. Materials

14.2.1. Reagents

1. Potato Dextrose Broth (PDB) medium.
2. GPS buffer: 0.1 M glycine, 0.3 M NaCl, 50 mM K_2HPO, pH 9.4.
3. 10% Sodium dodecyl sulfate (SDS).
4. 2-Mercaptoethanol.
5. Phenol.
6. Chloroform.
7. Liquid nitrogen.
8. EtOH.
9. CF-11 cellulose.
10. RNase-free H_2O or diethyl pyrocarbonate (DEPC) treated deionized H_2O (*see* **Note 1**).
11. RNase*Zap*® (Ambion, Inc.) or RNase Away® (Molecular BioProducts).
12. STE Buffer: 0.1 M NaCl, 0.05 M Tris–HCl, 0.001 M EDTA, pH 6.8.
13. Isoamyl alcohol.
14. Isopropyl alcohol.
15. RQ1 RNase-Free DNase (Promega).
16. 20× SSC: 3 M NaCl, 0.3 M Tri–sodium citrate, pH 7.0.
17. Glycogen or acrylamide gel.
18. 80% EtOH (DEPC treated and autoclaved).
19. *Taq* polymerase (Qiagen).
20. dNTPs (Promega).
21. 1× Tris–acetate–EDTA (TAE) electrophoresis buffer: 40 mM Tris–acetate, 1 mM EDTA; pH 8.0.

22. 0.5× Tris–borate–EDTA (TBE) electrophoresis buffer: 45 mM Tris base, 45 mM boric acid, 1 mM EDTA; pH 8.0.
23. Ethidium bromide (*see* **Note 2**).
24. 10× gel loading buffer.
25. Agarose.
26. 8 M LiCl (DEPC treated and autoclaved).
27. Oligonucleotide primers.
28. Superscript™ II Reverse Transcriptase (Invitrogen).
29. RNaseOUT™ (Invitrogen).
30. QIAquick® Gel Extraction Kit (Qiagen).

14.2.2. Equipment

1. Mortars and pestles (autoclaved and baked; pre-cooled at –20°C) (*see* **Note 1**).
2. 30-mL Corex® tubes.
3. BIO-RAD Econo-Pac® Chromatography Columns.
4. Gel electrophoresis equipment.

14.3. Methods

The methods described below outline *(1)* the preparation of the fungal cultures, *(2)* extraction of dsRNA and detection via gel electrophoresis, *(3)* mini-prep RNA extraction and detection via reverse transcription polymerase chain reaction (RT-PCR), *(4)* purification of amplified PCR products, and *(5)* sequencing of purified PCR products.

14.3.1. dsRNA Extraction

The steps in **Subheadings 14.3.1.1–14.3.1.3** describe the preparation of fungal cultures and the procedure for large-scale dsRNA extraction from *R. solani*. This procedure allows the detection of dsRNA and determination of their size.

14.3.1.1. Preparation of Fungal Isolates

Isolates of *R. solani* were grown in 100 mL of PDB in 1-L Erlenmeyer flasks by inoculating each flask with five, 5-mm diameter plugs. Isolates were grown until mycelial mats covered the surface of the PDB medium. The mycelial mats were collected on Miracloth, washed three times with sterile H_2O, squeezed lightly to remove water, frozen with liquid nitrogen, and stored at –80°C until use.

14.3.1.2. Large Scale dsRNA Extraction

The following is a modification of the procedure described by Morris and Dodds *(5)*. To prevent RNase contamination, wear gloves throughout the protocol (*see* **Note 1**).

1. Dispense the following in a 100-mL beaker with a magnetic stirring bar (RNase-free):
   - 20 mL GPS buffer
   - 2 mL 10% SDS
   - 0.2 mL 2-Mercaptoethanol
   - 20 mL Tris-saturated phenol containing 0.1% 8-hydroxyquinoline (*see* **Note 3**)
   - 20 mL Chloroform
2. Grind 10 g frozen mycelium in a pre-cooled mortar and pestle (placed on ice) to a fine powder with liquid nitrogen.
3. Dispense the powder with a spatula into the beaker with the reagents and stir for 30 min until thawed.
4. Dispense the emulsion into 30-mL Corex® tubes and centrifuge at 12,000 × *g* for 15 min.
5. Collect the upper aqueous phase into a clean beaker with stirring bar, leaving the interphase and the phenol:chloroform phase.

The following steps should be conducted at room temperature:

6. Add 0.176× volume of 100% EtOH (drop by drop, very slowly) to the beaker with constant stirring.
7. Add 1 g of CF-11 cellulose (autoclaved) with stirring.
8. Transfer the slurry to a sterile column. Let the cellulose settle while eluting the aqueous phase.
9. Wash the cellulose with 250 mL of STE buffer containing 15% ethanol.
10. Elute the dsRNA enriched sample with 1 mL of STE in DEPC treated H_2O, then with 4 mL of STE buffer and collect them in a 30-mL Corex® tube.
11. Add 0.1 volume of 3 M sodium acetate and an equal volume of isopropanol (or 2.5× volumes of EtOH), mix well and store at –20°C until use.

The following steps should be carried out at 4°C or on ice unless otherwise indicated:

12. Centrifuge at 12,000 × *g* for 15 min to pellet the dsRNA.
13. Wash the pellet carefully with 80% EtOH (prepared with DEPC treated H_2O); centrifuge for 5 min if the pellet is loose.
14. Repeat washing one more time with 80% EtOH.
15. Dry the pellet and dissolve in approximately 100 μL DEPC treated H_2O and transfer into 1.5-mL microcentrifuge tube.
16. Add 30 μL of 10× DNase buffer and 5 μL of Promega RQ1 RNase-Free DNAse; incubate at 37°C for 20 min.
17. Add 14 μL of 20× SSC buffer and RNase at 20 μg/mL final concentration. Mix and incubate at room temperature for

30 min (final 2× SSC). Alternatively, add 3 M sodium chloride to a final concentration of 0.3 M.

18. Extract once with phenol:chloroform (1:1, v/v) and once with chloroform:isoamyl alcohol (24:1). Centrifuge at 12,000 × *g* for 15 min. Add 1 μL of glycogen or linear acrylamide stock solution as provided by the manufacturer, and an equal volume of isopropanol or 2.5× volumes of EtOH, mix and incubate at −20°C overnight or until use.
19. Centrifuge at 12,000 × *g* for 15 min, wash the pellet with 80% DEPC treated cold EtOH (centrifuge if the pellet is loose), air dry for 10–15 min, and dissolve the pellet in DEPC treated H_2O.
20. Freeze the RNA at −80°C for short-term (one week) storage or precipitate again in sodium acetate–ethanol (*see* **step 11**) and store at −20°C for long-term storage.

14.3.1.3. Detection of dsRNA Via Gel Electrophoresis

The quality as well as the number and sizes of dsRNAs should be checked by agarose gel electrophoresis in 1× Tris–acetate–EDTA (TAE) buffer using dsRNA or DNA size standards (1-kb Plus DNA Ladder).

1. Weigh appropriate amount of agarose (1 g/100 mL of buffer) into an Erlenmeyer flask, and add the required volume of running buffer. The vessel should be no more than 20% filled. Weigh flask and dissolve the agarose by heating in a microwave or a steaming water bath for approximately 1–2 min. Weigh flask again and replace lost water if needed.
2. Cool the agarose solution to 60°C in a water bath or on the bench, and pour into a gel tray assembled in a submarine gel electrophoresis apparatus with a comb, to form the wells, and end-blocks.
3. When agarose has solidified, remove the comb by lubricating it around the immersed teeth with a small amount of running buffer, detach the end-blocks, place gel tray into the apparatus chamber and pour the running buffer to cover the gel.
4. Use a small volume (2–10 μL) from each dsRNA sample for electrophoretic analysis. Dilute dsRNA and DNA size standard aliquots with DEPC-treated H_2O (check agarose well capacity), and add 0.1 volume of 10× gel loading buffer, and apply all samples into separate wells using a micropipette. Place the submarine gel apparatus cover on the unit and connect electrode leads to the power source (black is connected to black and red is connected to red). Make sure that DNA will migrate from the black electrode (cathode) – closest to wells – to the red electrode (anode) – furthest from the wells. Turn on power source to apply current to the electrophoresis unit (15–20 V for an overnight run or 40 V for a 2–4 h run).

5. When electrophoresis has been completed, turn off the power supply, remove gel from the submarine apparatus, and place it into a sturdy container with an ethidium bromide solution (0.5 μg/mL or 25 μL of a 1 mg/mL ethidium bromide solution per 50 mL gel) in water or running buffer. Stain for 10–20 min with gentle agitation, decant staining solution, and destain gel in water or running buffer for 20 min.
6. Visualize nucleic acid bands by UV light (*see* **Note 4**).

14.3.2. Detection of dsRNA Via RT-PCR

The steps in **Subheadings 14.3.2.1–14.3.2.3** describe the preparation of the fungal cultures, the procedure for mini-prep RNA isolation of *R. solani* anastomosis group 3 (AG-3), followed by detection of a specific dsRNA element (M2) using RT-PCR.

14.3.2.1. Preparation of Fungal Isolates

An isolate of *R. solani* AG-3, Rs84, was grown in 25 mL of PDB in a 250-mL Erlenmeyer flask by inoculating with a 5-mm diameter plug. After incubation for 7–10 days at 24°C, the mycelial mat was washed three times with sterile deionized H_2O by decanting the PDB and rinsing the mycelial mat with sterile deionized H_2O. The mycelial mat was collected on no. 1 Whatman filter paper using a Büchner funnel by vacuum filtration following the third wash. The sample was freeze-dried (lyophilized) and stored at –20°C.

14.3.2.2. Mini-Prep RNA Extraction

Total genomic RNA was extracted by grinding lyophilized tissue to a fine powder in liquid nitrogen using a mortar and pestle (*see* **Note 1**). Total RNA was extracted using TRIZOL® Reagent (Invitrogen). Lithium chloride was used to improve RNA isolation by removing excessive amounts of carbohydrates *(4, 6)*.

The following steps should be conducted at room temperature unless otherwise indicated:

1. Homogenize 50 mg of fungal tissue in 1 mL of TRIZOL® Reagent (Invitrogen) by using a microcentrifuge pestle to mix the sample thoroughly (*see* **Note 5**). Incubate the samples for 5 min.
2. Add 200 μL of chloroform to the sample and shake vigorously by hand for 15 s. Incubate for 3 min. Centrifuge at 12,000 × *g* for 15 min.
3. Transfer upper aqueous phase to a 1.5-mL microcentrifuge tube.
4. Precipitate the RNA with 0.5 mL of isopropyl alcohol per 1 mL of TRIZOL® Reagent used. Incubate for 10 min. Centrifuge at 12,000 × *g* for 10 min.
5. Remove supernatant and wash the RNA pellet with 1 mL of 80% DEPC treated cold EtOH. Mix the sample with a vortex for 15 s and centrifuge at 7,500 × *g* for 5 min.

6. Remove supernatant and allow RNA pellet to air dry for 2–3 min.
7. Dissolve the pellet in 1 mL DEPC treated or RNase-free H_2O. Incubate at 55–60°C for 10 min. Centrifuge at 12,000 × *g* for 20 min.
8. Transfer aqueous phase to a 1.5-mL microcentrifuge tube. Add 333 µL of 8 M LiCl (DEPC treated and autoclaved).
9. Precipitate on ice or at 4°C overnight.
10. Centrifuge at 12,000 × *g* for 20 min.
11. Discard supernatant and wash RNA pellet with 0.5 mL of 80% DEPC treated EtOH. Air dry the RNA pellet for 2–3 min.
12. Resuspend the RNA pellet in 40 µL of DEPC treated or RNase-free H_2O.

14.3.2.3. Reverse-Transcription Polymerase Chain Reaction

Total RNA was denatured at 100°C for 1 min in the presence of 10 µM gene-specific primer (**Table 14.1**). Two negative control reactions were performed during the reverse transcription reaction (*see* **Note 6**). Twenty-base oligonucleotide primers, P40, P34N, and P36 were used in reverse-transcription of the total RNA for the 5′ region (bases 11–1,286), middle region (bases 1,264–2,454), and the 3′ region (bases 2,437–3,570), respectively for synthesis of complementary DNA (cDNA) (**Fig. 14.1**, **Table 14.1**).

Reverse transcription should be performed in a total volume of 20 µL:

Table 14.1
Oligonucleotide primers used for cDNA synthesis and amplification of three regions of the M2 double-stranded RNA element of *Rhizoctonia solani*

Primer	Sequence	Position (bp)	T_m (°C)
P39	5 -GTCCAATTAAGGACTCAGCT-3	11–30	51.9
P40	5 -TAAGTCGATCTGCAAGAGAG-3	1,305–1,286	51.3
P33[a]	5 -TTGCTCTCTTGCAGATCGAC-3	1,264–1,283	54.9
P34N[a]	5 -TAAGCTGCATGTAATGACGC-3	2,473–2,454	53.3
P35[a]	5 -GTCATTACATGCAGCTTACC-3	2,437–2,456	51.3
P36[a]	5 -GGGGCTTCTGGCGGAAAGAA-3	3,589–3,570	60.3

[a]Liu et al. (4)

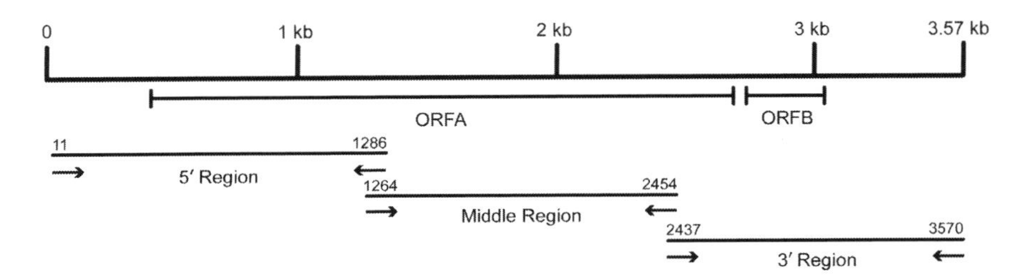

Fig. 14.1. Genome of the 3.57 kb M2 double-stranded RNA element of *Rhizoctonia solani* showing primer positions used to amplify three regions of the dsRNA. Locations and sizes for each region are approximate (see text for details). Open reading frames (ORFs A and B) are also shown.

1. Add the following components to a 1.5-mL microcentrifuge tube:
   - 1 μL 10 μM Gene-specific primer
   - 4 μL Total RNA
   - 3 μL RNase-free H_2O
2. Denature at 100°C for 1 min and quick chill on EtOH ice bath. Collect contents of the tube by brief centrifugation and add:
   - 4 μL 5× First-Strand Buffer
   - 2 μL 0.1 M DTT
   - 1 μL 10 mM dNTP mix
   - 0.5 μL RNase OUT™ (40 units/μL)
   - 3.5 μL RNase-free H_2O
3. Incubate at 42°C for 2 min.
4. Add 1 μL (200 units) of SuperScript™ II reverse transcriptase and mix this solution by gently pipetting up and down.
5. Incubate at 42°C for 90 min.
6. Heat inactivate at 70°C for 10 min.
7. Store cDNA at –20°C to use as a template for PCR amplification.

Oligonucleotide primers (Integrated DNA Technologies, Inc., Coralville, IA) (Table 14.1) were used for PCR to amplify the 3.57 kb M2 dsRNA (Fig. 14.1). Primers P33 and P34N were used to amplify a 1,151 bp fragment from the middle region of the M2 dsRNA (primer set #1). Primers P35 and P36 were used to amplify a 1,094 bp fragment from the 3′ end of the M2 dsRNA (primer set #2). Primers P39 and P40 were used to amplify a 1,236 bp fragment from the 5′ end of the M2 dsRNA (primer set #3).

PCR should be performed in a total volume of 50 μL containing the following components:

- 2-μL aliquot of undiluted cDNA template
- 5 U *Taq* polymerase (Qiagen Inc., USA)

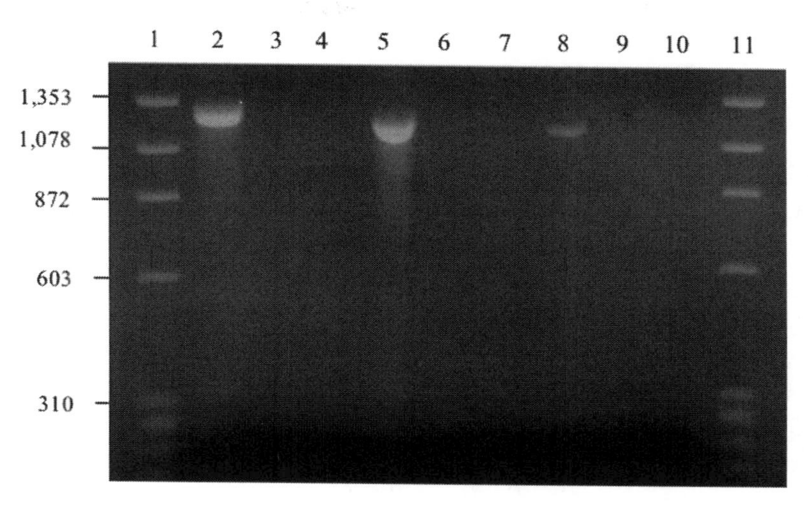

Fig. 14.2. Agarose gel analysis (1%) of the three amplified cDNA regions of the M2 dsRNA. *Lanes 1* and *11* show φX174 DNA digested with *Hae*III; sizes of bands (bp) are indicated on the *left*. *Lanes 2*, *5*, and *8* show amplification of the 5′ region, the middle region, and the 3′ region, respectively, of the M2 dsRNA from *Rhizoctonia solani* AG-3 (isolate Rs84). *Lanes 3*, *6*, and *9* show the cDNA synthesis in H_2O controls (replacing RNA template) in the presence of 10 μM primers, P40, P34N, and P36, respectively. *Lanes 4*, *7*, and *10* show the cDNA synthesis in H_2O controls in the absence of primers (RNA template was used in these reactions).

- 5 μL 10× reaction buffer containing 15 mM $MgCl_2$ (Qiagen Inc., USA)
- 1 μL 10 mM dNTP mix (Promega, Madison, WI)
- 1 μL of each 100 μM target-specific primer (Integrated DNA Technologies, Inc.)

The cycling parameters for PCR were 1 min at 94°C, 35 cycles of 1 min at 94°C, 45 s at 55°C, 1 min at 72°C, followed by a final synthesis step at 72°C for 7 min, and cooling at 4°C. The annealing temperatures in RT-PCR ranged from 51 to 55°C, depending on the T_m of the primers (e.g., 53°C for primer set #1; 55°C for primer set #2; and 51°C for primer set #3, **Table 14.1**). Amplified PCR products were analyzed by gel electrophoresis by using a 1% agarose gel in 0.5× Tris–borate–EDTA (TBE) buffer as described previously. cDNA fragments were visualized after staining with ethidium bromide on a UV transilluminator as described previously (**Fig. 14.2**).

14.3.3. Sequencing

In this section, procedures are presented for purifying and sequencing the amplified PCR products following RT-PCR. Purification of amplified products will remove non-target DNA, primers, polymerases, nucleotides, ethidium bromide, dyes, and salts.

14.3.3.1. Purification of PCR Products

DNA fragments amplified by each of the primer sets were recovered from a 1% SeaPlaque® GTG® Agarose Gel (Cambrex Bio Science Rockland, Inc.) and purified using the QIAquick® Gel Extraction Kit (Qiagen Inc., USA) according to manufacturer's instructions (*see* **Note 7**). This kit uses a spin-column that facilitates the recovery of DNA by capturing it on a membrane while eliminating contaminants and impurities that interfere with the sequencing reaction during the washing and elution steps.

14.3.3.2. cDNA Sequencing

Purified PCR products were sequenced using Big Dye Terminator Chemistry (version 3.1, Applied Biosystems). The sequencing reaction was performed in a total volume of 10 μL containing the following:

- 50–60 ng cDNA (*see* **Note 8**)
- 0.4 μL Big Dye reaction mix
- 1.8 μL 5× buffer
- 0.4 μL 10 μM Target-specific primer

The cycling parameters were 30 cycles of 10 s at 95°C, 5 s at 50°C, 4 min at 60°C, followed by cooling at 4°C. Sequencing reactions were cleaned to remove unincorporated dye terminators, residual salts, and other components that may interfere with sequencing (*see* **Note 9**). Samples were then analyzed on an ABI3700 DNA sequencer.

14.3.3.3. Analysis of Sequence Data

Forward and reverse chromatograms were aligned and adjusted by visual examination using Sequencher™ (Gene Codes Corporation, Ann Arbor, MI). Nucleotide sequences were compared using the National Center for Biotechnology Information (NCBI) database (http://www.ncbi.nlm.nih.gov). A nucleotide-protein Basic Local Alignment Search Tool (BLAST) search (blastx) was conducted to compare the nucleotide-translated query with the protein sequence database (**Fig. 14.3**). The results showed that the query sequence matches the dsRNA RNA-dependent RNA polymerase (*Thanatephorus cucumeris*, the sexual stage of *R. solani*). The 83% similarity represents the number of amino acids in the query sequence that match the number of amino acids in the database. Identities and conservative replacements resulted in a positive score of 60/66 with zero gaps.

```
>gi|4335684|gb|AAD17381.1|  dsRNA viral RNA-dependent RNA polymerase [Thanatephorus cucumeris]
Length=754

 Score =  120 bits (300),  Expect = le-25
 Identities = 55/66 (83%), Positives = 60/66 (90%), Gaps = 0/66 (0%)
 Frame = +2

Query  11   PQPYPFSKEWNREPKFLENEDKLASQLLHSALNKVIVLRDQYDWFIAKNASEMYFPFGYL  190
            PQ YPFSKEWNREPKFLENEDKLAS+LLHSALNKVI+LRDQY+WFIA+NASEMYFP  YL
Sbjct  687  PQLYPFSKEWNREPKFLENEDKLASRLLHSALNKVIILRDQYNWFIAENASEMYFPLSYL  746

Query  191  DNPKDK  208
            +N   K
Sbjct  747  NNSSAK  752
```

Fig. 14.3. The nucleotide-protein BLAST (blastx) result from NCBI for the nucleotide sequence of the 3′ end of the M2 dsRNA from *Rhizoctonia solani* AG-3 isolate Rs84.

14.4. Notes

1. To prevent RNase contamination, wear gloves at all times. Glassware must be baked at 250°C for approximately 5 h or overnight to ensure that it is RNase-free. Solutions (buffers) should be treated with 0.05% diethyl pyrocarbonate (DEPC) and incubated overnight. The solutions should then be autoclaved to remove any traces of DEPC.
2. Ethidium bromide is a potent mutagen. Gloves and safety glasses must be worn when handling ethidium bromide (and anything containing ethidium bromide). Please comply with the safety procedures regarding disposal of ethidium bromide waste documented by your facility.
3. Commercially available Tris–saturated phenol is stabilized with 0.1% 8-hydroxyquinoline.
4. UV-protective safety glasses should be worn when working with UV light.
5. RNase Away® (Molecular BioProducts) or RNase*Zap*® (Ambion, Inc.) may be used on glassware and plastic surfaces to remove RNase contamination.
6. One negative control should contain RNase-free H_2O in the presence of the 10 μM gene-specific primer and the second negative control should contain RNA template in the absence of primer.
7. Extraction of DNA fragments may be taken from standard agarose gels. The QIAquick® PCR Purification Kit may be used in place of the QIAquick® Gel Extraction Kit when a single band is present and direct purification of the amplified product is desired.
8. The optimal reaction composition for cDNA sequencing reactions requires 6 ng of DNA per 100 bp of amplified product. For each amplified product, one reaction must be prepared with the forward primer and one reaction must be prepared with the reverse primer.
9. A dye terminator removal product must be used to purify the DNA sequencing reactions prior to sequencing.

References

1. Castanho, B., and Butler, E.E. (1978) Rhizoctonia decline: a degenerative disease of *Rhizoctonia solani*. *Phytopathology* 68, 1505–1510.
2. Tavantzis, S.M., Lakshman, D.K., and Liu, C. (2002) Double-stranded RNA elements modulating virulence in *Rhizoctonia solani*, in *dsRNA Genetic Elements Concepts and Applications in Agriculture, Forestry and Medicine*. (Tavantzis, S.M., ed.), CRC Press, Boca Raton, FL, pp. 191–211.
3. Tavantzis, S.M., and Lakshman, D.K. (1995) Virus-like double-stranded RNA elements and hypovirulence in phytopathogenic fungi, in Pathogenesis and Host–Parasite

Specificity in Plant Disease: Histopathological, Biochemical, Genetic and Molecular Basis, Vol. III. (Kohmoto, K., Singh, U.S., Singh R.P., eds.), Elsevier (Pergamon) Press, Oxford, pp. 249–267.

4. Liu, C., Lakshman, D.K., and Tavantzis, S.M. (2003) Quinic acid induces hypovirulence and expression of a hypovirulence-associated double-stranded RNA in *Rhizoctonia solani*. *Curr. Genet.* 43, 103–111.
5. Morris, T.J., and Dodds, J.A. (1979) Isolation and analysis of double-stranded RNA from virus-infected plant and fungal tissue. *Phytopathology* 69, 854–858.
6. Salzman, R.A., Fujita, T., Zhu-Salzman, K., Hasegawa, P.M., and Bressan, R.A. (1999) An improved RNA isolation method for plant tissues containing high levels of phenolic compounds or carbohydrates. *Plant Mol. Biol. Rep.* 17, 11–17.

Chapter 15

Immunocapture-PCR for Plant Virus Detection

Vincent Mulholland

Summary

Immunocapture followed by the detection of viruses using polymerase chain reaction is a versatile, sensitive and robust diagnostic technique The application of this hybrid method of virus detection in plants is particularly useful in species or tissues containing inhibitory substances. In addition, antibody-mediated virus purification is usually simpler than other methods of isolation.

Key words: Immunocapture, PCR, RT-PCR, Reverse transcription, Virus, IC-PCR, IC-RT-PCR,

15.1 Introduction

Immunocapture-polymerase chain reaction (IC-PCR) combines two techniques, both of which are used frequently in diagnostics. The hybrid combination of these two methods into immunocapture-PCR has been a part of the armoury of plant virus detection methods for more than a decade *(1–9)*. IC exploits the specific binding capabilities of an antibody to recognise a molecule (or a group of similar molecules). This recognition is determined by the variable tips of the immunoglobulin molecule **(Fig. 15.1)**. Monoclonal antibodies are most frequently used in IC as they provide a stable source of reagent over an extended period. Polyclonal antibodies are produced in batches, so batch-to-batch variation of both antigen preparation and antibody becomes an issue. However, the requirements of an individual assay may mean that a polyclonal antibody preparation is the favoured option. In addition, one may use mixtures of antibodies to increase the range of viruses captured. The use of IC to purify the virus particles has the advantage that PCR-inhibitory compounds should not

R. Burns (ed.), *Methods in Molecular Biology, Plant Pathology, vol. 508*

DOI: 10.1007/978-1-59745-062-1_13

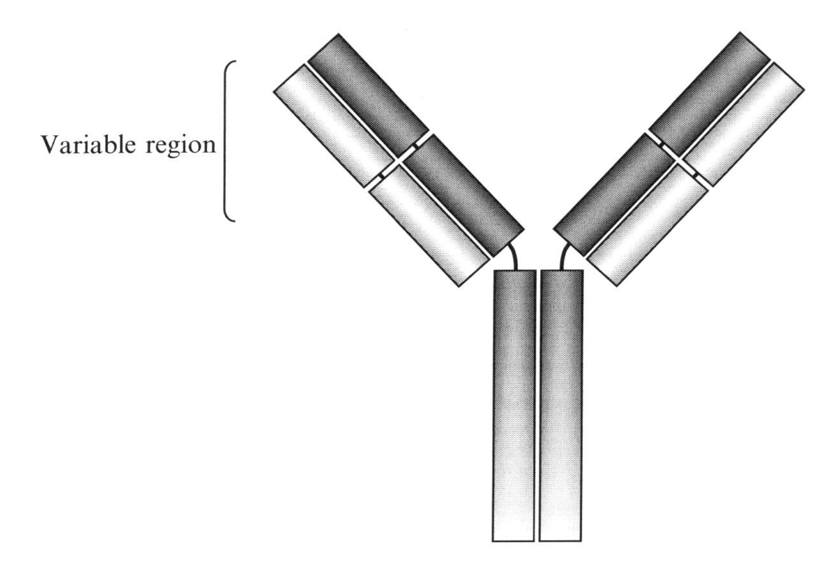

Fig. 15.1 Schematic representation of an antibody. The antibody consists of four polypeptides, two heavy chains, and two light chains. Antibody specificity is determined by the variable regions at the tips of the molecule.

co-purify *(1, 4, 10)*; these include polysaccharides, tannins and polyphenolics.

The PCR is an in vitro technique to amplify a DNA template to levels at which it can be detected. In many cases plant viruses contain RNA; RNA is not a suitable substrate for PCR, so must first be converted to complementary DNA (cDNA). This conversion is accomplished by the action of the enzyme reverse transcriptase (RT) an RNA-directed DNA polymerase, hence the abbreviation IC-RT-PCR. Following this conversion, the cDNA is amplified by the PCR. The primers used in the amplification process can be designed to be species-specific or could have a more relaxed specificity to allow detection of all viruses within a genus. This depends on the characteristics of the primer sites (conserved or variable) and on the primers themselves (degeneracy, use of modified nucleotides, etc.). In some instances, the presence of plant genomic sequences may give rise to false-positives – an example of this is the banana which has viral genes within the plant genome *(10, 11)*.

A judicious selection of antibody will allow the capture of viruses to be tailored to the assay requirements, ranging from subspecies-specific to genus-wide virus capture. This primary selection, based on antibody specificity, can then be supplemented with a second level of specificity at the PCR level. Much work on PCR detection methods, particularly for plant quarantine use, has focussed on the development of molecular detection assays for viruses with as wide a range as possible, to reduce the number of tests required during screening. Group-specific tests available include those for *Carlaviruses* *(12)*, *Potexvirus* *(13)*, *Potyvirus* *(14, 15)*, *Potexvirus* and *Tobamovirus* *(16)*, *Closterovirus* *(17, 18)*, *Bymovirus* *(19)*, *Tospovirus* *(20)*, *Cucumovirus* *(21)*, *Vitivirus* and *Trichovirus* *(22)* and *Begomovirus* *(23)*.

The method for IC-RT-PCR can be summarised as six steps: *(1)* tissue processing, *(2)* virus binding to an antibody-coated vessel, *(3)* release of viral RNA, *(4)* cDNA synthesis, *(5)* PCR amplification, and (6) detection. The two defining characteristics of the assay are the virus capture **(Fig. 15.2)** and the RT-PCR **(Fig. 15.3)** stages.

15.2 Materials

1. Dulbecco's phosphate-buffered saline without magnesium or calcium (PBS; pH 7.4): 8 g of NaCl, 0.15 g of KH_2PO_4, 2.9 g of $Na_2HPO_4 \cdot 12H_2O$, 0.2 g of KCl per liter.
2. Sample extraction buffer (SEB): PBS (pH 7.4) containing 0.05% Tween-20, 2% (w/v) polyvinylpyrrolidone 25000, 0.2% (w/v) ovalbumin, and 0.5% (w/v) bovine serum albumin.
3. Coating buffer (pH 9.6): 0.16% (w/v) Na_2CO_3, 0.29% (w/v) $NaHCO_3$.
4. Antibodies at 1 mg/mL, diluted to a working concentration of 1-in-100 in coating buffer (pH 9.6).
5. Washing buffer: PBS + 0.1% (v/v) Tween-20.
6. HOMEX 6 Plant Tissue Homogenizer (Bioreba AG).
7. Universal extraction bags 12 × 14 cm with synthetic intermediate layer (Bioreba AG).
8. Dispomix® Drive Unit (Medic Tools AG, Switzerland).
9. Dispomix® 25 S Tubes (Medic Tools AG, Switzerland).
10. PCR plates (96-well).
11. Plate seals (Dynatech Laboratories).
12. Enzyme solution: Cellulase "onozuka" RS and Macerozyme R-10 (Yakult Pharmaceuticals, also distributed by SERVA) 0.1 g of each enzyme in 1 mL of RNase-free water.
13. 1× RT buffer; 50 mM Tris–HCl (pH 8.3), 75 mM KCl, 3 mM $MgCl_2$.
14. 5× RT buffer; 250 mM Tris–HCl (pH 8.3), 375 mM KCl, 15 mM $MgCl_2$.
15. Moloney murine leukemia virus reverse transcriptase.
16. *Not*I poly dT primer; $GAATTCGCGGCCGCT_{18}$.
17. Anchored oligo dT primers; $T_{18}A$, $T_{18}C$ and $T_{18}G$.

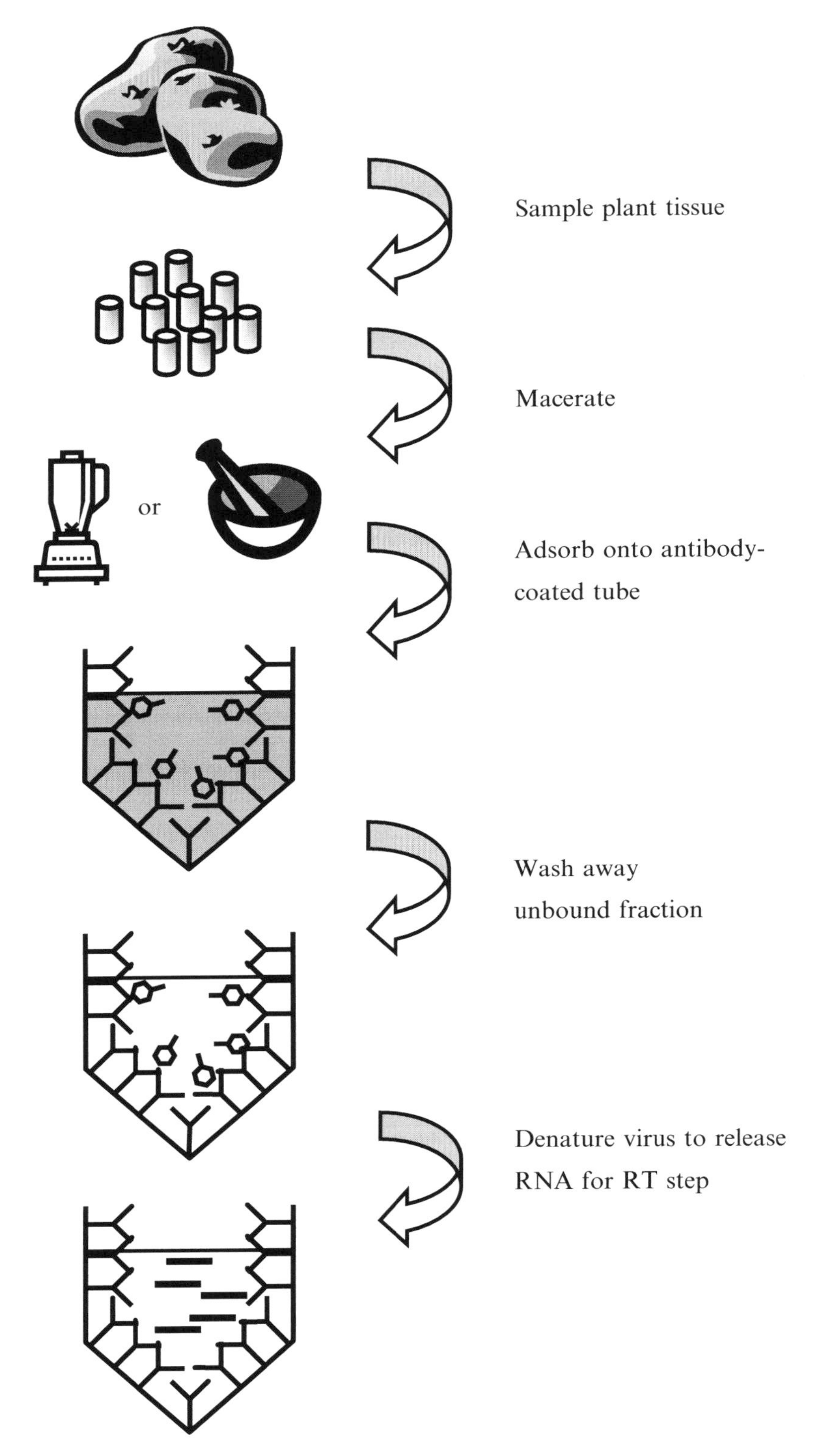

Fig. 15.2 Process-flow showing the major steps involved in tissue processing and viral RNA purification.

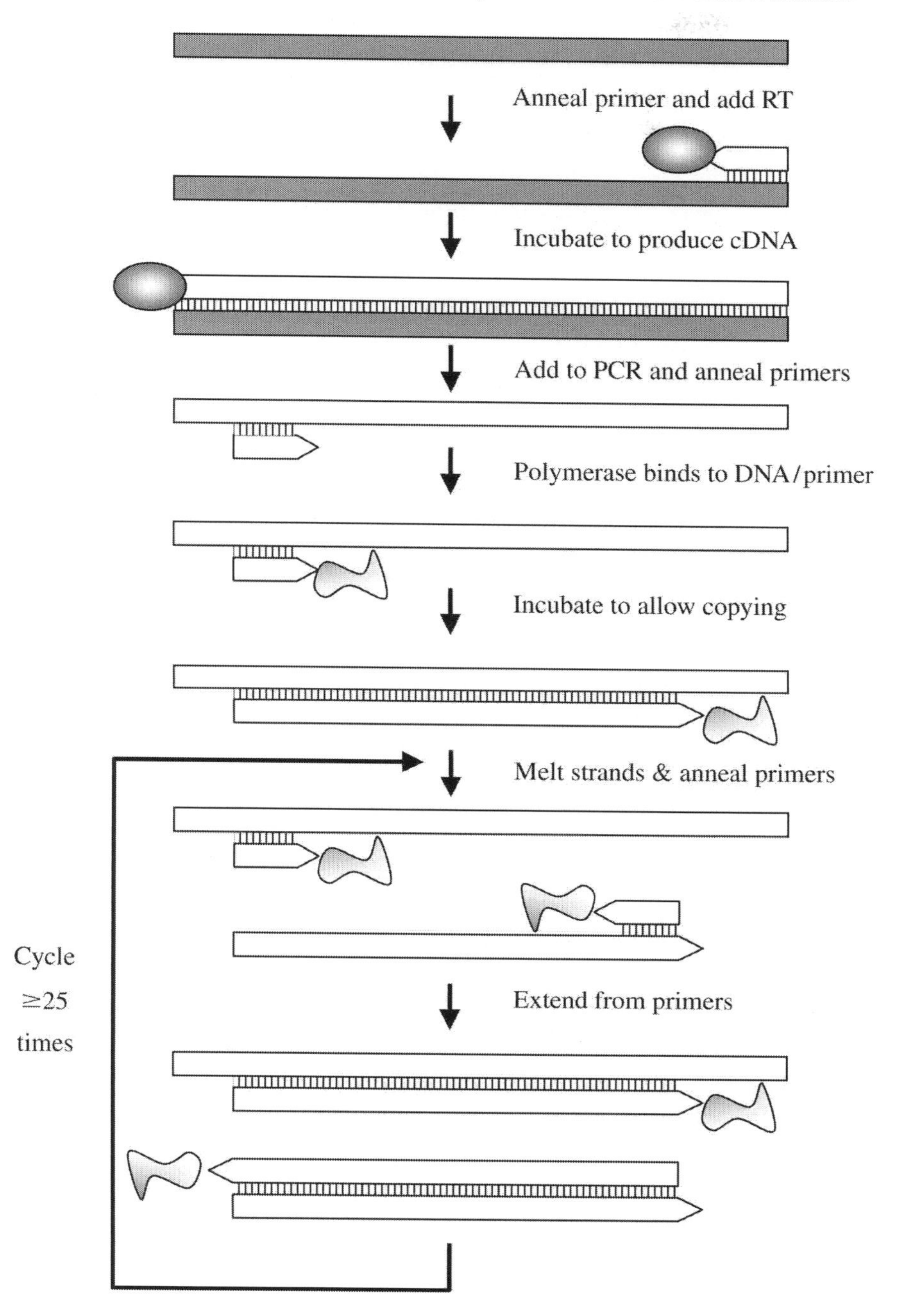

Fig. 15.3 A diagram showing steps in the RT-PCR process. RNA is converted to cDNA using the action reverse transcriptase from a primer (either a primer designed to a specific virus sequence, or an oligo-dT primer mainly used for mRNA-like molecules). The cDNA is added to a PCR and virus detection primers amplify a region from the virus.

18. Random hexamers; N_6.
19. dNTPs, 10 mM each with respect to dATP, dCTP, dGTP, and dTTP.
20. RNase-free H_2O.
21. 100 mM dithiothreitol (DTT).
22. Recombinant RNasin ribonuclease inhibitor (Promega).

15.3 Methods

The methods outlined below consist of (1) tube coating with the captured antibody, (2) sample extraction, (3) virus capture in the coated tubes, (4) cDNA synthesis from the captured viruses, and (5) PCR amplification of the cDNA. Tube coating requires an overnight incubation, so should be set up on the day prior to extractions.

15.3.1 Tube Coating

Coat 96-tube PCR plates with 50 mL of virus-captured antibody (1 mg/mL) diluted 1:100 in coating buffer pH 9.6, cover with a plate sealer and incubate overnight at 4°C. Antibody-coated plates may be prepared in advance and stored at –20°C for at least 4 weeks.

15.3.2 Sample Extection

15.3.2.1 Sample Extraction Using a HOMEX 6 Plant Tissue Homogenizer

1. Using a scalpel (*see* **Note 1**) sample approximately 1 g of tissue from the plant material, cut into thin slices (*see* **Note 2**).
2. Place the slices into Universal extraction bags, and add 1 mL of sample extraction buffer (SEB) to each bag.
3. Slot the polythene bags containing the tuber sample and SEB into the frame of a HOMEX 6 Plant Tissue Homogenizer, and grind.

15.3.2.2 Potato Tuber Extraction Using a Dispomix® (See Note 3)

1. Ten cores from tuber tissue are taken, giving approximately 1 g of plant material (*see* **Note 4**).
2. Place the cores into a Dispomix® 25 S tube (which has an internal strainer which holds back fibrous material). Add 1 mL SEB to the cores and seal the tube.
3. Place the tube onto the Dispomix® Drive Unit.
4. Run the extraction program (*see* **Note 5**).

15.3.3 Virus Capture

1. After antibody coating, wash the PCR plate wells three times with 200 mL of washing buffer, then block with 100 mL of SEB for 1.5 h at 37°C.
2. Knock the SEB out onto a paper tissue pad and wash the plates twice with 200 mL of wash buffer.

3. Add 50 mL of tuber extract to individual tubes on the plate.
4. To this extract, add 50 mL of enzyme solution diluted in SEB (*see* **Note 6**). Incubate the plates overnight at 4°C (*see* **Note 7**).
5. Remove the extracts from the tubes by knocking the plates on an absorbent pad, then wash three times with 200 mL of wash buffer. Should the sample adhere to the wells, soak the plates for 3 min during the wash stage.
6. Rinse with 200 mL of 1× RT buffer and flick out the remaining liquid before proceeding with the cDNA synthesis step.

15.3.4. cDNA Synthesis

15.3.4.1. First-Strand Primer Binding

For each reaction required, add the appropriate multiples of the following into a master mix. An appropriate primer should be chosen from the following list; *Not*I poly dT primer, anchored poly dT primer, virus-specific primer, random hexamers or a mixture of two or more primers.

Dispense 30 mL of the master mix into each tube. Incubate plates (in a PCR machine) at 65°C for 5 min, then 24°C for 30 min to allow primer annealing.

Ingredient	Volume per sample
5× First-strand buffer	6 µL
DTT solution (100 mM)	3 µL
RNasin (40 U/µL)	0.15 µL
Primer (1 µg/µL)	3 µL
H_2O	17.85 µL

15.3.4.2. First-Strand Synthesis

For each reaction required, add the appropriate multiples of the following into a master mix:

Ingredient	Volume per sample (µL)
5× First-strand buffer	2
dNTPs (10 mM)	2
DTT solution (100 mM)	1
RNasin (40 U/µL)	0.15
M-MLV RT (200 U/µL)	0.5
H_2O	4.45

Dispense 10 mL of the master mix into each tube. Incubate plates in a PCR machine at 37°C (*see* **Note 8**) for 2 h for RT, 95°C for 5 min to stop the reaction and a final 4°C hold.

15.3.5 PCR Amplification

There are numerous PCR assays for plant viruses which may be performed using the cDNA produced as described above, including group-specific PCRs and real-time PCRs. As a first approximation, use 2.5 mL of the cDNA in each PCR.

15.4 Notes

1. Several scalpels should be used in rotation. The scalpel blades are decontaminated by successive washes, of 1 min duration, in the following solutions; water, 0.2 M NaOH, water, 96% ethanol, and then briefly flamed to remove the ethanol. Caution – plastic-handled disposable scalpels should not be flamed; use metal-handled scalpels.
2. Plant tissue is sampled on a disposable surface (such as a polythene bag) on top of a ceramic tile, giving a clean cutting surface for each tuber.
3. A unit similar in design to the Dispomix® has recently become available; the Ultra Turrax® Tube Drive (IKA). The cost of this system is lower, but the speed is constant and the direction of rotation of the blades is not programmable. Also, the tubes do not have a sieve. The utility of all such machines should be determined empirically for particular sample types.
4. Cores are taken using a cork borer-type sampler. The cylindrical cores should be approximately 5 mm long with a diameter of 5 mm. The location of core sampling may affect the efficiency of detection for individual viral species – this factor should be determined during development of an assay.
5. The potato tuber extraction program for the Dispomix® is as follows, showing rpm for each second of the program:
   - 3,600; 3,800; 3,800; –1,500; 3,900; 3,900; 3,900; 3,900; –1,500; 3,800; 3,800; 3,500; 2,800; –200; 0; 3,600; 3,800; 3,800; –1,500; 3,900; 3,900; 3,900; 3,900; –1,500; 3,800; 3,800; 3,500; 2,800; –200; 0; 3,600; 3,800; 3,800; –1,500; 3,900; 3,900; 3,900; 3,900; –1,500; 3,800; 3,800; 3,500; 2,800; –200; 0.

   The numbers should be entered into a 45-s profile in the Dispomix® profile editor.
6. For 150 samples, 2× 1 mL of enzyme solutions are made in 1.5-mL microfuge tubes and mixed on a culture wheel

at room temperature until dissolved (the enzyme solution should be made in advance as it takes some time to dissolve). Before the addition to the tuber extract, 2 mL of the enzyme solution is mixed with 8 mL of SEB in a pipette trough, to give the correct dilution.

7. The action of the enzymes may be improved by incubation at room temperature for 2 h prior to the overnight incubation.

8. Some viruses have inherent secondary structure in their genome which can inhibit the production of cDNA by early termination of the synthesis reaction. To overcome this problem one can raise the temperature of incubation used in first-strand synthesis to 42°C or higher. This may reduce secondary structures, but will also reduce the effective half-life of the reverse transcriptase. AMV reverse transcriptase may be used instead, because it has an optimal temperature of 42°C. Unfortunately, AMV RT has more endogenous RNaseH activity than M-MLV RT, thus on average AMV RT produces shorter cDNA fragments. RNaseH-deficient RT enzymes are also available (e.g., the Superscript enzymes from Invitrogen), and there is some evidence that these may be a more sensitive type of RT enzymes for PCR assays. The RT conditions required for the efficient detection of individual viruses can only be determined empirically.

References

1. Wetzel, T., Candresse, T., Macquaire, G., Ravelonandro, M., and Dunez, J. (1992) A highly sensitive immunocapture polymerase chain reaction method for plum pox potyvirus detection. *J. Virol. Methods* 39, 27–37.
2. Nolasco, G., de Bias, C., Torres, V., and Ponz, F. (1993) A method combining immunocapture and PCR amplification in a microtiter plate for the detection of plant viruses and subviral pathogens. *J. Virol. Methods* 45, 201–218.
3. Kokko, H. I., Kivineva, M., and Karenlampi, S. O. (1996) Single-step immunocapture RT-PCR in the detection of raspberry bushy dwarf virus. *Biotechniques* 20, 842–846.
4. Mumford, R. A., and Seal, S. E. (1997) Rapid single-tube immunocapture RT-PCR for the detection of two yam potyviruses. *J. Virol. Methods* 69, 73–79.
5. Jacobi, V., Bachand, G. D., Hamelin, R. C., and Castello, J. D. (1998) Development of a multiplex immunocapture RT-PCR assay for detection and differentiation of tomato and tobacco mosaic tobamo-viruses. *J. Virol. Methods* 74, 167–178.
6. Pasquini, G., Simeone, A. M., Conte, L., and Barba, M. (1998) Detection of plum pox virus in apricot seeds. *Acta Virol.* 42, 260–263.
7. Sefc, K. M., Leonhardt, W., and Steinkellner, H. (2000) Partial sequence identification of grapevine-leafroll-associated virus-1 and development of a highly sensitive IC-RT-PCR detection method. *J. Virol. Methods* 86, 101–106.
8. Sharman, M., Thomas, J. E., and Dietzgen, R. G. (2000) Development of a multiplex immunocapture PCR with colourimetric detection for viruses of banana. *J. Virol. Methods* 89, 75–88.
9. Helguera, P. R., Taborda, R., Docampo, D. M., and Ducasse, D. A. (2001) Immunocapture reverse transcription-polymerase chain reaction combined with nested PCR

greatly increases the detection of Prunus necrotic ring spot virus in the peach. *J. Virol. Methods* 95, 93–100.

10. Harper, G., Dahal, G., Thottappilly, G., and Hull, R. (1999) Detection of episomal banana streak badnavirus by IC-PCR. *J. Virol. Methods* 79, 1–8.
11. Harper, G., Osuji, J. O., Heslop-Harrison, J. S., and Hull, R. (1999) Integration of Banana streak badnavirus into the *Musa* genome: molecular and cytogenetic evidence. *Virology* 255, 207–213.
12. Badge, J., Brunt, A., Carson, R., Dagless, E., Karamagioli, M., Philips, S., et-al. (1996) A carlavirus-specific PCR primer and partial nucleotide sequence provides further evidence for the recognition of cowpea mild mottle mosaic virus as a whitefly-transmitted carlavirus. *Eur. J. Plant Pathol.* 102, 305–310.
13. van der Vlugt, R. A. A., and Berendsen, M. (2002) Development of a general potexvirus cDNA and PCR primer set. *Eur. J. Plant Pathol.* 108, 367–371.
14. Gibbs, A., and Mackenzie, A. (1997) A primer pair for amplifying part of the genome of all potyvirids by RT-PCR. *J. Virol. Methods* 63, 9–16.
15. Chen, J., Chen, J., and Adams, M. J. (2001) A universal PCR primer to detect members of the *Potyviridae* and its use to examine the taxonomic status of several members of the family. *Arch. Virol.* 146, 757–766.
16. Gibbs, A., Armstrong, J., Mackenzie, A. M., and Weiller, G. F. (1998) The GPRIME package: computer programs for identifying the best regions of aligned genes to target in nucleic acid hybridisation-based diagnostic tests, and their use with plant viruses. *J. Virol. Methods* 74, 67–76.
17. Karasev, A. V., Nikolaeva, O. V., Koonin, E. V., Gumpf, D. J., and Garnsey, S. M. (1994) Screening of the closterovirus genome by degenerate primer-mediated polymerase chain reaction. *J. Gen. Virol.* 75, 1415–1422.
18. Tian, T., Klaasen, V. A., Soong, J., Wisler, G., Duffus, J. E., and Falk, B. W. (1996) Generation of cDNAs specific to lettuce infections yellows closterovirus and other whitefly-transmitted viruses by RT-PCR and degenerate oligonucleotide primers corresponding to the closterovirus gene encoding the heat shock protein homolog. *Phytopathology* 86, 1167–1173.
19. Monger, W. A., Clover, G. R. G., and Foster, G. D. (2001) Molecular identification of *Oat mosaic virus* as a *Bymovirus. Eur. J. Plant Pathol.* 107, 661–666.
20. Chu, F.-H., Chao, C.-H., Chung, M.-H., Chen, C.-C., and Yeh, S.-D. (2001) Completion of the genome sequence of Watermelon silver mottle virus and utilization of degenerate primers for detecting Tospoviruses in five serogroups. *Phytopathology* **91**, 361-368.
21. Choi, S. K., Choi, J. K., Park, W. M., and Ryu, K. H. (1999) RT-PCR detection and identification of three species of cucumoviruses with a genus-specific single pair of primers. *J. Virol. Methods* 83, 67–73.
22. Saldarelli, P., Rowhani, A., Routh, G., Minafra, A., and Digiaro, M. (1998) Use of degenerate primers in a RT-PCR assay for the identification and analysis of some filamentous viruses, with special reference to clostero- and vitiviruses of the grapevine. *Eur. J. Plant Pathol.* 104, 945–950.
23. Brown, J. K. (2000) Molecular markers for the identification and global tracking of whitefly vector–*Begomovirus* complexes. *Virus Res.* 71, 233–260.

Chapter 16

Multiplex Polymerase Chain Reaction (PCR) and Real-Time Multiplex PCR for the Simultaneous Detection of Plant Viruses

V. Pallas, J. Sanchez-Navarro, A. Varga, F. Aparicio, and D. James

Summary

Multiplex Polymerase Chain Reaction (PCR) can be used for the simultaneous detection of plant viruses. Multiple primer pairs or polyvalent primer pairs can be used to detect and identify several viruses in a single PCR.

Key words: Multiplex, Polymerase chain reaction, DNA, Plant viruses, Primers.

16.1. Introduction

Traditional plant-virus detection techniques have been based on bioassays or serology. These are relatively inexpensive, but of limited sensitivity. During the last years, incorporation of molecular detection techniques (e.g. molecular hybridization, RT-PCR, NASBA, etc.) has increased the spectrum of available techniques, and probes, for the detection of plant pathogens, including viruses. Detection methods based on nucleic acid analysis, alone or in combination with serological methods (e.g. molecular hybridization, RT-PCR, nested RT-PCR, IC-PCR, cooperative-PCR, etc.) offer an alternative to the worldwide use of serological methods. This is due to their high specificity and low detection limits, specifically with techniques that use the PCR amplification approach *(1–7)*. The incorporation of molecular probes or chemicals that allows the detection of the amplified PCR products during the reaction (real-time PCR) has proved to be a powerful and sensitive tool for rapid reliable detection, characterization, and quantification of pathogens, including plant viruses *(8)*.

R. Burns (ed.), *Methods in Molecular Biology, Plant Pathology, vol. 508*

DOI: 10.1007/978-1-59745-062-1_16

Real-time PCR has been found to be more sensitive than conventional PCR for the detection of some plant viruses *(9, 10)*. Accurate quantification of DNA targets may be achieved, provided a suitable standard curve can be created. Real-time PCRs usually take a shorter time to complete, compared to conventional PCR, and electrophoretic analysis is not required. See reviews by James et al. *(8)*, Mackay et al. *(11)*, and Wilhelm and Pingoud *(12)* for detailed information about the process of real-time PCR.

A valuable feature of the molecular technologies is the possibility to detect several pathogens in a single reaction. This makes these approaches very attractive for routine diagnosis by reducing time, labour, and cost. Several strategies have been successfully employed that allow the simultaneous detection and (or) identification of several viruses in a single PCR by using either multiple primer pairs (multiplex RT-PCR; *(7, 13–18)* or a polyvalent primer pair (polyvalent PCR) that is able to drive the amplification of several targets *(3, 19)* (see (8) for a review)). The possibility to simultaneously amplify two or more targets has several obvious advantages: (a) different regions of the target may be amplified to confirm and verify detection, (b) detection may be combined with specific identification such as strain typing or identification of variants, (c) several organisms may be identified in a single assay, and (d) oligonucleotide primers that target host endogenous genes or spiked templates may be included as assay monitors to avoid false negative results. In this chapter we describe two protocols to perform multiplex PCR analysis using either the standard one-step RT-PCR with specific primers for each virus or the real-time RT-PCR using universal primers and melting curve analysis.

16.2. Materials

16.2.1. Multiplex One-Step RT-PCR

1. Extraction Buffer: 6 M guanidine hydrochloride, 0.2 M sodium acetate (pH 5.2), 25 mM EDTA, 1 M potassium acetate, 2.5% polyvinylpyrrolidone (PVP), 1% Β-mercaptoethanol (add just before use). Adjust to pH 5.6–5.8 with acetic acid.
2. Washing buffer: 10 mM Tris–HCl (pH 7.5), 0.5 mM EDTA, 50 mM sodium chloride (NaCl), 50% ethanol.
3. NaI Solution: 6 M sodium iodide (NaI), 1.87% sodium sulphite (Na_2SO_3). Autoclave. Store in the dark.
4. Silica solution (Sigma-Aldrich Cat. S5631). Dissolve desired Silicon dioxide powder in DEPC-water at 1 g/10 V ratio. Store 24 h at 4°C (Silica powder should be compacted at the bottom of the recipient). Discard supernatant and dissolve

the pellet at 1 g/10 V of DEPC-water. Store 5 h at 4°C. Discard supernatant, and dissolve the pellet with DEPC-water at 1 g/1 V ratio. Adjust to pH 2.0 with HCl and store in a dark bottle at 4°C (this solution can be stored at 4°C up to 1 year).

5. SuperScript III™ One-Step RT-PCR System with Platinum Taq DNA Polymerase (Invitrogen).
6. Ribonuclease inhibitor (HPRI).
7. Absolute ethanol.
8. Diethylpyrocarbonate-treated water (DEPC water).
9. Tri Reagent (Sigma-Aldrich Cat. T9424).
10. Chloroform.
11. Isopropanol.
12. Liquid nitrogen.
13. Pellet pestle polypropylene.
14. Centrifuge.
15. Water bath.
16. 1.5-ml Eppendorf tubes.
17. Plastic bags.

16.2.2. Multiplex Real-Time RT-PCR

1. QIAGEN RNeasy Total RNA Kit (QIAGEN Inc, Mississauga, Ontario).
2. FastPrep FP 120A Instrument (Cat# 6001-120, Qbiogene, Inc., Carlsbad, CA).
3. 2.0-ml microfuge tubes, flat bottomed and screw capped.
4. Glass beads (Cat# 18000-46, Fine Science Tools, Vancouver, BC).
5. PPV-P1 primer (10 μM stock solution).
6. Nad5-R primer (5 μM stock for herbaceous hosts, and 10 μM stock for woody hosts).
7. 5× First Strand buffer (Invitrogen Corp., Burlington, Ontario).
8. 0.1 M DTT.
9. 10 mM dNTP Mix (Prepared from 100 mM dNTP set, Cat# 10297-018, Invitrogen).
10. RNaseOUT™ (40 U/μl, Invitrogen).
11. SUPERSCRIPT™ II (Invitrogen).
12. Diethylpyrocarbonate-treated (DEPC) H_2O.
13. Stratagene's RoboCycler thermocycler (La Jolla, California).
14. PPV-U primer (10 μM stock solution).
15. PPV-FM primer (10 μM stock solution).

16. PPV-FD primer (10 μM stock solution).
17. PPV-RR primer (10 μM stock solution).
18. Nad5-F primer (10 μM stock solution).
19. Nad5-R primer (5 μM stock for herbaceous hosts, and 10 μM stock for woody hosts).
20. 10× Karsai buffer (see *29*).
21. 50 mM $MgCl_2$.
22. 10 mM dNTP Mix.
23. Platinum® Taq DNA polymerase High Fidelity (Invitrogen).
24. SYBR Green I (Cat# S-9430, Sigma, Oakville), diluted 1:5,000 in TE, pH 7.5 as per *(29)*.
25. Sterile PCR-grade Millipore® H_2O.
26. Agarose (BioRad).
27. Tris–Borate–EDTA (TBE) buffer, 10× concentrate (Sigma, Cat# T4415-4L).
28. Ethidium bromide (10 mg/ml stock).
29. SmartCycler® II Thermal Cycler (Cepheid, Sunnyvale, CA) with data interpretation software SmartCycler® Software version 2.0d.

16.3. Methods

16.3.1. Multiplex One-Step RT-PCR

Multiplex RT-PCR using specific primers has been used successfully for routine diagnosis of plant viruses and viroids *(7, 15–17)* allowing simultaneous detection and identification of at least eight plant viruses plus an internal control *(18)*. We describe a protocol to perform standard one-step RT-PCR with special emphasis on the best RNA extraction protocol observed, together with the recommended amplification kit and primer considerations. Finally, we present an example of multiplex one-step RT-PCR applied to the analysis of Apple mosaic virus (ApMV), Prunus necrotic ringspot virus (PNRSV), Prune dwarf virus (PDV), American plum line pattern virus (APLPV), Plum pox virus (PPV), Apple chlorotic leaf spot virus (ACLSV), Apricot latent virus (ApLV), and Plum bark necrosis stem pitting associated virus (PBNSPaV) in different host, plus an internal control (**Fig. 16.1**; **Table 16.1**).

The methods described below outline two different RNA extraction procedures (Silica method and Tris Reagent method, Sigma) plus a sensitive and reliable one-step RT-PCR method to simultaneously detect and differentiate several viruses.

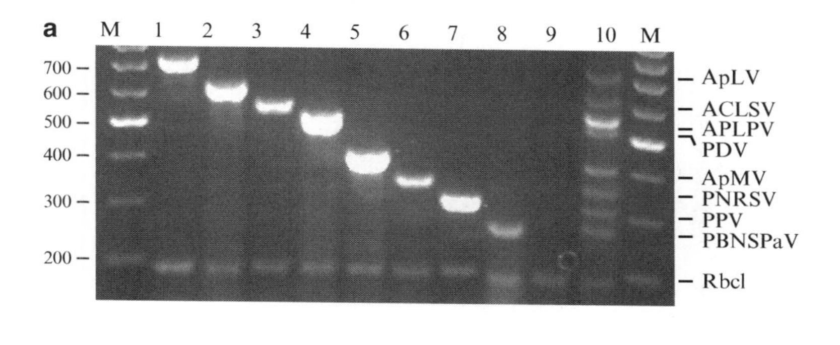

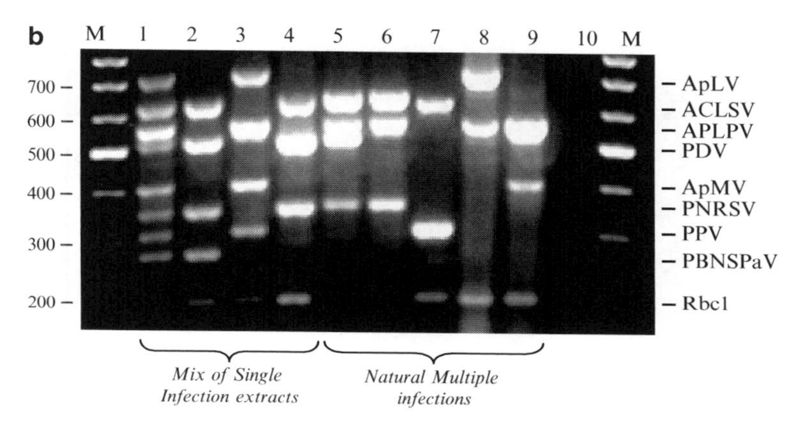

Fig. 16.1. (a) One-step multiplex RT-PCR analysis of single (*lanes 1–8*) or simulated multiple (*lane 10*) infected stone fruit trees. *Lane 1* GF305-ApLV; *lane 2* plum-ACLSV; *lane 3* plum-APLPV; *lane 4* peach-PDV; *lane 5* apricot-ApMV; *lane 6* cherry-PNRSV; *lane 7* GF-305-PPV; *lane 8* plum-PBNSPaV; *lane 9* healthy plum; *lane 10* mix of total nucleic acids analysed in *lanes 1–8*. (**b**) Multiple RT-PCR analysis of simulated (*lanes 1–4*) or natural (*lanes 5–9*) multiple infections. *Lane 1* ApLV, ACLSV, APLPV, PDV, ApMV, PNRSV, PPV, and PBNSPaV; *lane 2* ACLSV, PDV, PNRSV, and PBNSPaV; *lane 3* ApLV, APLPV, ApMV, and PPV; *lane 4* ACLSV, PDV, and PNRSV; *lane 5* ACLSV, APLPV, PDV, and PNRSV; *lane 6* ACLSV, APLPV, and PNRSV; *lane 7* ACLSV and PPV; *lane 8* ApLV and APLPV; *lane 9*: APLPV and ApMV; *lane 10* healthy cherry. *M* 300 ng of 100-bp molecular weight marker. The expected PCR fragment for each virus is indicated (Figure reproduced from Sanchez-Navarro et al. (18) with permission from Springer.

16.3.1.1. Total RNA Extraction Using the Silica Method

This protocol described by Foissac et al. *(20)* with some modifications is based on the use of Silica chemical. This procedure has the advantage of avoiding the use of phenol (*see* **Note 1**).

1. Homogenize 100 mg of plant tissue in 2 ml of extraction buffer (1:20, w/v relation) in a plastic bag.
2. Transfer 500 µl of the sap extract sample to a 1.5-ml Eppendorf tube.
3. Add 100 µl of 10% PVP and incubate 10 min at 70°C (mix intermittently).
4. Transfer tubes to ice and incubate for 5 min.
5. Centrifuge at 12,000 × *g* for 10 min at 4°C.

Table 16.1
Oligonucleotide primers used in the one-step multiplex RT-PCR

Virus		Sequence 5′–3′	Product size (nt)	Location	Acc. No.
ApLV	s	GGGAATAGAGCCCCAAGAAG	717	(RNA 3) 667–1,393	AF057035
	a	TGCCGTCCCGATTAGGTTG			
ACLSV	s	CCATCTTCGCGAACATAGC	632	6,902–7,537	NC001409
	a	GTCTACAGGCTATTTATTATAAG			
APLPV	s	GGTCGTCAAGGGAGAGGC	563	(RNA 3) 1,490–2,053	NC003453
	a	GGCCCCTAAGGGTCATTTC			
PDV	s	CAACGTAGGAAGTTCACAG	517	(RNA 3) 1,610–2,129	L28145
	a	GCATCCCTTAAAGGGGCATC			
ApMV	s	CGTGAGGAAGTTTAGGTTG	417	(RNA 3) 1,638–2,056	U15608
	a	GCCTCCTAATCGGGGCATCAA			
PNRSV	s	GAACCTCCTTCCGATTTAG	346	(RNA 4) 527–884	AJ133208
	a	GCTTCCCTAACGGGGCATCCAC			
PPV	s	CAATAAAGCCATTGTTGGATC	313	9,196–9,506	M92280
	a	CTCTGTGTCCTCTTCTTGTG			
PBNSPaV	s	TACCGAAGAGGGTTTGGATG	270	149–421	AF195501
	a	TAGTCCGCTGGTACGCTACA			
Rbcl gene (internal control)	s	TACTTGAACGCTACTGCAG	186	685–868	AF206813
	a	CTGCATGCATTGCACGGTG			

s sense primer; a antisense primer

6. Transfer 300 μl of supernatant to a fresh Eppendorf tube and add 150 μl of absolute ethanol, 300 μl of NaI solution and 50 μl of Silica-gel solution.
7. Incubate at room temperature with gentle shaking for 30 min.
8. Centrifuge at 7,500 × *g* for 1 min at room temperature. Discard supernatant.
9. Wash the pellet by adding 1 ml of washing buffer. Centrifuge at 7,500 × *g* for 1 min at room temperature.
10. Repeat wash step at least twice.

11. Dry pellet by air drying for 5–10 min and re-suspend in 150 μl of DEPC water. Incubate at 60°C for 4 min to facilitate the complete dissolution of the RNA from the silica reagent.
12. Centrifuge at 7,500 × *g* for 1 min at room temperature and transfer the supernatant containing purified RNA to a new tube. Store at −20°C for immediate analysis (few days), or −80°C for long storage periods.

16.3.1.2. Total RNA Extraction Using the Tri Reagent Method

This procedure is based on the one described by the manufacturer (Sigma) (*see* **Note 2**).

1. Put 100 mg of plant tissue in an Eppendorf tube. Freeze it in liquid nitrogen and grind it using a polypropylene pestle. Avoid thawing of the samples during the homogenization process.
2. Add 1 ml of Tri reagent and vortex vigorously.
3. Let the sample stand for 5 min at room temperature.
4. Add 200 μl of chloroform and vortex vigorously.
5. Allow the sample to stand for 5 min at room temperature.
6. Centrifuge at 12,000 × *g* for 10 min at 4°C.
7. Transfer the aqueous phase to a fresh Eppendorf tube and add 0.8 V of isopropanol.
8. Incubate at room temperature for 10 min. Centrifuge at 12,000 × *g* for 10 min at 4°C. Discard supernatant.
9. Air dry pellet for 5–10 min and re-suspend in 100 μl of DEPC water and incubate at 60°C for 4 min to facilitate the complete dissolution of the RNA pellet.
10. Centrifuge at 12,000 × *g* for 1 min at room temperature and transfer the RNA-containing supernatant to a new tube. Store at −20°C for immediate analysis (few days), or −80°C for long storage periods.

16.3.1.3. One-Step Multiplex RT-PCR

Multiplex reverse transcriptase polymerase chain reaction (RT-PCR) has been used successfully for simultaneous detection and identification of several viruses in a single reaction *(7, 15, 18, 21, 22)*. In this section a protocol is described to detect diverse RNA viruses present in a plant sample, using SuperScript III™ One-Step RT-PCR System with Platinum *Taq* DNA Polymerase (Invitrogen), according to the manufacturer's recommendations. In this procedure all the reactions, e.g. cDNA synthesis followed immediately by PCR amplification, are carried out in the same tube (*see* **Note 3**).

Before addressing multiplex detection by RT-PCR several factors should be taken into account in designing the specific primers present in the reaction cocktail:

1. It is recommendable to design primers that would amplify a region of the viral RNA that accumulates at a high concentration in the infected tissue (usually the coat protein gene), selecting regions that are conserved in all the isolates. To carry out this analysis we must confirm that the designed primers are able to prime amplification of all isolates by performing a BLASTN analysis (http://www.ncbi.nlm.nih.gov/BLAST/). This procedure will confirm the capacity of the primers to amplify at least all isolates present in the database.
2. Primer size around 20 nucleotides in length is sufficient to confer good specificity in the reaction. Moreover, it is desirable that all primers have a similar melting temperature in order to avoid "asymmetric" amplification of the most efficiently primed product. Another important consideration is related to the 3′ terminus of the primer. Perfect matching of the 3′-terminal region is critical to ensure good start of the polymerase or the reverse transcriptase enzymes. To ensure efficiency of priming it is recommendable that primers should end (3′) in a G or C, or CG, or GC nucleotides. Finally, self-complementary primers should be avoided.
3. To avoid asymmetric amplification of the smallest amplicons it is recommendable to design the primers in order to generate PCR products of similar length but sufficient to be discriminated by size in agarose or polyacrylamide gels (*see* **Note 4**).
4. To amplify simultaneously all desired viruses in one RT-PCR all primers must be mixed in a 10× cocktail primer. Together with the pathogen-specific primers it is recommended that a pair of primers be included that would amplify a host RNA target to be used as an internal positive control (*see* **Note 5**). Although it may have to be further optimized for each pair of primers, a concentration of 1.5–5 μM for each primer is recommended as starting 10× cocktail, except for the internal control primers that would be around ten times less concentrated (*see* **Note 6**).

The following recipe has been optimized to perform one-step RT-PCR in a final volume of 10 μl (*see* **Note 7**) using 0.2-ml nuclease-free and thin wallet PCR tubes. It is important to keep PCR tubes on ice until its placement in the preheated thermal cycler.

1. Mix in the PCR tube:
   - 5 μl of 2× reaction Mix (Stratagene).
   - 0.5 μl total RNA.
   - 1 μl 10× cocktail primers.
   - 1 μl PVP 10% (*see* **Note 8**).
   - 10 U of Ribonuclease inhibitor (HPRI).

- 0.4 μl of SuperScript™ III RT/Platinum *Taq* Mix (Stratagene).
- Autoclaved distilled water to 10 μl.

2. Centrifuge briefly, if necessary, to collect all the components at the bottom of the tube.
3. Depending on the thermal cycler used, overlay with mineral oil.
4. Place PCR tubes in the preheated thermal cycler and perform the multiplex RT-PCR with the following cycling conditions: 1 cycle of 45–60°C for 30 min (cDNA synthesis step); 1 cycle of 94°C for 2 min (denaturation step); 40 cycles: 94°C for 15 s, 50–60°C for 30 s, 68°C for 1 min/kb (choose the extension time according to the length of the longest viral target to be amplified); 1 cycle of 68°C for 5 min.

16.3.1.4. Analysis of PCR Products in Agarose or Acrylamide Gels

Depending on the size of the amplicons generated agarose or acrylamide gels can be used. Efficient range separation of PCR products between 0.1 and 2.5 or 0.08 and 0.5 kbp can be performed using 2% agarose or 5% polyacrylamide gels, respectively.

16.3.2. Multiplex Real-Time RT-PCR

In this protocol a relatively simple and rapid real-time multiplex PCR technique is described for identification of the two most common strains of Plum pox virus (PPV), D and M (Fig. 16.2). This approach utilizes: (a) SYBR Green I with melting curve analysis, with unique melting temperatures for PPV-D (T_m 84.3–84.43°C) and PPV-M (T_m: 85.34–86.11°C); (b) universal primers that detect all strains of PPV; and (c) control primers that

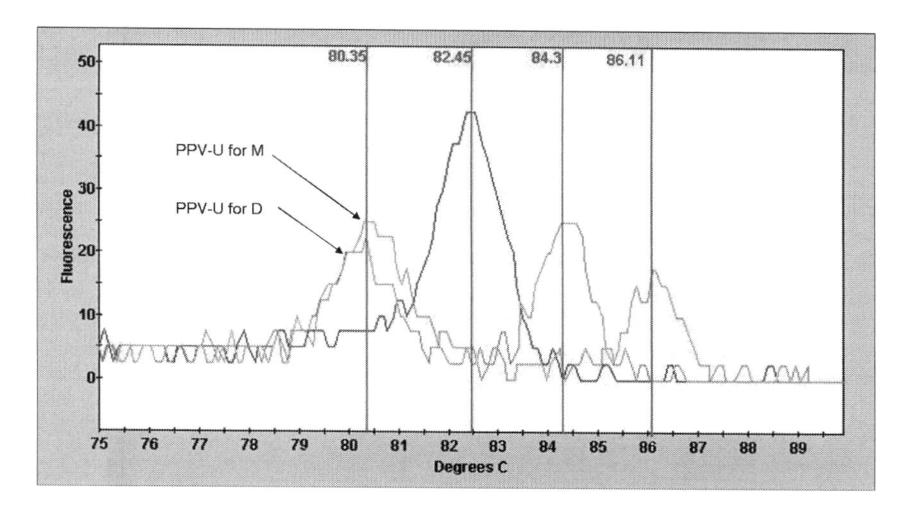

Fig. 16.2. Multiplex melting curves using SYBR Green I for the detection of Nad5, PPV D (D), and PPV M (M) fragments amplified with specific primers, showing strain identification in infected *N. benthamiana*. The T_ms associated with the fragments amplified simultaneously using the universal PPV primers, PPV-U, are also shown (PPV-U for M and PPV-U for D) (Figure reproduced from Varga and James (23), with permission from Elsevier).

target the plant endogenous Nad5 gene *(23, 24)*. See **Table 16.2** for the primer sequences. The Nad5 primers were described by Menzel et al. *(5)* and were designed to span two exon regions, making it mRNA specific. The assay utilizes a universal reverse primer P1 designed by Wetzel et al. *(25)*. Upstream of this primer another universal primer (PPV-RR) was designed to allow amplification of a universal PPV fragment (74 bp), a D-specific fragment (114 bp), and an M-specific fragment (380 bp) (**Table 16.3**). The techniques described have the capacity to confirm the presence of all PPV isolates with amplification of the universal PPV-specific fragment (74 bp). Similar multiplex techniques may

Table 16.2
Oligonucleotide primers used in the multiplex real-time PCR

Primer	Sequence (5 –3)	Size	Target a	Position
PPV-P1 b	ACCGAGACCACTACACTCCC	20	PPV	9,560–9,579[c]
PPV-U	TGAAGGCAGCAGCATTGAGA	20	PPV	9,424–9,243[c]
PPV-FD	TCAACGACACCCGTACGGGC	20	D	9,384–9,403l[c]
PPV-FM	GGTGCATCGAAAACGGAACG	20	M	9,118–9,137[c]
Nad5-F	GATGCTTCTTGGGGCTTCTTGTT	23	Plant	Menzel et al. (5)[d]
PPV-RR	CTCTTCTTGTGTTCCGACGTTTC	23	PPV	9,475–9,497[c]
Nad5-R	CTCCAGTCACCAACATTGGCATAA	24	Plant	Menzel et al. (5)[d]

[a] Primers were designed to detect all PPV strains (PPV-U, and PPV-RR), D-specific (primer PPV-FD), M-specific (primer PPV-FM), or plant-specific (primers Nad5-F and Nad5-R)
[b] PPV-P1 primer *(25)* was used for cDNA synthesis
[c] Relative position on *Plum pox virus* accessions D (X16415) and M (M92280)
[d] Relative position of the Nad5 primers on the nad5 gene for D37958 *(5)*. The primers are mRNA specific as they overlap two exons

Table 16.3
Expected amplicon size with each primer pair used in the multiplex polymerase chain reactions

Forward primer	Reverse primer	Target	Expected size (bp)
PPV-U	PPV-RR	All strains	74
PPV-FD	PPV-RR	D strain	114
PPV-FM	PPV-RR	M strain	380
Nad5-F	Nad5-R	Plant nad5 gene	181

be useful for differentiating strains of other plant viruses, as well as simultaneous detection of different viruses.

16.3.2.1. Total RNA Extraction

1. Place 0.75 mg of glass beads in each 2.0-ml microfuge tube (*see* **Note 9**).
2. Add 100 mg of tissue to each tube with glass beads, and add 500 μl (for herbaceous hosts) or 650 μl (for woody hosts) of RLT buffer (QIAGEN RNeasy Total RNA kit).
3. Use the FastPrep FP 120A Instrument to grind the samples.
4. Centrifuge the microfuge tubes containing ground samples at 5,000 × *g* for 5 min and collect supernatant.
5. Add supernatant to QIAGEN QIAShredder spin column and complete the extraction procedure using QIAGEN RNeasy Total RNA Kit as described by the kit's supplier.
6. Elute total RNA with 35 μl of DEPC H_2O.

16.3.2.2. Two-Tube System of Real-Time Multiplex RT-PCR

cDNA Synthesis

1. Combine 5 μl total RNA (4 μg) suspended in DEPC H_2O, with 2 μl of each antisense primer stock (PPV-P1 – 10 μM (*see* **Note 10**); Nad5-R – 5 μM or 10 μM for herbaceous or woody tissue, respectively).
2. Incubate at 72°C for 5 min, then place on ice immediately.
3. Perform RT in a final reaction volume of 20 μl by adding; 4 μl 5× First strand buffer, 2 μl 0.1 M DTT (*see* **Note 11**), 1 μl 10 mM dNTP Mix, 0.5 μl RNaseOUT™, 1 μl SUPER-SCRIPT™ II, and 2.5 μl DEPC H_2O. Alternatively an RT Master mix may be prepared and 11 μl added to each tube. In calculating reagent volumes for preparing the master mix the number of samples is usually increased by 1–2 samples to ensure that there is enough of the mix for all samples.
4. Carry out RT by incubating RT Mix at 42°C for 60 min, followed by incubation at 99°C for 5 min.

Real-Time PCR

1. Combine 1.0 μl PPV-U primer, 0.875 μl PPV-FM primer, 0.375 μl PPV-FD primer, 0.25 μl PPV-RR primer, 0.25 μl Nad5-F primer, 0.25 μl Nad5-R primer, 2.5 μl 10× Karsai buffer, 0.5 μl 10 mM $MgCl_2$, 0.1 μl Platinum® *Taq* DNA polymerase High Fidelity, 3 μl 1:5,000 SYBR Green I, and sterile PCR-grade Millipore® H_2O to adjust the final volume to 24 μl (*see* **Note 12**). Alternatively a PCR Master mix may be prepared and 24 μl added to each tube. In calculating reagent volumes for preparing the master mix the number of samples is usually increased by 1–2 samples to ensure that there is enough of the mix for all samples.
2. Add 1.0 μl of cDNA diluted with sterile PCR-grade Millipore® H_2O (diluted 1:4 or 1:5 for herbaceous or woody samples, respectively, *see* **Note 13**).

3. Perform real-time PCR using the SmartCycler® II Thermal Cycler with the following cycling conditions: 2-min incubation at 95°C, followed by threshold-dependent cycling for 15 s at 95°C, and 60 s at 60°C. Advance cycling to melt stage once fluorescence passes a threshold of 30 (set manually) plus an extra nine cycles.
4. Take fluorescence readings during anneal/extension step (60°C incubation).
5. Perform melting from 60°C to 95°C, at 0.1°C/s, and with a smooth curve setting averaging 1 point.
6. Visualize the melting peak by plotting the first derivative of fluorescence units against the melting temperature. The melting temperature (T_m) is defined as the peak of the curve, and if the highest point is a plateau, use the mid-point of this plateau to define the T_m.
7. Electrophoretic analysis to identify the PCR products may be carried out using a 1.5% agarose gel in TBE buffer, at 80 V for 60 min, with ethidium bromide staining. **Table 16.3** provides a list of the amplified products and their sizes.

16.3.2.3. One-Tube System of Real-Time Multiplex RT-PCR

1. Dilute total RNA (from both herbaceous and woody tissue) with DEPC H_2O (1/10), and in the PCR tube combine 2.5 μl of the diluted total RNA with; 2.5 μl 10× Karsai buffer, 0.5 μl 5 μM PPV-U primer, 0.5 μl 5 μM PPV-RR primer or 0.5 μl 5 μM PPV-P1 primer, 0.5 μl 5 μM Nad5-F primer, 0.5 μl 5 μM Nad5-R primer, 0.5 μl 10 mM dNTP, 1.0 μl 50 mM $MgCl_2$, 0.2 μl RNaseOUT™, 0.1 μl SUPERSCRIPT™ III, 0.1 μl Platinum® Taq DNA polymerase High Fidelity, 1.0 μl 1:5,000 SYBR Green I, and 15.1 μl DEPC H_2O. Alternatively a Master mix may be prepared that excludes the total RNA. Aliquots (22.5 μl) of the master mix are then added to each tube, followed by the addition of 2.5 μl of the diluted total RNA. In calculating reagent volumes for preparing the master mix the number of samples is usually increased by 1–2 samples to ensure that there is enough of the mix for all samples.
2. Perform real-time PCR using the SmartCycler® II Thermal Cycler with the following cycling conditions: 10 min incubation at 50°C, followed by a two-step thermal cycle of 2 min at 95°C, followed by threshold-dependent cycling for 15 s at 95°C and 60 s at 60°C. Advance cycling to melt stage once fluorescence passes a threshold of 20 (set manually) plus an extra nine cycles.
3. Take fluorescence readings during anneal/extension step (60°C incubation).
4. Perform melting from 60 to 95°C, at 0.1°C/s, and with a smooth curve setting averaging 1 point.

5. Visualize the melting peak by plotting the first derivative of fluorescence units against the melting temperature. The melting temperature (T_m) is defined as the peak of the curve, and if the highest point is a plateau, use the mid-point of this plateau to define the T_m.
6. Electrophoretic analysis to identify the PCR products may be carried out using a 1.5% agarose gel in TBE buffer, at 80 V for 60 min, with ethidium bromide staining. Table 16.3 provides a list of the amplified products and their sizes.

16.4. Notes

1. Silica method is an inexpensive procedure that renders high quality of total RNA isolation with any kind of plant material including necrotic leaves or tissues containing high amounts of phenolic compounds (e.g. stone fruit leaves, tomato plants). From the point of view of quality and yield Silica extraction is comparable to the RNeasy Kit (Quiagen) *(18)*.
2. TRI Reagent renders high amounts of total RNA. However, we only recommend this procedure for herbaceous hosts.
3. We do not recommend using the standard two-step RT-PCR (reverse transcriptation and subsequent PCR amplification) to perform the multiplex RT-PCR analysis. Although this procedure has been successfully used for the simultaneous detection of three stone fruit viruses *(7)*, it does not work well when four or more viruses are to be detected. Unlike the two-step RT-PCR, the one-step RT-PCR works satisfactory at least for the simultaneous detection of eight plant viruses plus an internal control *(18)*. In addition, in order to avoid premature termination of cDNA synthesis due to structured RNAs or high percentage of "G" or "C" residues the use of thermo-stable reverse transcriptases (60°C) is desirable. Comparative analysis of the sensitivity between standard single-target PCR and multiplex PCR indicates that, in general, both techniques show similar sensitivity *(5, 7, 26)*. In many cases, however, a significant reduction in sensitivity is observed under some conditions *(15, 18)*. Competition between different PCR products during amplification *(15)* or for primers between targets could explain this lost of sensitivity. In any case, we observed a decrease in sensitivity when the primer cocktail contained more than five pairs of primers *(18)*. The loss of sensitivity was attributed to the number of different primers rather than the concentration of primers present in the cocktail.

4. Nine different amplicons ranging from 200 to 700 bp length are well amplified simultaneously, with approximately equal amounts. In this range of lengths, PCR products that differ by 50–100 bp are well separated in 2% Agarose gel (**Fig. 16.1**).
5. In order to discriminate potential false negative results (e.g. degraded RNA, RT-PCR inhibitors, etc.), an internal amplification control should be included. This control consists of the inclusion in the RT-PCR of a pair of primers targeting highly conserved host plant RNA that would be co-amplified simultaneously with the viral sequences. Rubisco or NADH dehydrogenase subunits genes have been successfully used as conserved internal control targets *(18, 21)*. To ensure amplification from the mRNA instead of contaminating genomic DNA, the specific sense primer must overlap two continuous exons *(21)*. Finally, it is important to confirm experimentally that the internal control will be amplified in all hosts to be analysed.
6. The optimal final primer concentration in the reaction should be between 0.15 and 0.5 μM, although it could be increased up to 1 μM. In the case of the internal control, we must reduce the final concentration to 0.05 μM to ensure sufficient PCR product without competing with the other amplicons. To select the appropriate primer concentration, start with a cocktail of primers of equal concentration (e.g. 2.5 μM final concentration of each primer in the 10× cocktail) and check it experimentally by amplifying all target viruses in a single reaction, including the internal control. In the case that one of the viruses is not amplified, the concentration of the corresponding primers should be increased 0.5 μM in the 10× cocktail and tested again experimentally.
7. Normally, the one-step RT-PCR is designed to be performed in a final volume of 50 μl. However, satisfactory results are observed using a final volume of 10 μl. In this last case no more than 0.5 μl of purified total RNA must be used. An excess of volume could inhibit the RT-PCR.
8. The presence of PVP in the reaction enhances amplification. This component is used to capture putative inhibitors of the RT-PCR.
9. Glass beads may be added to tubes and stored for use eventually. This is also an ideal system for sample storage as 100-mg samples of tissue may be added to tubes with beads, and stored at –80°C for processing at some later time.
10. Use of the P1 primer designed by Wetzel et al. *(25)* for cDNA synthesis, followed by use of an internal reverse primer (PPV-RR) for real-time PCR improved the reliability and sensitivity of the assay.

11. Dithiothreitol (DTT) is not used in the one-tube system of real-time multiplex RT-PCR. DTT can negatively affect RT-PCR using SYBR Green I *(27, 28)*.
12. It is always convenient to ensure that reagents are mixed to achieve uniform distribution. A brief period of low-speed centrifugation after mixing will ensure that all the mixture is at the bottom of the PCR tube.
13. Higher dilutions of the cDNA from infected woody samples (1:5) compared to infected herbaceous samples (1:4) were required for the best melting curve results. This may indicate the presence of inhibitors in cDNA derived from infected woody hosts.

References

1. Sanchez-Navarro, J. A., Aparicio, F., Rowhani, A., and Pallas, V. (1998) Comparative analysis of ELISA, nonradioactive molecular hybridization and PCR for the detection of prunus necrotic ringspot virus in herbaceous and Prunus hosts. *Plant Pathol.* 47, 780–786.
2. Pallas, V., Mas, P., and Sanchez-Navarro, J. A. (1998) Detection of plant RNA viruses by nonisotopic dot-blot hybridization. *Methods Mol. Biol.* 81, 461–468.
3. Foissac, X., Svanella-Dumas, L., Gentit, P., Dulucq, M. J., Marais, A., and Candresse, T. (2005) Polyvalent degenerate oligonucleotides reverse transcription-polymerase chain reaction: a polyvalent detection and characterization tool for trichoviruses, capilloviruses, and foveaviruses. *Phytopathology* 95, 617–625.
4. Marini, D. B., Zhang, Y. P., Rowhani, A., and Uyemoto, J. K. (2002) Etiology and host range of a Closterovirus associated with plum bark necrosis-stem pitting disease. *Plant Dis.* 86, 415–417.
5. Menzel, W., Jelkmann, W., and Maiss, E. (2002) Detection of four apple viruses by multiplex RT-PCR assays with coamplification of plant mRNA as internal control. *J. Virol. Methods* 99, 81–92.
6. Olmos, A., Bertolini, E., and Cambra, M. (2002) Simultaneous and co-operational amplification (Co-PCR): a new concept for detection of plant viruses. *J. Virol. Methods* 106, 51–59.
7. Saade, M., Aparicio, F., Sanchez-Navarro, J. A., Herranz, M. C., Myrta, A., Di Terlizzi, B., and Pallas, V. (2000) Simultaneous detection of the three ilarviruses affecting stone fruit trees by nonisotopic molecular hybridization and multiplex reverse-transcription polymerase chain reaction. *Phytopathology* 90, 1330–1336.
8. James, D., Varga, A., Pallas, V., and Candresse, T. (2006) Strategies for simultaneous detection of multiple plant viruses. *Can. J. Plant Pathol.* 28, 16–29.
9. Olmos, A., Bertolini, E., Gil, M., and Cambra, M. (2005) Real-time assay for quantitative detection of non-persistently transmitted Plum pox virus RNA targets in single aphids. *J. Virol. Methods* 128, 151–155.
10. Schneider, W. L., Sherman, D. J., Stone, A. L., Damsteegt, V. D., and Frederick, R. D. (2004) Specific detection and quantification of Plum pox virus by real-time fluorescent reverse transcription-PCR. *J. Virol. Methods* 120, 97–105.
11. Mackay, I. M., Arden, K. E., and Nitsche, A. (2002) Real-time PCR in virology. *Nucleic Acids Res.* 30, 1292–1305.
12. Wilhelm, J. and Pingoud, A. (2003) Real-time polymerase chain reaction. *Chembiochem.* 4, 1120–1128.
13. Henegariu, O., Heerema, N. A., Dlouhy, S. R., Vance, G. H., and Vogt, P. H. (1997) Multiplex PCR: critical parameters and step-by-step protocol. *Biotechniques* 23, 504–511.
14. James, D. (1999) A simple and reliable protocol for the detection of apple stem grooving virus by RT-PCR and in a multiplex PCR assay. *J. Virol. Methods* 83, 1–9.
15. Bertolini, E., Olmos, A., Martinez, M. C., Gorris, M. T., and Cambra, M. (2001) Single-step multiplex RT-PCR for simultaneous and colourimetric detection of six RNA viruses in olive trees. *J. Virol. Methods* 96, 33–41.
16. Ito, T., Ieki, H., and Ozaki, K. (2002) Simultaneous detection of six citrus viroids and Apple stem grooving virus from citrus plants by multiplex reverse transcription

polymerase chain reaction. *J. Virol. Methods* 106, 235–239.

17. Thompson, J. R., Wetzel, S., Klerks, M. M., Vaskova, D., Schoen, C. D., Spak, J., and Jelkmann, W. (2003) Multiplex RT-PCR detection of four aphid-borne strawberry viruses in *Fragaria* spp. in combination with a plant mRNA specific internal control. *J. Virol. Methods* 111, 85–93.
18. Sanchez-Navarro, J. A., Aparicio, F., Herranz, M. C., Minafra, A., Myrta, A., and Pallas, V. (2005) Simultaneous detection and identification of eight stone fruit viruses by one-step RT-PCR. *Eur. J. Plant Pathol.* 111, 77–84.
19. James, D. and Upton, C. (1999) Single primer pair designs that facilitate simultaneous detection and differentiation of peach mosaic virus and cherry mottle leaf virus. *J. Virol. Methods* 83, 103–111.
20. Foissac, X., Svanella-Dumas, L., Dulucq, M. J., Candresse, T., and Gentit, P. (2001) Polyvalent detection of fruit tree tricho, capillo and foveaviruses by nested RT-PCR using degenerated and inosine containing primers (PDO RT-PCR). *Acta Hortic.* 550, 37–43.
21. Navarro, J. A., Botella, F., Maruhenda, A., Sastre, P., Sanchez-Pina, M. A., and Pallas, V. (2004) Comparative infection progress analysis of Lettuce big-vein virus and Mirafiori lettuce virus in lettuce crops by developed molecular diagnosis techniques. *Phytopathology* 94, 470–477.
22. Hassan, M., Myrta, A., and Polak, J. (2006) Simultaneous detection and identification of four pome fruit viruses by one-tube pentaplex RT-PCR. *J. Virol. Methods* 133, 124–129.
23. Varga, A. and James, D. (2005) Detection and differentiation of Plum pox virus using real-time multiplex PCR with SYBR Green and melting curve analysis: a rapid method for strain typing. *J. Virol. Methods* 123, 213–220.
24. Varga, A. and James, D. (2006) Real-time RT-PCR and SYBR Green I melting curve analysis for the identification of Plum pox virus strains C, EA, and W: Effect of amplicon size, melt rate, and dye translocation. *J. Virol. Methods* 132, 146–153.
25. Wetzel, T., Candresse, T., Ravelonandro, M., and Dunez, J. (1991) A polymerase chain-reaction assay adapted to Plum pox potyvirus detection. *J. Virol. Methods* 33, 355–365.
26. Nassuth, A., Pollari, E., Helmeczy, K., Stewart, S., and Kofalvi, S. A. (2000) Improved RNA extraction and one-tube RT-PCR assay for simultaneous detection of control plant RNA plus several viruses in plant extracts. *J. Virol. Methods* 90, 37–49.
27. Deprez, R. H. L., Fijnvandraat, A. C., Ruijter, J. M., and Moorman, A. F. M. (2002) Sensitivity and accuracy of quantitative real-time polymerase chain reaction using SYBR green I depends on cDNA synthesis conditions. *Anal. Biochem.* 307, 63–69.
28. Pierce, K. E., Rice, J. E., Sanchez, J. A., and Wangh, L. J. (2002) QuantiLyse (TM): reliable DNA amplification front single cells. *Biotechniques* 32, 1106–1111.
29. Karsai, A., Muller, S., Platz, S., and Hauser, M. T. (2002) Evaluation of a home-made SYBR Green I reaction mixture for real-time PCR quantification of gene expression. *Biotechniques* 32, 790–796.

Chapter 17

Fluorescent-Based Techniques for Viral Detection, Quantification, and Characterization

Mathieu Rolland, Agnès Delaunay, and Emmanuel Jacquot

Summary

Fluorescent-based technologies offer opportunities for developing new assays for detection, quantification, and characterization of viral isolates. According to the intrinsic characteristics of fluorescent-based tools (high specificity, sensitivity, and reliability), such type of molecular assays makes possible investigations on original studies such as evolutionary processes (including fitness measurement of isolates), quantitative epidemiology, or the analysis of synergism and antagonism between closely related isolates. The development of these tools is very simple and requires, in complement to basic molecular knowledge such as extraction, cloning, and (RT)-PCR procedures, only the identification of short specific sequence(s) in the targeted viral genome. The Single Nucleotide Polymorphism (SNP) and the 'real-time' RT-PCR assays are proposed as fluorescent-based tools for qualitative and quantitative viral detection, respectively. Moreover, the SNaPshot technology is described as method for isolate characterization.

Key words: Real-time PCR, Single nucleotide polymorphism, SNaPshot, TaqMan® probes.

17.1. Introduction

As any virus-based research programme would not be possible without efficient and reliable assay for detection and characterization of the pathogen, numerous tools allowing identification of viruses in tested samples have been developed and published. Viral particles are constituted by the association of nucleic acids and proteins *(1)*. In consequence, the biochemical properties of these two types of molecules have been extensively used, according to the steady upgrade of molecular technologies, to design detection tools. During the last 20 years, the time- and space-consuming biologically based methods for viral characterization have been efficiently substituted for serological technologies

R. Burns (ed.), *Methods in Molecular Biology, Plant Pathology, vol. 508*

DOI: 10.1007/978-1-59745-062-1_17

(targeting antigenic properties of viral coat proteins) *(2)* and/or for molecular approaches (targeting nucleotide sequences of the viral genome) *(3)*. The progressive increased specificity, sensitivity, and reliability offered by each new technology have justified their use for the improvement of viral detection tools. One of the last described innovations in molecular technologies applied to pathogen detection procedures is the use of fluorescent molecules *(4)*. Covalently linked to free nucleotides (fluorescent deoxyNTPs or dideoxyNTPs) or to primers (5′- and/or 3′-end labelled primers), the specific wavelength produced by each of these chemical compounds makes possible to monitor the recruitment of the labelled molecules during or after the molecular reaction(s) that targets the pathogen. Such properties have been used to develop series of new detection tools *(5–10)*. Among them, the fluorescent-based single nucleotide polymorphism (SNP) detection *(11)*, the quantitative 'real-time' RT-PCR amplification *(12)*, and the multiple SNP interrogation methods *(13)* offer new opportunities for the development of innovative detection, quantification, and characterization assays. In order to illustrate the use of these fluorescent-based technologies in viral detection, the technical set up of such procedures for Potato virus Y (PVY, *Potyvirus*, single-stranded positive sense RNA virus infecting *Solanaceous* and being transmitted by aphids) detection, quantification, and characterization is described. **Figure 17.1** illustrates the genetic organization of PVY and includes the targeted sequences (located within the HC-Pro/P3 genes) used to design probes and primers. Nucleotides A/G_{2213} and A/C_{2271} correspond to the polymorphic nucleotides used to discriminate PVY^N (using isolate 605; *(14)* and PVY^O (using isolate 139;*(15)* isolates in the presented techniques. **Figure 17.2** is a schematic representation of the three presented fluorescent-based detection, quantification, and characterization assays.

17.2. Materials

17.2.1. Required Materials, Products, and Chemicals

1. Sterile plastic wares (microtubes, cones, etc.).
2. Sterile razor blade.

Fig. 17.1. Genomic organization of *Potato virus Y* and nucleotide sequence (from nt 2161 to nt 2278) of isolates N-605 and O-139. All primers and probes requested for the described fluorescent-based techniques are presented. *Black boxes* correspond to positions of the targeted single nucleotide polymorphisms (A/G_{2213} and A/C_{2271}) in the PVY sequence. FAM, Vic and MGB dyes are presented by *grey box*, *circle* and *star*, respectively. *Kpn*I and *Sac*I restriction sites, used to clone the viral sequence in pBluescript plasmid, are presented in *bold uppercase letters* in HCf and P3r primers. [a]Nucleotide positions are according to Jakab et al. *(14)*.

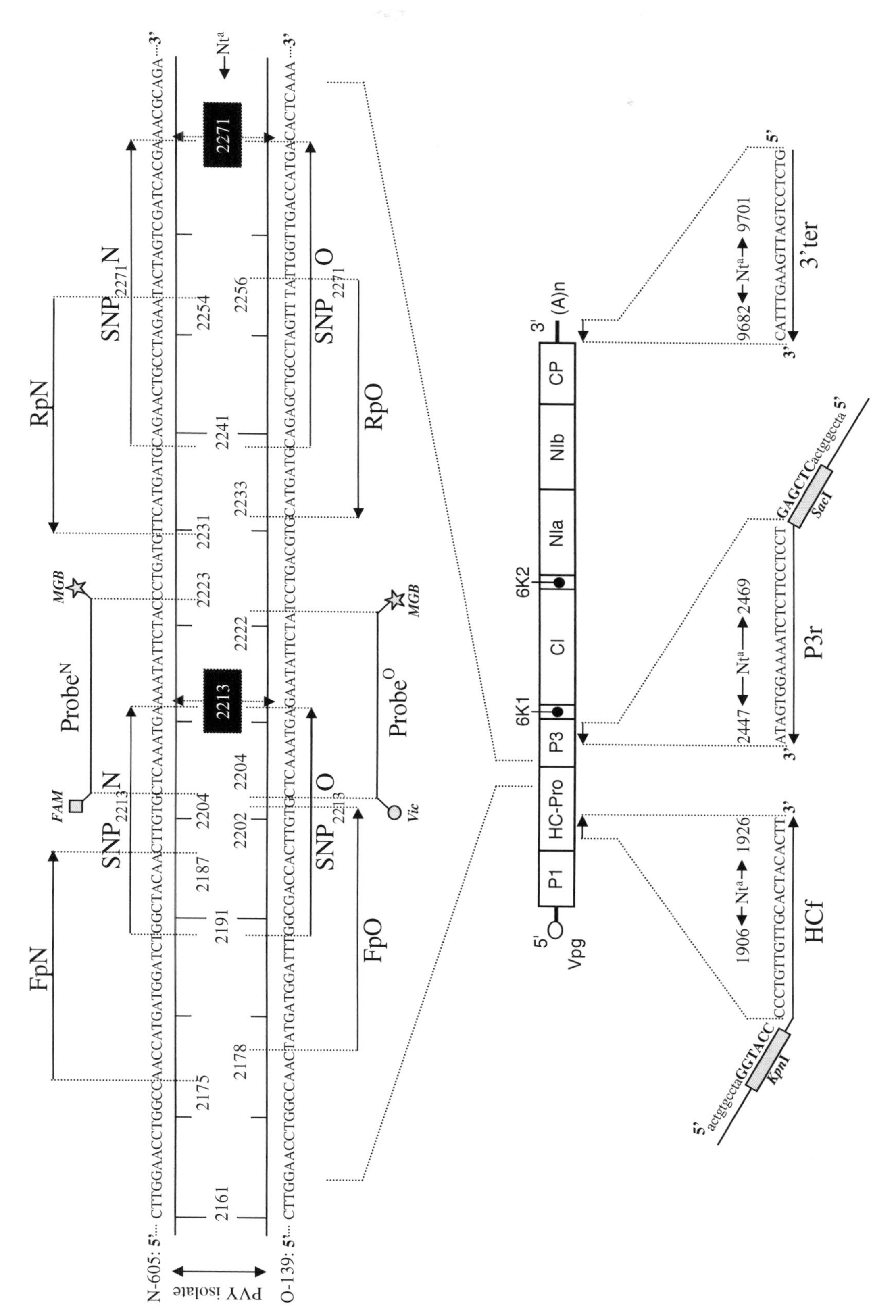

FpN
RpN
Probe^N
SNP2213N
SNP2271N
FAM
MGB
2213
2271
Nt^a
N-605: 5'... CTTGGAACCTGGCCAACCATGATGGATCTGGCTACAACTTGTGCTCAAATGAAAATATTCTACCCTGATGTTCATGATGCAGAACTGCCTAGAATACTAGTCGATCACGAAACGCAGA ...3'
O-139: 5'... CTTGGAACCTGGCCAACTATGATGGATTTGGCGACCACTTGTGCTCAAATGAGAATATTCTATCCTGACGTGCATGATGCAGAGCTGCCTAGTTTATTGGTTGACCATGACACTCAAA ...3'
PVY isolate
FpO
RpO
Probe^O
SNP2213O
SNP2271O
Vic
MGB
2161
2175
2178
2187
2191
2202
2204
2204
2222
2223
2231
2233
2241
2254
2256
5'
Vpg
P1
HC-Pro
P3
6K1
CI
6K2
NIa
NIb
CP
3'
(A)n
5' actgtgcctaGGTACC
KpnI
CCCTGTTGTTGCACTACACTT 3'
1906 Nt^a 1926
HCf
3' ATAGTGGAAAATCTCTTCCTCCT GAGCTCactgtgccta 5'
SacI
2447 Nt^a 2469
P3r
3' CATTTGAAGTTAGTCCTCTG 5'
9682 Nt^a 9701
3'ter

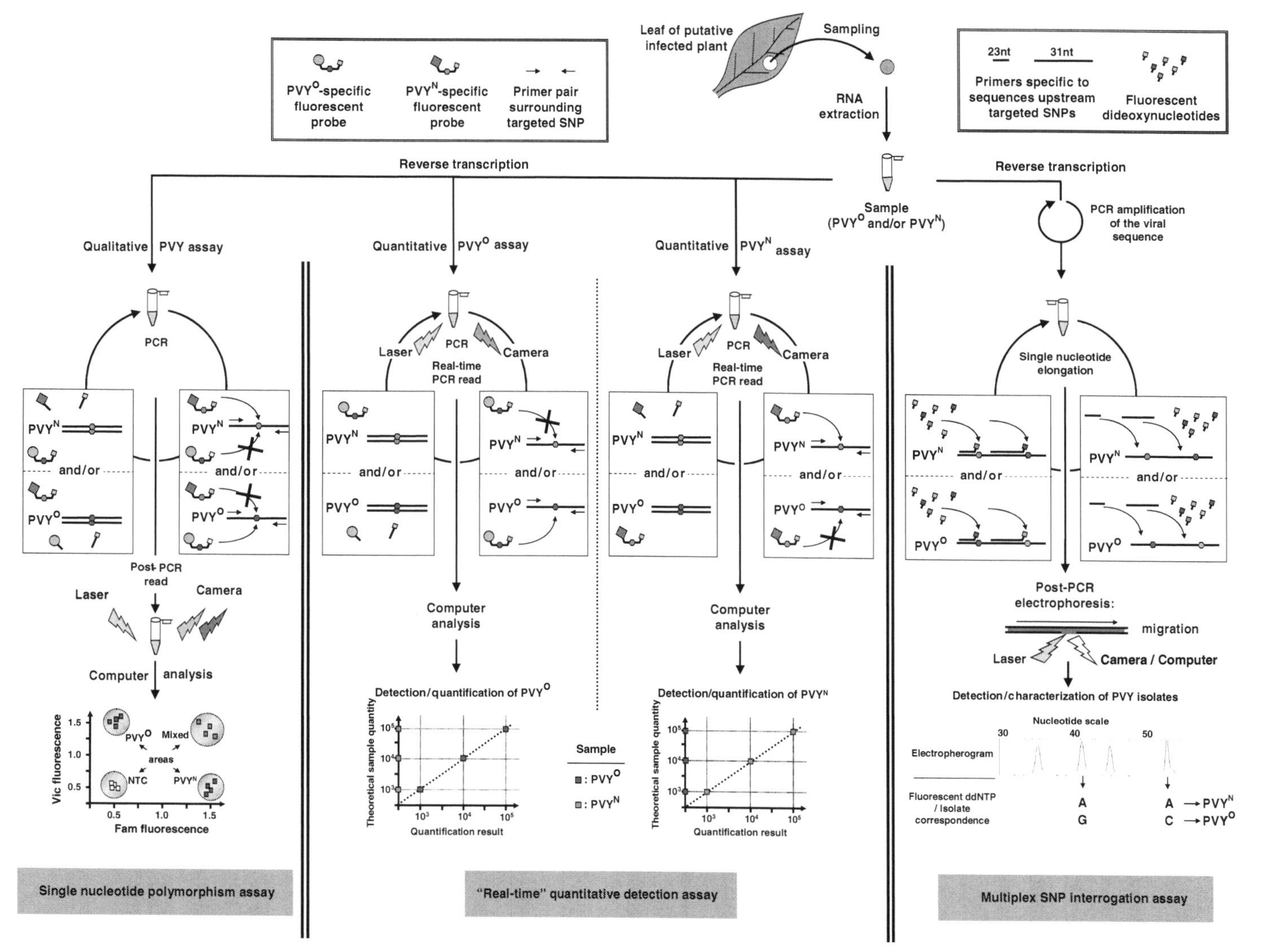

Fig. 17.2. Schematic representation of the procedures used in single nucleotide polymorphism, 'real-time' quantitative detection and multiplex SNP interrogation assays.

3. Sterile plastic pestle.
4. Sterile toothpick.
5. Multiscreen plate (Millipore, http://www.millipore.com/).
6. Wet ice.
7. Deionized water.
8. RNase-free water.
9. 10 mg/ml BSA.
10. 200 mg/ml Ampicillin.
11. Agarose.
12. Phenol/chloroform/isoamyl alcohol (24/24/2, v/v/v).
13. Chloroform/isoamyl alcohol (24/1, v/v).
14. Hi-Di Formamide.
15. Absolute ethanol and 70% ethanol solution.
16. 5 mM Dithiotreitol.
17. 3 M NaOAc.
18. Parafilm®.
19. Sephadex G50.
20. RNasine® (Promega, http://www.promega.com/).
21. AMV Reverse Transcriptase (Promega).
22. Taq Polymerase with buffer and $MgCl_2$ solution (Promega).
23. Restriction enzymes (*Sac*I and *Kpn*I) (New England Biolabs, http://www.neb.com/).
24. T3 RNA polymerase (Promega).
25. RNase-free DNaseI (Promega).
26. Calf Intestinal Phosphatase (Promega).
27. 10 mM dNTPs.
28. 10 mM rNTPs.
29. Primers and TaqMan® probes (according to the matrix sequence, *see* **Fig. 17.1**).
30. T3 (5′-ATTAACCCTCACTAAAGGGA-3′) and T7 (5′-TAA TACGACTCACTATAGGG-3′) primers.
31. SybrSafe™ (Invitrogen, http://www.invitrogen.com/).
32. DNA Smart ladder (Eurogentec, http://www.eurogentec.be/).
33. Escherichia coli DH5α cells (Invitrogen).
34. pBluescript plasmid (Stratagene, http://www.stratagene.com/).
35. PVY Polyclonal antibody.

36. Thermocycler.
37. Agarose gel device.
38. Power supply.
39. Heated block.
40. Imaging system.
41. Centrifuge.
42. Storage/incubating devices for −80, −18, +4, +37, and +42°C.
43. Spectrophotometer apparatus.
44. Electroporator system for bacterial transformation.
45. SV RNA extraction kit®, (Promega).
46. GenElute PCR Clean-up kit (Sigma, http://www.sigmaaldrich.com/).
47. QIAprep® miniprep kit (Qiagen, http://www1.qiagen.com/).
48. One-Step RT-PCR Master Mix Kit® (Applied Biosystems, http://www.appliedbiosystems.com/).
49. SNaPshot® kit (Applied Biosystems).
50. LIZ-120® ladder (Applied Biosystems).
51. ABI7700 apparatus (Applied Biosystems).
52. 3130 Genetic analyser (Applied Biosystems).
53. Allelic discrimination software® (applied biosystems).
54. GeneMapper® (Applied Biosystems).

All Home-Made Buffers, Solutions, and Mixtures Have to Be Prepared with Deionized Water

17.2.2. Standard Solutions and Buffers

1. 1× TBE buffer pH 8.0: 10.78 g Tris, 5.50 g boric acid, 0.58 g EDTA, deionized water to prepare 1 L (*see* **Note 1**).
2. 3 M NaOAc, pH 5.5: 36 ml acetic acid, 114 ml sterile water, 24 g NaOH (*see* **Note 2**). Complete to 200 ml with sterile water.
3. Solid LB media: 10 g bactotrypton, 5 g yeast extract, 5 g NaCl, 20 g agar, deionized water to prepare 1 L. Do not use Agar to prepare liquid LB media.
4. PBS buffer: 8 g NaCl, 2.9 g $Na_2HPO_4 \cdot 12H_2O$, 0.2 g KCl, 0.2 g KH_2PO_4, deionized water to prepare 1 L.
5. Carbonate buffer: 1.59 g Na_2CO_3, 2.93 g $NaHCO_3$, deionized water to prepare 1 L.
6. Extraction buffer: PBS, 0.05% (v/v) Tween20.
7. Ethanol 70% solution must be prepared using absolute ethanol.

17.3. Methods

The methods described below outline *(1)* the total nucleic acid extraction procedures, *(2)* the production of standards, *(3)* the qualitative single nucleotide polymorphism detection assay, *(4)* the 'real-time' quantitative detection assay, and *(5)* the multiplex SNP interrogation assay.

17.3.1. Nucleic Acid Extraction

The first required step to perform any molecular-based detection procedure is to extract, from the original sample to be tested, a fraction containing nucleic acids. Different procedures make possible such extraction step. Among them, leaf soak, immunocapture, and phenol/chloroform extraction procedures are described. As an alternative to these extraction procedures, commercial kits for nucleic acid extractions from plant tissues exist and can be used. The benefit of these kits is presented.

17.3.1.1. Leaf Soak Extraction

The leaf soak extraction is the simplest way to produce crude fractions containing nucleic acids *(16)*. Due to the simplicity of the extraction protocol, the produced fractions contain, in complement to nucleic acids, numerous plant compounds that could impair the use of the leaf soak fractions in molecular technologies such as cloning or sequencing. However, the described fluorescent-based procedures can generally be applied to fractions resulting from leaf soak extractions.

1. Collect small parts (±1 cm^2) of leaf tissue from a plant to be sampled using sterile razor blade (*see* **Note 3**)
2. Cut the collected leaf fragments in thin slices and place slices in a sterile microtube
3. Add 100 µl of sterile extraction buffer
4. Heat the sample for 15 min at 95°C and place the heated sample in wet ice for 10 min
5. Centrifuge at 8,000 × *g* for 5 min and transfer the resulting supernatant in new sterile microtube
6. Dilute the supernatant 10–100 times in sterile extraction buffers
7. Store leaf soak extracts at –20°C until used

17.3.1.2. Immunocapture

The immunocapture corresponds to the first step of standard ELISA procedure *(17–18)*. This does not constitute a real nucleic acid extraction procedure. It allows the specific attachment of viral particles on antibody-coated wells. Particles fixed on antibody-coated wells contain viral genome that can be used as target in the next steps of the presented procedure.

1. Collect, in a sterile microtube, leaf fragments (0.2 g) from plant to be sampled using sterile razor blade
2. Grind collected sample using a sterile plastic pestle in the presence of 200 μl of sterile extraction buffer to produce crude sap extract
3. Coat microtitre plates with 100 μl of PVY polyclonal antibody (1 μg/ml) in carbonate buffer for 2 h at 37°C (*see* **Note 4**)
4. Wash plates three times with sterile extraction buffer
5. Fill wells with 100 μl of crude sap extract and incubate overnight at 4°C
6. Wash plates three times with sterile extraction buffer and discard the buffer after the last wash
7. Store Parafilm®-covered plates at –20°C until used

17.3.1.3. Phenol/ Chloroform

The classical phenol/chloroform procedure allows the production of purified nucleic acid fractions

1. Collect, in a plastic bag, leaf fragments (1 g) from the plant to be sampled using sterile razor blade
2. Grind collected sample in the presence of 1 ml of sterile extraction buffer to produce crude sap extract
3. Mix, in a sterile microtube, 100 μl of produced crude sap extract and 100 μl of cold phenol/chloroform/isoamyl alcohol (49/49/2, v/v/v) and vortex vigorously for 15 s
4. Centrifuge at 10,000 × *g* for 10 min at 4°C and transfer the upper layer in a new sterile microtube
5. Add 100 μl of cold chloroform/isoamyl alcohol (24/1, v/v), vortex vigorously for 15 s, and centrifuge at 10,000 × *g* for 10 min at 4°C
6. Transfer the upper layer in a new sterile microtube containing 200 μl of cold ethanol (100%) and 10 μl of 3 M NaOAc (pH 5.7)
7. Mix gently and place the tube at –20°C for 2 h (*see* **Note 5**)
8. Centrifuge at 10,000 × *g* for 10 min at 4°C, discard supernatant, and add 100 μl of ethanol (70%)
9. Mix gently and centrifuge at 10,000 × *g* for 10 min at 4°C
10. Discard supernatant and dry the pellets at room temperature or using vacuum device
11. Suspend the pellets in 100 μl of sterile nuclease free water
12. Store at –20°C until used

17.3.1.4. Commercial Extraction Kits

Numerous commercial extraction kits are available to prepare purified nucleic acid fractions from plant tissues. Some propose

DNA extraction procedures (e.g. DNeasy plant Kit® (Qiagen)); other are optimized for RNA extraction (e.g. SV RNA extraction kit®, (Promega)). All these kits are based on successive steps corresponding to grinding, lysis, clarifying, nucleic acid-matrix interaction, washing, and elution. A DNase RNase-free activity or an RNase DNase-free activity is included either within one of the previously listed steps or as a supplementary step. Differences between kits correspond to the time required to perform the complete extraction protocol, the requirement of centrifugation or vacuum devices, the maximal nucleic acid yield, and the purity of eluted fraction. They must be tested and compared for each plant/virus application to find the more adapted one.

17.3.2. Cloning and Production of Standard Fractions Containing Viral Transcripts

The described fluorescent-based techniques require the use of calibrated standard fractions containing known quantities of the targeted molecular sequence. Such fractions can be obtained using either purified virus or cloned genomic viral sequence overlapping the targeted molecular markers. As for almost all viruses a specific purification protocol has already been published (e.g. purification of PVY, *(19)*, the following description will focus on the cloning strategy for production of the calibrated standard.

The complete procedure for production of standards is illustrated in **Fig. 17.3**. RNA and DNA manipulation were performed by standard recombinant DNA methods *(20)* to insert viral sequence into pBluescript KS(+) plasmid (*see* **Note 6**).

17.3.2.1. Reverse Transcriptase

Viral cDNA corresponding to viral genome was produced using AMV reverse transcriptase (RT) and a reverse primer complementary to the 3′ end of the viral sequence (**Fig. 17.1**, 3′ter primer).

1. Prepare the following RT mixture (20 μl/reaction)
   - 5× buffer (provided with enzyme), 4 μl
   - 10 mM dNTPs, 2 μl
   - 10 μM RT primer (3′ter), 2 μl
   - Sterile RNase-free water, 4.3 μl
   - 40 U/μl RNasine®, 0.5 μl
   - 10 U/μl AMV Reverse Transcriptase (Promega), 0.2 μl
   - 25 mM $MgCl_2$ 2μl
   - Total nucleic acid extract[1], 5 μl
2. Incubate 1 h at 42°C
3. Store at –20°C until used for PCR

[1]For samples prepared using the immunocapture procedure, sample corresponds to fixed material in ELISA plate well. The mix must include 5 μl of nuclease free water as substitute for the volume of nucleic acid extract. RT reaction occurs directly in the well of microtitration plates.

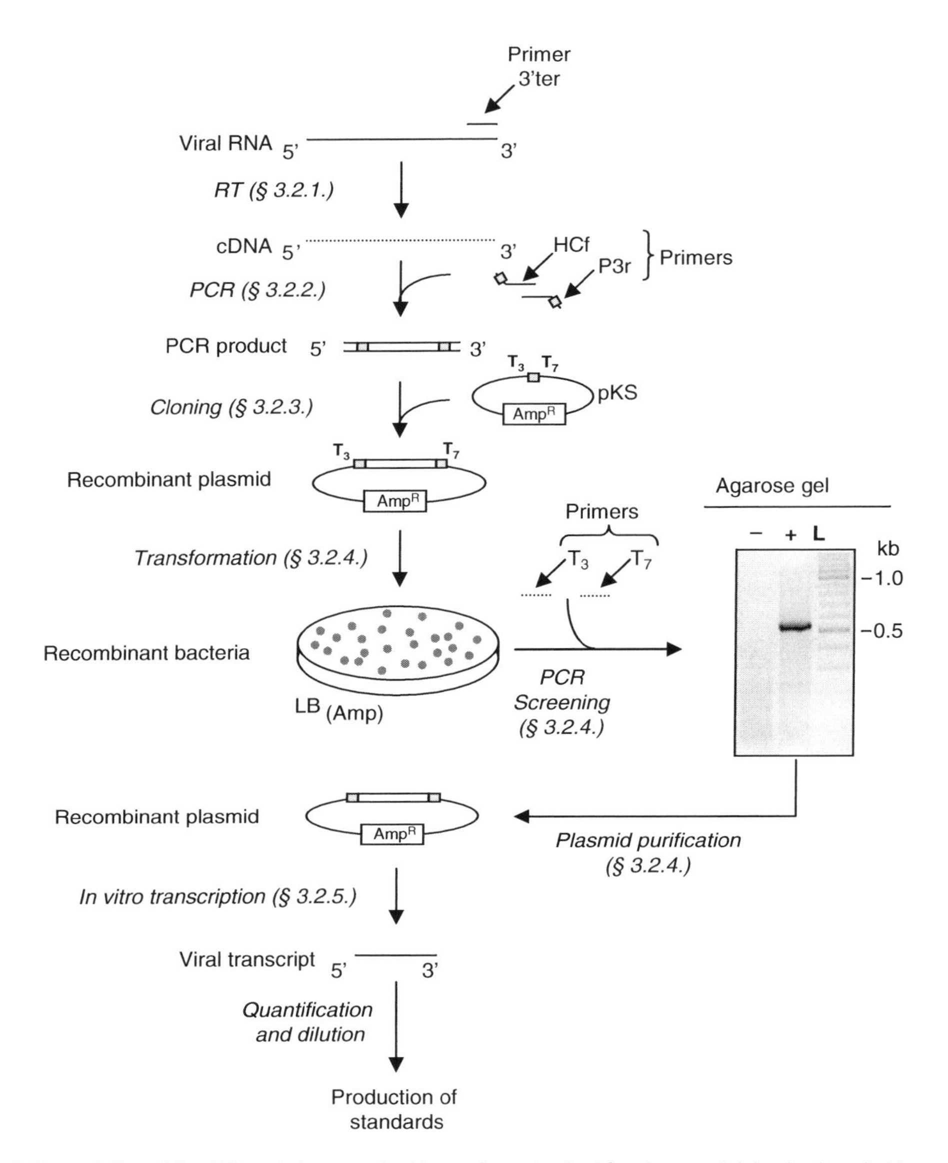

Fig. 17.3. Presentation of the different steps required to produce standard fractions containing in vitro viral transcripts. The information required to perform the procedure are described in the listed paragraphs.

17.3.2.2. PCR Amplification of Targeted Viral Sequence

The cDNA region overlapping the targeted sequence was amplified using appropriate forward and reverse primers (**Fig. 17.1**, HCf and P3r). The latter have extra viral sequences including restriction sites (*see* **Note 7**) that make possible to clone amplified PCR fragments into *Kpn*I and *Sac*I restriction sites of the pBluescript polylinker.

1. Prepare the following PCR mixture (25 μl/reaction)
   - 10× buffer (provided with enzyme), 2.5 μl

- 25 mM $MgCl_2$ (provided with enzyme), 2 μl
- 10 mM dNTPs, 1 μl
- 10 μM HCf primer, 1 μl
- 10 μM P3r primer, 1 μl
- Sterile water, 12.25 μl
- 5 U/μl Taq polymerase, 0.25 μl
- cDNA (RT product), 5 μl

2. Run PCR programme. For the selected HC-Pro/P3 PVY region, amplification conditions correspond to 5 min at 94°C followed by 40 cycles of 1 min at 94°C, 1 min at 52°C, and 1 min at 72°C. A final elongation step is added (10 min, 72°C) after the last PCR cycle
3. Run one-tenth of the PCR product on 1% agarose gel in 1× TBE in the presence of 0.6 mg/ml SybrSafe® (*see* **Note 8**)
4. Visualize nucleic acids under ultraviolet light
5. Check the viral sequence amplification efficiency
6. Store PCR product at −20°C until used for restriction enzyme digestions

17.3.2.3. Restriction Enzyme Digestion of PCR Fragments

The resulting PCR products are extracted from the PCR mix reaction using a phenol/chloroform procedure or using appropriate PCR product extraction kit (*see* **Note 9**). Cleaned amplified sequence is digested using appropriate restriction enzymes (*Kpn*I and *Sac*I) according to the following procedure.

1. Prepare the digestion mix (50 μl/reaction) using the following components
   - 10× buffer (provided with enzyme(s)), 5 μl
   - BSA (10 mg/ml), 5 μl
   - Sterile water, 29 μl
   - 10 U/μl appropriate restriction enzyme 1[2], 0.5 μl
   - 10 U/μl appropriate restriction enzyme 2 (see footnote 2), 0.5 μl
   - DNA (PCR product), 10 μl
2. Incubate at optimal temperature for enzyme activity (*see* **footnote 2**) (generally 37°C) for 1 h
3. Extract DNA fragments using phenol/chloroform procedure or appropriate DNA purification kit (e.g. GenElute PCR Clean-up (Sigma))

[2]The simultaneous use of two enzymes is performed according to their buffer and temperature compatibility. If common buffer/temperature is not available then perform two digestion steps, and after **step 2** go **step 3**; otherwise skip **steps 3–4**.

4. Repeat **steps 1–3** with the second requested restriction enzyme
5. Run one-tenth of the digested products on 1% agarose gel in 1× TBE in the presence of 0.6 mg/ml SybrSafe®
6. Visualize digested DNA fragment under ultraviolet light
7. Store DNA fragments at –20°C until used for cloning into pBluescript plasmid

17.3.2.4. Cloning and Selection of Recombinant Bacteria

After digestion with appropriate restriction enzymes, DNA extraction, and ligation in linearized pBluescript plasmid, the recombinant vector was transformed into *Escherichia coli* DH5α cells by standard methods *(20)*. The DH5α cells were then plated on LB media supplemented with ampicillin (200 μg/ml) and incubated overnight at 37°C. Single colonies were selected and screened using polymerase chain reaction for the presence of recombinant plasmids.

1. Select and collect a single colony using sterile toothpick
2. Make a print on new plate and put remaining cells into sterile 0.2 -μl microtube
3. Place the new printed plate at 37°C for 16–20 h
4. Add the appropriate compounds to perform standard PCR using T3 and T7 primers and the cells as DNA matrix. Adjust water to maintain the concentration of components in the PCR mix
5. Run PCR programme. Amplification conditions correspond to 5 min at 94°C followed by 40 cycles of 1 min at 94°C, 1 min at 52°C, and 1 min at 72°C. A final elongation step is added (10 min, 72°C) after the last PCR cycle
6. Load one-tenth PCR products on 1% agarose gel in 1× TBE in the presence of 0.6 mg/ml Sybrsafe®
7. Visualize fragment under ultraviolet light
8. Identify transformed cells containing recombinant plasmid according to the presence of amplified fragment (around 500 bases) corresponding to the inserted viral sequence in pBluescript vector between T3 and T7 hybridization sites

17.3.2.5. In Vitro Production of Viral Transcripts

According to the presence of recombinant plasmid, select transformed *E. coli* cells from the printed plate stored at 4°C after the 16- to 20-h period at 37°C. Selected cells were grown overnight at 37°C in LB liquid media supplemented with ampicillin (200 μg/ml). The recombinant plasmid DNA was isolated (QIAprep® miniprep (Qiagen)), quantified at 260 nm, and used as matrix for the in vitro transcription of viral RNA.

1. Linearize, for 1 h at 37°C, 1 μg of recombinant plasmid by *Sac*I

2. Purify linearized DNA using phenol/chloroform procedure
3. Resuspend DNA in 5 µl of nuclease-free water
4. Mix DNA plasmid (1 µg; 5 µl), T3 RNA polymerase (15 U), rNTPs (10 mM), and dithiotreitol (5 mM)
5. Incubate the mixture for 3 h at 37°C
6. Add RNase-free DNaseI (5 U) to the mixture and incubate for 15 min at 37°C
7. Purify in vitro transcripts using phenol/chloroform procedure
8. Resuspend viral RNA transcripts in 100 µl nuclease-free water
9. Determine RNA concentration at 260 nm
10. According to the theoretical length (563 bases) of the produced in vitro transcript, calculate the RNA copy number and dilute, in sterile deionized water, viral RNA in order to produce tenfold serially diluted fractions containing viral RNA copies number in the range 10^7 to 10^3 copies/µl

17.3.3. Qualitative SNP Detection Assay

The SNP technology, usually applied for allelic discrimination assays, gene segregation studies, chromosomal mapping, and cloning of genetic determinant(s) of diploid organisms (for a review, see *(21)*, is a simple method allowing the detection and the characterization of a nucleotide within a known polymorphic sequence. Developed to characterize the homo- or heterozygous status of a tested diploid individual for a selected gene, this technique can be applied to polymorphic haploid organisms (such as single-stranded RNA viruses) to identify samples containing only one (visualized as homozygous) or a combination of two variants (considered by the assays as heterozygous). Targeting a well-defined polymorphic sequence in the viral genome, the SNP technique could be appropriate to detect and discriminate viral isolates.

The fluorescent-based SNP detection assay requires the use of primer pairs that frame the targeted polymorphic nucleotide and two fluorescent probes that specifically hybrid the targeted polymorphic sequence.

The presented technique is based on the use of probes with differential (intact or cleaved) fluorescent characteristics. Indeed, the TaqMan® probes (**Fig. 17.4** deal with the fluorescent resonance energy transfer (FRET, *(22)* that allows to transfer energy from a reporter dye to a quencher dye. When these two dyes are linked to the same molecule (primer), excitation of the reporter induces the transfer of its energy to the quencher. The latter produces a fluorescent signal that could be recorded by an appropriate camera. However, the cleavage of the probe separates the

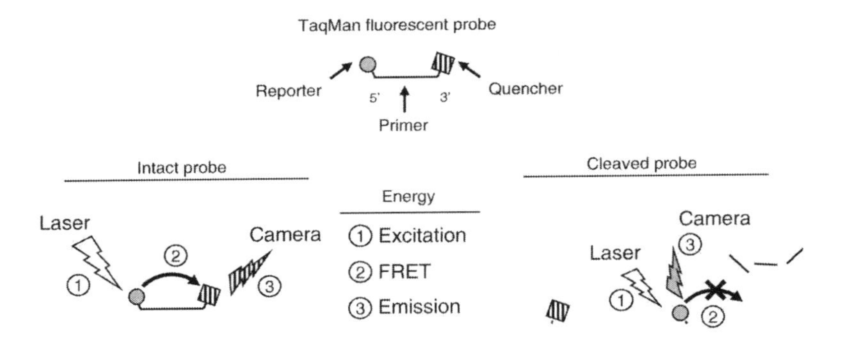

Fig. 17.4. Characteristics of the TaqMan® florescent probe and schematization of the fluorescence resonance energy transfer (FRET).

dyes and prevents the FRET process. Then, the recorded fluorescence in the presence of cleaved probe corresponds to the reporter signal. Such cleavage could occur during the synthesis of new strands according to the 5′–3′ exonuclease activity of the used DNA polymerase (*see* **Note 10**).

17.3.3.1. Design of Probes

Probes correspond to primers that generally do not exceed 30 bases in length. To design probes, use the following guidelines.

1. Design probes with melting temperature (T_m) in the range 65–67°C
2. Fit the G/C content of probes in the range 20–80%
3. Do not select sequence with runs of four or more identical nucleotide especially true for Guanine
4. Do not define primer with a Guanine at the 5′ end
5. Try to place the polymorphic nucleotide in the middle of the probe

According to these guidelines, two probes (**Fig. 17.1**, $Probe^N$ and $Probe^O$) were designed to target polymorphism of PVY nucleotide 2213.

17.3.3.2. Design of Probes

The designed probes must be linked to a reporter dye at the 5′ end and to a quencher at the 3′ end (*see* **Note 11**). In the presented procedure, probes were linked to 6-FAM and Vic reporters for $Probe^N$ and $Probe^O$, respectively (**Fig. 17.1**). Moreover, MGB molecules were selected as quencher in the developed assays.

17.3.3.3. Design of Primer Pair

Primer pair should be designed to allow amplification of small region that does not exceed (if possible) 150 bases in length. The amplified region includes sequence targeted by the probes.

To design primer pair, use the following guidelines.

1. Design primers with melting temperature (T_m) in the range 58–60°C
2. Fit the G/C content of primers in the range 30–80%
3. Do not select sequence with runs of four or more identical nucleotide especially true for Guanine
4. Select sequence allowing the presence of more than two adenine and/or thiamine at the 3′ end of the primers
5. Define primers and probes as close as possible to each other (without overlapping)

According to these guidelines, two primer pairs that frame the targeted SNP_{2213} were designed (**Fig. 17.1**, primers FpN, FpO, RpN, and RpO).

17.3.3.4. Reverse Transcription and Polymerase Chain Reaction

Prior to being analysed, samples (nucleic acid extracts containing or not viral RNA molecules) are submitted to reverse transcription and to polymerase chain reaction in the presence of appropriate enzymes, primers, probes, free dNTP, and buffer. Commercial kits (e.g. TaqMan one-Step RT-PCR Master Mix Kit® (Applied Biosystems)) make possible to perform these two successive steps in a single reaction. The following procedure is described for the use of the RT-PCR kit listed.

1. Prepare the following one-step RT-PCR mixture
   - 2× mix buffer[3] (provided in the kit), 12.5 µl
   - 20 µM FpN primer, 1 µl
   - 20 µM RpN primer, 1 µl
   - 20 µM FpO primer, 1 µl
   - 20 µM RpO primer, 1 µl
   - 20 µM $Probe^N$, 0.25 µl
   - 20 µM $Probe^O$, 0.25 µl
   - Sterile water, 4.875 µl
   - 40× enzyme mixture[4] (provided in the kit), 0.625 µl
   - Sample[5], 2.5 µl
2. Run one-step RT-PCR programme. Amplification conditions correspond to 30 min at 48°C, 10 min at 95°C, and 40 cycles of 15 s at 95°C and 1 min at 60°C

[3]Includes dNTP and hot-start Taq polymerase (with 5′–3′ exonuclease activity).

[4]Includes reverse transcriptase and RNase inhibitors.

[5]Correspond to crude nucleic acid extracts, calibrated in vitro viral transcripts or deionized RNA-free water.

3. At the end of the run, perform laser excitation of each sample in order to collect fluorescent signal (in the range 500–660 nm) produced by the dyes present in the samples (*see* **Note 12**)

   Collected signals are analysed by computer (e.g. allelic discrimination software (Applied Biosystems)) in order to extract from the crude fluorescent records the presence of the reporter – and quencher (when fluorescent quencher is used) – specific wavelengths according to the nature of the used dyes.

17.3.3.5. Requested Control Samples

In complement with unknown samples to be tested, series of control samples should be run. Such samples correspond to two sets of fractions (in 4–8 replicates each), each set containing calibrated quantity (10^6 copies) of one type of the targeted polymorphic sequences. Moreover, no template control (NTC; nucleic acid-free water) should be added to the list of the tested control samples.

17.3.3.6. Data Analysis

Crude fluorescent signal (Fs) corresponding to the specific wavelength of used dyes was recorded by a camera for each sample (**Fig. 17.5a**). Background fluorescence signal associated with NTC and fluorescence data associated with pure PVY^N or PVY^O

a

		Sample type		
		NTC	N-605	O-139
Fs	FAM	0.522+/– 0.024	1.535+/– 0.118	0.538+/– 0.007
	Vic	0.484+/– 0 .008	0.464+/– 0.026	1.277+/– 0.040

b

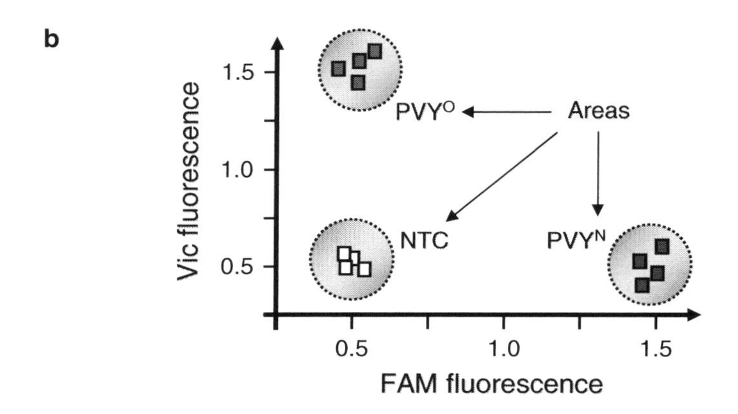

Fig. 17.5. Crude fluorescent signals (Fs) produced by the single nucleotide polymorphism assay (**a**) and graphic illustration of collected Fs data sets (**b**). Mean fluorescence and standard error associated to FAM- and Vic-labelled probes are presented for no template control (NTC), PVY^N-605 and PVY^O-139. Areas for PVY^O, PVY^N and no viral (CTC) detection are presented.

controls were used to define areas corresponding to each type of samples (**Fig. 17.5b**).

Any PVY isolate, used as unknown sample in such SNP assay, is associated with an Fs data set. The latter makes possible the assignment of detected PVY isolate in its corresponding group (PVY^N or PVY^O) according to the identity of the targeted SNP. Samples corresponding to mixture of PVY^N and PVY^O produce FS data with significant level of both FAM- and Vic-specific fluorescence (e.g. mixed samples are associated to FAM and Vic Fs values close to 1.5 and 1.2, respectively).

17.3.4. The 'Real-Time' Quantitative Detection Assay

The 'real-time' fluorescent-based technology makes possible the detection and the quantification of a molecular target (RNA or DNA) in not yet characterized samples. This technique is based on a procedure closely related to the one described for the qualitative SNP detection assay. However, some differences, such as the use of only one fluorescent probe in the mix and the requirement of a range of calibrated (quantified) standards, exist between SNP and 'real-time' procedures. To quantify the target copy number in tested samples, the amplification of the targeted sequence must be monitored in 'real time' during the PCR. Such monitoring produces a description of the increase of fluorescent signals (reporter dyes) during the exponential amplification of the target sequence. Then, comparisons between fluorescent signals associated with calibrated and unknown samples make possible the estimation of the number of targets in the tested sample.

The procedure is described for specific PVY^N detection and quantification. The specificity of this method is supported by the sequence of the designed primers and probe (**Fig. 17.1**, FpN, RpN, and $Probe^N$). The use of PVY^O primers and probe (**Fig. 17.1**, FpO, RpO and $Probe^O$) changes the specificity of the 'real-time' assay to PVY^O isolates.

17.3.4.1. Parameters of the Quantitative One-Step RT-PCR

The RT-PCR is performed using parameters defined for the qualitative SNP procedure except for the primers and probe concentrations. Indeed, the latter should be optimized in order to allow the precise duplication of the matrix at each PCR cycle.

Primers and probes could be used in the range of 50–900 and 50–200 nM, respectively. Numerous studies have demonstrated that optimal concentrations for 'real-time' assays are close to 300 and 200 nM for primers and probes, respectively. However, concentrations should be tested to find the optimal primers and probe concentrations to be used for each new developed assay. For each tested primer and probe concentration, test samples containing 10^6 copies of targeted sequence.

1. Prepare the following one-step RT-PCR mixture (25 µl/ reaction)

- 2× mix buffer[6] (provided in the kit), 12.5 µl
- 20 µM FpN primer[7] variable
- 20 µM RpN primer (see footnote 7) variable
- 20 µM Probe^{N} (see footnote 7) variable
- Sterile water variable
- 40× enzyme mixture[8] (provided in the kit), 0.625 µl
- Sample (10^6 copies of targeted sequence), 2.5 µl

2. Run one-step RT-PCR programme. Amplification conditions correspond to 30 min at 48°C, 10 min at 95°C, and 40 cycles of 15 s at 95°C and 1 min at 60°C
3. At each PCR cycle, perform laser excitation of each sample in order to collect fluorescent signal (in the range 500–660 nm) produced by the different dyes
4. Perform computer analysis of the collected data (e.g. SDS® software (Applied Biosystems)) to draw, for each tested sample, the cycle-dependent increase of fluorescent signal (**Fig. 17.6**)

The collected data made possible to draw the cycle-dependent variation of the reporter fluorescence signal for each tested primer and probe concentration. The optimal concentrations are defined to obtain the maximal increase of reporter fluorescence as soon as possible during the PCR. In the present assay, primers were used at 300 nM and probe at 200 nM.

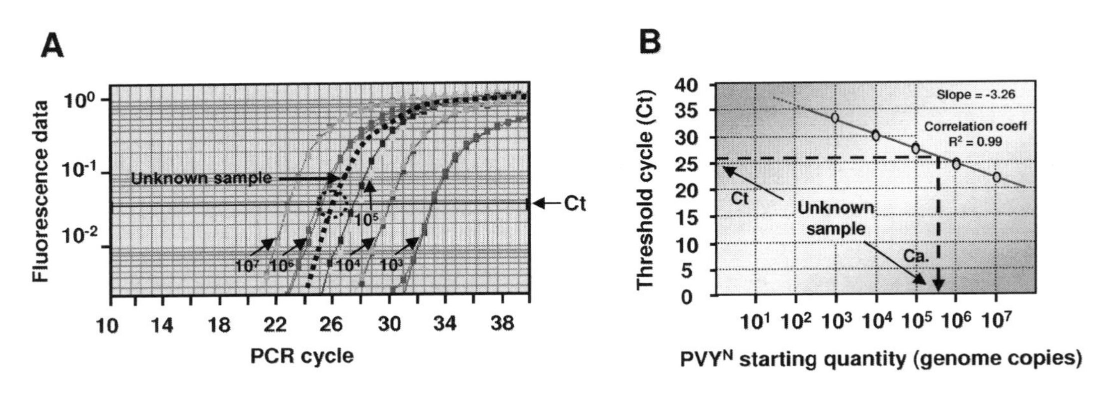

Fig. 17.6. 'Real-time' RT-PCR data for PVY^N detection and quantification (**a**) and standard curve for quantification of unknown samples (**b**). Amplification plots of tested samples corresponding to a tenfold dilution series containing PVY^N viral transcripts in the range 10^7–10^3 copies/reaction. Each fraction was tested in duplicate. The threshold cycle (C_t) used for standard curve calculation is represented with a *black horizontal line*. Standard curves associated to PVY^N 'real-time' RT-PCR assay is illustrated in *plain line*. The standard curve slope and correlation coefficient are listed in the graph. Unknown sample, associated to $C_t = 26$, contains a calculated 6×10^5 targeted RNA copies.

[6]Includes dNTP and hot-start Taq polymerase (with 5′–3′ exonuclease activity).

[7]Reaction presented for PVY^N sequence detection/quantification, for PVY^O application replace primers and probe by FpO, RpO, and Probe^O.

[8]Includes reverse transcriptase and RNase inhibitors.

17.3.4.2. The Use of Standards

In complement with unknown samples to be tested, series of standard samples should be run. The standards correspond to in vitro PVY^N or PVY^O viral transcripts. Fractions containing from 10^7 to 10^3 copies of viral RNA were used. The resulting cycle-dependent fluorescence data (**Fig. 17.6a**) was used to define a fluorescent cycle threshold limit (C_t) within the linear part (using log scale) of the increase of fluorescence data associated to each calibrated sample. The number of PCR cycles required for these samples to reach the defined C_t was calculated and used to draw the standard curve (**Fig. 17.6b**). The slope of this curve describes the efficiency of the quantitative assay. Indeed, the reliable quantification of unknown samples requires the duplication of the target at each PCR cycle. A standard curve characterized by a −3.32 slope illustrates the exact duplication of the matrix and reflects the accuracy of the quantification of unknown samples.

17.3.4.3. The Use of 'Real-Time' Controls

To prevent interpretation of non-specific cleavage of the fluorescent TaqMan® probes, two extra controls must be prepared. They correspond to the non-template control (NTC) and non-amplification control (NAC). These two controls test the non-specific cleavage of the probe and the specific amplification of non-targeted DNA sequence. The NAC, a specific control for reverse transcription step, is requested only for RNA quantification assays.

1. Prepare the NTC one-step RT-PCR mixture (*see* **Subheading 17.3.4.1, step 1**) with sterile water as sample
2. Prepare the NAC one-step RT-PCR mixture (*see* **Subheading 17.3.4.1, step 1**) with sample corresponding to 10^6 copies of viral transcripts and without the 40× enzyme mixture
3. Run real-time PCR. Amplification conditions correspond to 30 min at 48°C, 10 min at 95°C and 40 cycles of 15 s at 95°C and 1 min at 60°C
4. At each PCR cycle, perform laser excitation of each sample in order to collect fluorescent signal (in the range 500–660 nm) produced by the different dyes
5. Perform computer analysis of the collected data (e.g. SDS® software (Applied Biosystems)) to draw, for each tested sample, the cycle-dependent increase of fluorescent signal

Fluorescent data associated with the NTC and NAC samples describe the non-specific cleavage of the used probes during the assay. Such non-specific cleavage, corresponding to the background fluorescence in the assay, should not reach the defined fluorescent threshold (C_t) before data associated with the most diluted control sample to allow reliable target quantification. Actually, in optimal conditions, control samples should be associated with fluorescent data that do not reach the C_t value even at the 40th PCR cycle.

17.3.4.4. Quantification of Unknown Samples

Data corresponding to collected fluorescent signals associated with each unknown sample were plotted to determine the number of PCR cycles required to reach the defined C_t fluorescence level (**Fig. 17.6a**, dotted line). The resulting C_t corresponds to the 'y' value to be used in standard curve equation to calculate the RNA copy number of the tested sample (**Fig. 17.6b**, dotted line in standard curve graph).

17.3.5. Multiplex SNP Interrogation Assay

The multiplex SNP is a fluorescent-based technique that allows the simultaneous identification of several (up to ten) nucleotides in nucleic acid sequence(s). This molecular technique allows detection and characterization of multiple polymorphic sequences of a pathogen's genome. Applied to RNA viruses, the corresponding procedure requires a RT-PCR amplification of part(s) of the viral sequence, the selection of polymorphic nucleotides and the production of labelled primers. The multiplex SNP data correspond to electropherogram produced using a capillary electrophoresis step. To perform such assay, a primer pair that frames the targeted polymorphic nucleotide(s) and primers of various lengths (SNaPshot primers) that hybridize the sequences immediately upstream of the interrogated SNP sites should be designed. The presented procedure is described to target the polymorphism of nucleotides 2213 and 2271 of the PVY sequence (**Fig. 17.1**).

17.3.5.1. Design of Primer Pair and Snapshot Primers

The primer pair required for the described multiplex SNP interrogation assay could correspond to the already designed HCf and P3r primers. Indeed, the viral sequence amplified using these primers includes the two selected SNP sites. In complement to primer pair, the multiplex SNP interrogation assay requires the design of SNaPshot primer(s) that specifically hybrid viral sequence one nucleotide upstream selected SNP sites. To design SNaPshot primers use the following guidelines:

1. Fix the melting temperature (T_m) of the primers above 50°C
2. The difference in length between primers should be 6–8 nucleotides for primers shorter than 36 nucleotides and 4–6 nucleotides for others (*see* **Note 13**)
3. Use only HPCL-purified primers

According to these guidelines, four primers were designed (**Fig. 17.1**, primers $SNP_{2213}N$, $SNP_{2213}O$, $SNP_{2271}N$ and $SNP_{2271}O$).

17.3.5.2. Production of Amplified Viral Sequence

After nucleic acid extraction, samples are submitted to reverse-transcription and to polymerase chain reaction. For the described multiplex SNP interrogation assay, the SNP sites are located close to each other. Then, a simplex PCR allows the amplification of a region containing the two SNP sites. However, when targeted SNPs are distant or located on different nucleic acid templates, they cannot be amplified during a single PCR. Then, several simplex PCRs or multiplex PCR should be performed.

1. Prepare RT mix and perform RT in the presence of PVY 3′ter primer
2. Prepare PCR mix and perform PCR in the presence of HCf and P3r primers
3. Run one-tenth of the PCR product on 1% agarose gel in 1× TBE in the presence of 0.6 mg/ml SybrSafe® (Invitrogen)
4. Visualize nucleic acids under ultraviolet light and check the viral sequence amplification efficiency
5. Prepare, in TBE buffer, a Sephadex G50 solution
6. Place 400 µl of the Sephadex G50 solution in each well (according to the number of sample to test) of the 96-well multiscreen plate (Millipore)
7. Place a collecting 96-well plate under the multiscreen plate
8. Centrifuge 1,500 × *g* for 3 min to dry the Sephadex matrix
9. Discard the buffer and place a new sterile 96-well plate under the multiscreen plate
10. Load each PCR product on the dried Sephadex G50 and centrifuge 1,500 × *g* for 3 min to perform a filtration step of the post-PCR fraction through the Sephadex G50
11. Discard the plate containing Sephadex G50
12. Store the cleaned post-PCR fraction at –20°C until use for Primer Extension step

17.3.5.3. Primer Extension Procedure

The cleaned amplified viral sequence is used as matrix in the primer extension (PE) procedure. The latter is performed during a cycling reaction in the presence of a DNA polymerase, the SNaPshot primers and the four fluorescent dideoxynucleotides (ddNTPs).

1. Prepare the PE mixture (5 µl/reaction)
   - 2.5× SNaPshot ready mix[9] 2 µl
   - 10 µM $SNP_{2213}N$ 0.1 µl
   - 10 µM $SNP_{2213}O$ 0.1 µl
   - 10 µM $SNP_{2271}N$ 0.1 µl
   - 10 µM $SNP_{2271}O$ 0.1 µl
   - Sterile water 0.6 µl
   - DNA (cleaned PCR product) 2 µl
2. Run PE programme in thermocycler for 25 cycles of 10 s at 96°C, 5 s at 50°C and 30 s at 60°C

[9]Includes the four labelled ddNTPs and Taq polymerase.

3. At the end of the last cycle, add 1 U of Calf Intestinal Phosphatase and incubate at 37°C for 1 h
4. Store at –20°C until used

During the PE step, a single base (ddNTP) is added at the 3′ end of each hybridized primer. In consequence, the size of the resulting products corresponds to the +1 size (linked dideoxynucleotide) of the used primers. Moreover, the fluorescence of the labelled primer indicates the nature of the linked ddNTP. This nucleotide selection occurred according to the identity of nucleotide at the targeted SNP site on the DNA matrix.

17.3.5.4. Analysis of Labelled Primers Using Capillary Electrophoresis

Capillary electrophoresis (CE) allows to differentiate labelled primers present in samples according to their length and fluorescence. To precisely describe each labelled SNaPshot primer, samples are analysed in the presence of a mixture of standards labelled (with a specific dye) molecules. Such commercial mixture (e.g. Liz-120® (Applied Biosystems)) corresponds to a fluorescent ladder and makes possible the identification of the length of each labelled SNaPshot primer present in the test. Products are run on a 3130 Genetic analyser (Applied Biosystems) and analysed using GeneMapper® (Applied Biosystems).

1. Prepare the CE mixture (10 μl/reaction) in a microtube
   - Hi-Di Formamide 8.5 μl
   - Liz-120 Size standard 0.5 μl
   - Sample (labelled primer post PE) 1 μl
2. Heat the mixture at 95°C for 5 min
3. Place immediately heated samples in wet-ice for 1 min
4. Place the microtube on the tray of the 3130 Genetic analyser (Applied Biosystems)
5. Run electrophoresis (15 kV) for 24 min
6. Collect fluorescent signals during the run
7. Analyse crude data using GeneMapper software

17.3.5.5. Data Analysis

During capillary electrophoresis, fluorescence data were recorded as a function of migration time. The software calculates the size of observed fluorescent fragments according to migration time of the standards labelled nucleic acids (**Fig. 17.7**). In the described assay, the lengths of the used primers are 23 and 31 nucleotides. However fluorescence peaks observed for the labelled primers correspond to 29 and 37 nucleotides (*see* **Note 14**).

Multiple SNP results can be manually scored according to labelled primers length and to the corresponding fluorescent signal. However, the GeneMapper® software (Applied Biosystems) can be calibrated to automatically assign each peak of the crude result and to combine all assigned peak to define the range of genotypes in the tested samples.

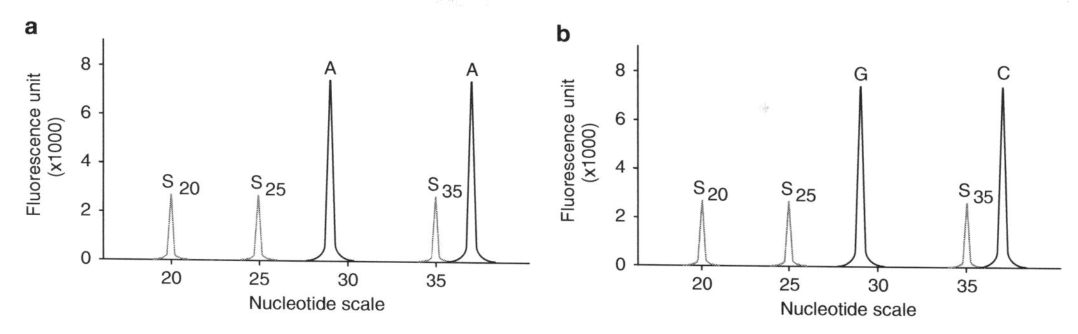

Fig. 17.7. SNaPshot electropherogram obtained for (**a**) PVY^N-605 and (**b**) PVY^O-139. The nucleotide scale is calculated according to migration time of the labelled nucleic acid standards (S_{20}, S_{25}, and S_{35}).

PVY^N genotype is characterized by two fluorescent peaks corresponding to ddA-labelled primers. Such results are in agreement with nature of the targeted SNPs on PVY^N genome (A_{2213} and A_{2271}). PVY^O isolates present G_{2213} and C_{2271} sequence identity. Corresponding genotypes are characterized by peak associated to ddG- and ddC-labelled primers.

17.4. Notes

1. The preparation of TBE buffer is easy. However, long term storage of home-made TBE buffers could induce precipitation of EDTA. Such phenomenon is rarely observed on ready-to-use TBE buffers. Moreover, commercial TBE buffer is not expensive and available from numerous manufacturers.
2. Addition of NaOH during preparation of NaOAc solution produces thermal exogenous reaction. This solution must be prepared in wet ice.
3. As alternative to razor blade, sampling of leaves can be performed using a sterile microtube cap as perforating device.
4. In the described procedure, the time for coating and sap incubation can be reduced to 30–60 min each. However, short incubation time could impair immunocapture efficiency especially for low concentrated samples.
5. Nucleic acids precipitation is improved at low temperature. Use cold ethanol solution (stored at –20°C). The two hours storage at –20°C of nucleic acids during the precipitation step can be replace by 15 min at –80°C.

6. Many other commercial plasmids can be use to clone the viral sequence. Requirements are: the presence of cloning sites and functional system for in vitro transcription of the inserted sequence.
7. The use of the presented degenerated primers in PCR procedure produces fragments with restriction sites that could be easily cloned in plasmids. However, digestion of any sequence located at the edge of a double-stranded DNA fragment requires the presence of some (from 2 to 8) extra nucleotides. Indeed, restriction enzymes could not efficiently identify (and cleave) a targeted sequence located too close to the edge of a PCR fragment. Add extra nucleotides at the 5′ end of used primers to use PCR products in such cloning strategy.
8. SyBr safe™ is a new chemical that substitutes Ethidium Bromide for visualization of nucleic acid after agarose gel electrophoresis.
9. Check the efficiency of nucleic acid extraction by agarose gel electrophoresis to adjust volumes in the next steps (digestion, PCR,...) of the procedure.
10. In complement to TaqMan® probes, other dually labelled primers (Scorpio®, Molecular beacon®...) are available to be used as probes.
11. The most frequently used fluorescent dyes correspond to 6-FAM and Vic. However, other molecules (Tet, Ned, Hex, etc.) exist and could be used. The quencher generally corresponds to the fluorescent Tamra dye. However, the latter produces a fluorescent signal that could slightly affect the measurement of the reporter's fluorescence during the assay. In order to improve the fluorescence monitoring, non-fluorescent quenchers have been developed (e.g. Minor Groove Binding® molecule; MGB (Applied Biosystems)). Moreover, the latter increases the fixation of the probe on the targeted sequence. In consequence, the melting temperature of the primer decreases in the presence of such quencher molecules. Then, the length of probe required to reach the appropriate Tm is lowered.
12. The described assay was designed using the ABI7700 apparatus (Applied Biosystems). The use of other 'real-time' apparatus could require some changes in the presented procedure (e.g. PCR parameters).
13. If necessary, increase the length of primers by the addition of a non-homologous poly (dT), poly (dA), poly (dC) or poly (dGACT) at the 5′ end.
14. Due to the structural characteristics of fluorescent dyes, the length of short labelled primers is generally overestimated

by the computer analysis. Indeed, the presence of the dye induces a delay in migration equivalent to few nucleotides. Such bias on primer length assignment modifies the crude results. However, the differences in length (from 4 to 8 nucleotides) between SNaPshot primers were fixed to avoid consequences of such a bias

Acknowledgements

The authors wish to thank Michel Tribodet, Pierre Lefeuvre, Flora Croizat, Valérie Balme-Sinibaldi, Laurent Glais and Camille Kerlan for their help during the set up of the different fluorescent-based PVY detection tools.

References

1. Brunt, A.A., Crabtree, K., Dallwitz, M.J., Gibbs, A.J., Watson, L., Zurcher, E.J. (eds.) (1996 onwards). Plant Viruses Online: Descriptions and Lists from the VIDE Database. Version: 20th August 1996. URL http://biology.anu.edu.au/Groups/MES/vide/
2. Clark, M.F., Adams, A.N. (1977). Characteristics of the microplate method of enzyme-linked immunosorbent assay for the detection of plant viruses. *Journal of General Virology* 34, 475–483.
3. Henson, J.M., French, R. (1993). The polymerase chain reaction and plant disease diagnosis. *Annual Review of Phytopathology* 31, 81–109.
4. Heid, C.A., Stevens, J., Livak, K.J.,Williams, P.M. (1996). Real time quantitative PCR. *Genome Research* 6, 986–994.
5. Leone, G., van Schijndel, H., van Gemen, B., Kramer, F.R., Schoen, C.D. (1998). Molecular beacon probes combined with amplification by NASBA enable homogeneous, real-time detection of RNA. *Nucleic Acids Research* 26, 2150–2155.
6. Walsh, K., North, J., Barker, I., Boonham, N. (2001). Detection of different strains of Potato virus Y and their mixed infections using competitive fluorescent RT-PCR. *Journal of Virological Methods* 91, 167–173.
7. Salmon, M.A., Vendrame, M., Kummert, J., Lepoivre, P. (2002). Detection of apple chlorotic leaf spot virus using a 5 nuclease assay with a fluorescent 3' minor groove binder-DNA probe. *Journal of Virological Methods* 104, 99–106.
8. Klerks, M.M., Leone, G.O., Verbeek, M., van den Heuvel, J.F., Schoen, C.D. (2001). Development of a multiplex AmpliDet RNA for the simultaneous detection of Potato leafroll virus and Potato virus Y in potato tubers. *Journal of Virological Methods* 93, 115–125.
9. Schnell, R.J., Olano, C.T., Kuhn, D.N. (2001). Detection of avocado sunblotch viroid variants using fluorescent single-strand conformation polymorphism analysis. *Electrophoresis* 22, 427–432.
10. Boonham, N., Walsh, K., Smith, P., Madagan, K., Graham, I., Barker, I. (2003). Detection of potato viruses using microarray technology: towards a generic method for plant viral disease diagnosis. *Journal of Virological Methods* 108, 181–187.
11. Jacquot, E., Tribodet, M., Croizat, F., Balme-Sinibaldi, V., Kerlan, C. (2005). A single nucleotide polymorphism-based technique for specific characterization of YO and YN isolates of Potato virus Y (PVY). *Journal of Virological Methods* 125, 83–93.
12. Balme-Sinibaldi, V., Tribodet, M., Croizat, F., Lefeuvre, P., Kerlan, C., Jacquot, E. (2006). Improvement of Potato virus Y (PVY) detection and quantitation using PVYN- and PVYO-specific real-time RT-PCR assays. *Journal of Virological Methods* 134, 261–266.

13. M., Rolland, L., Glais, C., Kerlan, E. Jacquot, (2008). A multiple single nucleotide polymorphisms interrogation assay (SNaPshot) for reliable Potato virus Y (PVY) group and variant characterization. Journal of Virological Methods 147, 108–117.
14. Jakab, G., Droz, E., Brigneti, G., Baulcombe, D., Malnoe, P. (1997). Infectious in vivo and in vitro transcripts from a full-length cDNA clone of PVY-N605, a Swiss necrotic isolate of potato virus Y. *Journal of General Virology* 78(12), 3141–3145.
15. Singh, M., Singh, R.P. (1996). Nucleotide sequence and genome organization of a Canadian isolate of the common strain of potato virus Y (PVYO). *Canadian Journal of Plant Pathology* 18, 209–224.
16. Roberts, C.A., Dietzgen, R.G., Heelan, L.A., Maclean, D.J. (2000). Real-time RT-PCR fluorescent detection of tomato spotted wilt virus. *Journal of Virological Methods* 88, 1–8.
17. Ptacek, J., Skopek, J., Dedic, P., Matousek, J. (2002). Immunocapture RT-PCR probing of potato virus Y isolates. *Acta Virologica* 46, 63–68.
18. Yu, C., Wu, J., Zhou, X. (2005). Detection and subgrouping of Cucumber mosaic virus isolates by TAS-ELISA and immunocapture RT-PCR. *Journal of Virological Methods* 123, 155–161.
19. Leiser, R.M., Richter, J. (1978). Purification and some characteristics of potato virus Y. *Archiv fur Phytopathologie und Pflanzenschutz* 14, 337–350.
20. Sambrook, J., Fritsch, E.F., Maniatis, T. (1989). Molecular cloning – a laboratory manual. 2nd ed. Cold Spring Harbor, New York.
21. Oefner, P.J. (2002). Sequence variation and the biological function of genes: methodological and biological considerations. *Journal of Chromatography B* 782, 3–25.
22. Marras, S.A., Kramer, F.R., Tyagi, S. (2002). Efficiencies of fluorescence resonance energy transfer and contact-mediated quenching in oligonucleotide probes. *Nucleic Acids Research* 30(21), e122.

Chapter 18

Analysis of Population Structures of Viral Isolates Using Single-Strand Conformation Polymorphism Method

Agnès Delaunay, Mathieu Rolland, and Emmanuel Jacquot

Summary

The analysis of viral populations requires the use of techniques that describe characteristics of individuals. The single-strand conformation polymorphism (SSCP) makes possible the identification of genetic differences between viral sequences and constitutes an alternative to the expensive and time-consuming cloning and sequencing strategies. Applied to small genomic regions (from 100 to 500 bases in length), SSCP patterns could describe, under appropriate experimental conditions, single nucleotide variations in the studied sequence. The different steps of a complete SSCP procedure, from sampling to pattern analysis (including nucleic acid extraction, RT-PCR amplification, double-stranded DNA quantification, polyacrylamide gel preparation, electrophoresis conditions, and staining procedures), are described using a region (500 bases) of the barley yellow dwarf virus-PAV (BYDV-PAV, *Luteovirus*) genome as molecular target.

Key words: Mutation detection, Viral adaptation, Band intensity, SSCP pattern, Polyacrylamide gel electrophoresis.

18.1. Introduction

Many research fields require the analysis of structure and evolution of pathogen populations at the molecular level. In most cases, the requested data could be obtained using sequencing procedures *(1)* applied to the whole or to part(s) of the genome of each member of the analyzed population. However, such widespread method is expensive (especially when used for population analyses) and sometimes inappropriate to the defined scientific aims. Indeed, the description of every single point mutation within the pathogen genomic sequence is not absolutely necessary to describe

R. Burns (ed.), *Methods in Molecular Biology, Plant Pathology, vol. 508*

DOI: 10.1007/978-1-59745-062-1_18

the diversity of a population. According to the expected results, alternative techniques such as single-strand conformation polymorphism (SSCP) can be used to highlight genetic differences of individuals within a defined population of a pathogen *(2–7)*. This method, applied to PCR products, makes possible the rapid identification of any modification of the nucleic acids that affects tri-dimensional structure of the corresponding single-strand (ss) DNA molecules. Based on the differential migration of ssDNA in a nondenaturant one-dimensional polyacrylamide gel, the SSCP procedure allows the simultaneous process of numerous samples and the description of their genetic diversity. However, quality of postelectrophoresis produced patterns and the acrylamide gel resolution depend on many parameters (temperature, sample preparation, gel characteristics, etc.) *(8–10)*. Moreover, conformational polymorphisms underlined by SSCP analyses depend on intrinsic parameters of the analyzed nucleic acid sequence such as the length of the ssDNA fragment, the number of nucleotide changes, their locations in the sequence, and the four bases ratio in the analyzed nucleotide region. In SSCP patterns, each polymorphic signal (band) between two samples reflects a variation within the analyzed population, but all together these bands represent only part of the complete population diversity as SSCP is not able to detect sequence modifications that do not impact on secondary and tertiary structure of the single strands. Then, optimization steps of SSCP protocols are requested to reach the most accurate description of the population variability. When optimized, the SSCP procedures are appropriate for several purposes including description of a single nucleotide change in the analyzed sequence *(11–13)*, identification of multicopy gene polymorphisms *(14)*, or monitoring modifications of the structure of a population *(6, 9)*.

Diversity of viral populations can be analyzed at different scales. Genetic comparisons could be performed between isolates of a viral species collected from different locations *(15)*, host species *(16)*, or plants *(17)*. Moreover, viral population analyses could also be carried out at a single infected plant level *(6)*. Indeed, most viruses (especially RNA viruses) produce their progeny using a polymerase devoid of proof reading activity. The resulting progeny obtained in each infected plant is considered as a viral population according to the genomic differences (mutation, deletion, or insertion) between molecular entities *(6)*. Characterization of such genetic diversity is a key step to understand processes (emergence, maintenance, and spread of new variants) involved in viral evolution. Viral population structures can be monitored using standard procedures including PCR, cloning steps, and sequencing clones. However, SSCP is an appropriate alternative method to describe such population structures.

In order to illustrate the use of SSCP in viral population analyses, complete SSCP procedure applied to *Barley Yellow Dwarf Virus*-PAV (*Luteoviridae* family, type-member of *Luteovirus* genus) is presented. BYDV-PAV population of isolates collected from different host species was monitored.

18.2. Materials

18.2.1. Required Materials, Products, and Chemicals

1. Sterile plastic wares (microtubes, cones, 96-well plate, etc.).
2. Syringe and needles.
3. Cellophane sheets (VWR, http://www.vwr.com/).
4. Storage/incubating devices for −18, 4, and 42°C.
5. Thermocycler.
6. Stirring machine.
7. Agarose gel device.
8. Acrylamide gel device.
9. Imaging system.
10. Image J software (http://rsb.info.nih.gov/ij/).
11. Power supply.
12. Heated Bloc.
13. Drying support.
14. Wet ice.
15. Deionized water.
16. RNase-free water.
17. Calibrated DNA smart ladder (Eurogentec, http://www.eurogentec.be/).
18. SybrSafe™ (Invitrogen, http://www.invitrogen.com/).
19. RNasine® (Promega, http://www.promega.com/).
20. Reverse Transcriptase.
21. Taq Polymerase with buffer and $MgCl_2$ solution.
22. Primers (according to the matrix sequence, *see* **Fig. 18.1**).
23. 10 mM dNTPs.
24. 1× TBE buffer.
25. 5 M EDTA.
26. Xylene cyanol and bromophenol blue.
27. 40% Acrylamide/bis-acrylamide (19/1).
28. 10% (w/v) Ammonium persulfate (APS).

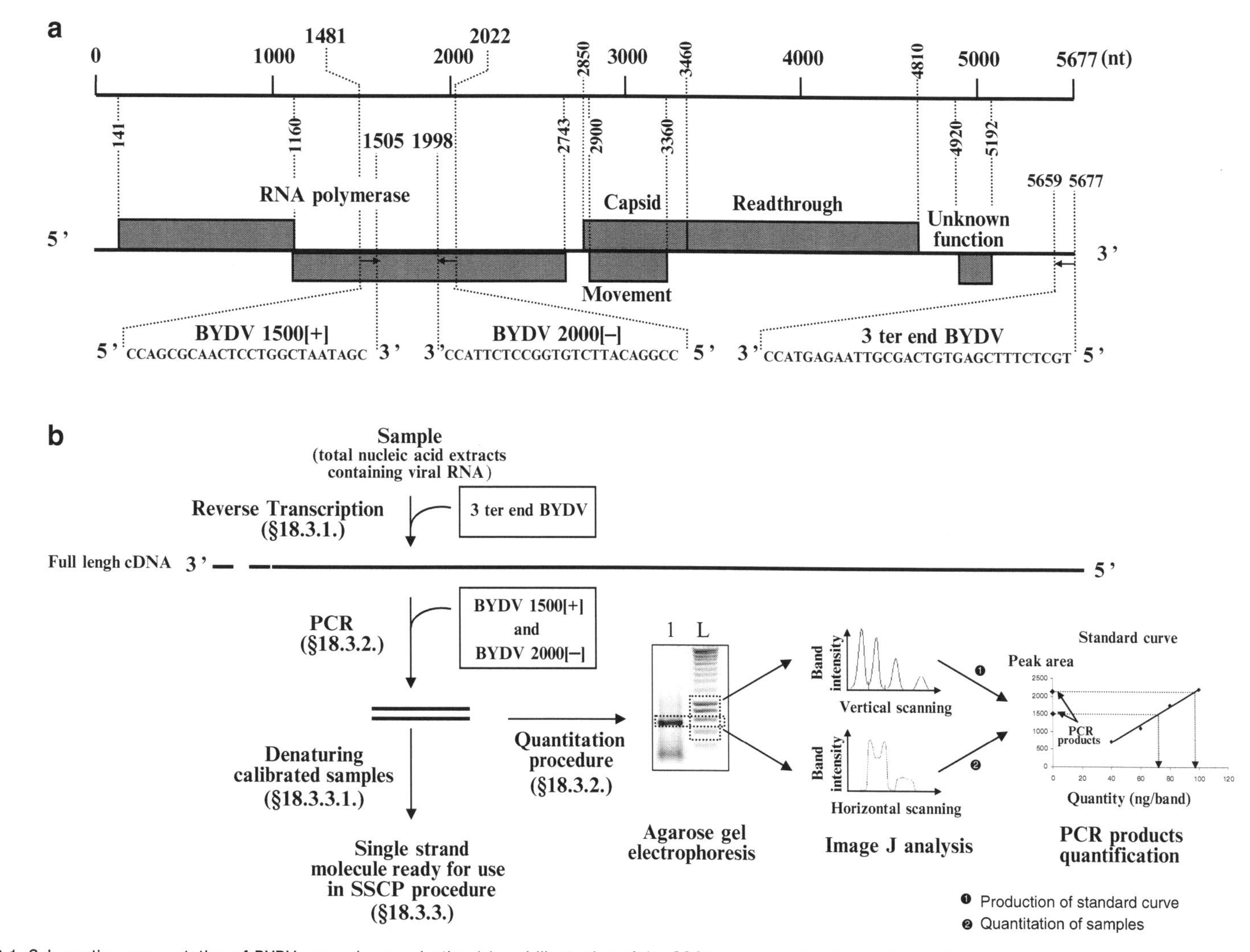

Fig. 18.1. Schematic representation of BYDV genomic organization (**a**) and illustration of the SSCP procedure (**b**). Nucleotide scale and denoted positions for open reading frames (*gray boxes*) and primers (3 ter end BYDV, BYDV 1500[+] and BYDV 2000[–]) are according to Miller et al. (*18*). The known functions of the ORFs are listed. The information required to perform the steps of the procedure are described in the listed paragraphs of this chapter.

29. TEMED® (Bio-rad laboratories, http://www.bio-rad.com/).
30. 100 and 70% Ethanol solutions.
31. 10% (w/v) Nitric acid.
32. 1 g/l Silver nitrate.
33. 36% (v/v) Formaldehyde.
34. 30 g/l Sodium carbonate anhydrous.
35. 10% (v/v) Acetic acid.

All home-made buffers, solutions, and mixtures have to be prepared with deionized water as well as all washing steps for silver nitrate staining procedure.

18.2.2. Reverse Transcription

1. Prepare reverse transcription (RT) mix (10 μl/reaction) according to the number of samples to be tested. Place tubes (especially enzymes and dNTPs) in wet ice.
2. Final concentrations in the mix are 2× buffer, 0.2 mM of each dNTP, 2 U/μl RNasine, 1 U/μl Reverse Transcriptase, and 1 mM appropriate primer (**Fig. 18.1**) (Vol/reaction)

   5× buffer (provided with enzyme), 4 μl.

   10 mM dNTPs, 2 μl.

   5 mM primer (3 ter end BYDV), 2 μl.

   Sterile RNase-free water, 0.5 μl.

   40 U/μl RNasine®, 0.5 μl.

   10 U/μl Reverse Transcriptase, 1 μl.

18.2.3. Amplification of Viral Sequence (Polymerase Chain Reaction)

1. Prepare PCR mix (23 μl/reaction) according to the number of samples to be tested. Place tubes (especially for enzymes and dNTPs) in wet ice.
2. Including the 2 μl of sample added to this mixture, final concentrations are 1× buffer, 1.5 mM $MgCl_2$, 0.1 mM of each dNTP, 2 U/μl Taq polymerase, and 0.8 mM of appropriate primer pair (**Fig. 18.1**) (Vol/reaction)

   10× buffer (provided with enzyme), 2.5 μl.

   10 mM $MgCl_2$ (provided with enzyme), 1.5 μl.

   10 mM dNTPs, 1 μl.

   5 mM forward primer (BYDV 1500[+]), 4 μl.

   5 mM reverse primer (BYDV 2000[–]), 4 μl.

   Sterile water, 9.5 μl.

   5 U/μl Taq polymerase, 0.5 μl.

18.2.4. Standard Solutions

1. 1× TBE buffer pH 8.0: 10.78 g Tris, 5.50 g boric acid, 0.58 g EDTA, deionized water to prepare 1 L. However, ready-to-use 10× TBE buffer is available from numerous manufacturers.

2. Denaturing solution: formamide (9.5 ml), 5 M EDTA (0.5 ml), xylene cyanol (5 mg), and bromophenol blue (5 mg).
3. Ammonium persulfate 10% (w/v) in deionized water should be prepared in small quantity as this solution should not be conserved more than a month at 4°C or aliquots could be stored at −20°C for 6 months, avoiding freeze–thaw cycles.
4. Prepare 70 and 10% ethanol solutions using absolute ethanol.
5. The ssDNA-staining process described in the presented procedure requires different solutions such as sodium carbonate (30 g/l), silver nitrate (1 g/l), nitric acid 1% (v/v), and acetic acid 10% (v/v). Silver nitrate and sodium carbonate solutions must be prepared extemporaneously and silver nitrate solution must be kept in the dark. All solutions can be stored at 4°C, especially recommended for silver nitrate and sodium carbonate solutions.

18.2.5. Polyacrylamide Gel

Solutions with different acrylamide/bis-acrylamide ratios are available to prepare gels. SSCP procedures are commonly performed on polyacrylamide gel prepared using "19:1" (acrylamide/bis-acrylamide) ratio. Commercial premixed solutions correspond to acrylamide concentration in the range 30–40%. Final acrylamide concentration in the gel must be defined to obtain optimal resolution of the analyzed fragments. Optimal final acrylamide percentage must be in the range 3–5% for 100–500 base fragments and in the range 8–12% for smaller fragments (*see* **Note 1**).

Always be very careful when handling acrylamide; liquid form of acrylamide is a potent neurotoxin. Use gloves, glasses, and lab coat when handling.

Prepare a 3.5% acrylamide premix. Mix gently (to avoid bubbles) the following components:

10× TBE buffer, 2 ml.

Sterile water, 14.4 ml.

40% Acrylamide/bis-acrylamide (19:1), 3.65 ml.

Indicated volumes allow to prepare four 8 × 7 × 0.75 cm (w × l × d) gels. Such premix could be prepared in larger volume and stored at 4°C until 6 months. Polymerization starts when TEMED and ammonium persulfate are added in the mix.

18.3. Methods

The BYDV-PAV genome is a single-stranded RNA molecule of 5.677 bases in length (**Fig. 18.1**). The presented SSCP procedure was designed to analyze ssDNA of 541 bases. Standard procedures for reverse transcription and PCR amplification of viral RNA are described. The first step of the procedure (i.e., extraction of the viral nucleic acids) was performed according to **Subheading 17.3.1** in Chapter 17.

18.3.1. Reverse Transcription

1. Mix 10 µl RT mix including the 3ter end BYDV primer (**Fig. 18.1**) and 10 µl of total nucleic acid extracts containing viral RNA
2. Incubate 1 h at 42°C
3. Store produced cDNA at –20°C until used for PCR

18.3.2. Viral Sequence Amplification and DNA Quantitation

The nucleotide sequence targeted to analyze genetic diversity of the tested samples should correspond to a 100–500 base region. In optimal conditions, SSCP is appropriate to detect a single point mutation in an ssDNA fragment of 500 bases *(19)*. However, in some specific applications, sequence variations in longer fragments could be recorded with adapted SSCP running conditions *(20, 21)*. The selected sequence (nucleotides: 1481–2022, **Fig. 18.1**) has to be amplified by PCR prior being loaded on acrylamide gel.

1. Mix 2 µl RT product (cDNA) with 23 µl PCR mix including BYDV 1500[+] and BYDV 2000[–] primers (**Fig. 18.1**)
2. Run PCR program. For the selected BYDV-PAV region, amplification conditions correspond to 5 min at 94°c followed by 40 cycles of 45 s at 94°C, 45 s at 55°C, and 1 min at 72°C. A final elongation step is added (10 min, 72°C) after the last PCR cycle
3. Migrate 5 µl of each PCR product and 5 µl of calibrated DNA smart ladder on 1% agarose gel in 1× TBE, in the presence of 0.6 mg/ml SybrSafe. Running conditions are usually 30 min migration under 10 V/cm
4. Visualize nucleic acids under ultraviolet light (**Fig. 18.1**) and integrate, using appropriate recording device, an optimal fluorescence signal for the detection of both PCR-amplified fragment and DNA ladder components. Do not saturate recorded signal during integration process
5. Use Image J software to produce a "fluorescent vs. concentration" calibration curve based on data associated with each fragment present in the used DNA ladder
6. Calculate the concentration (ng/ml) of the 541 base DNA fragments in tested samples

18.3.3. Polyacrylamide Gel Electrophoresis

18.3.3.1. Preparation of Samples

This is a crucial step in the SSCP procedure. Indeed, quantitation of nucleic acids and production of ssDNA molecules have strong impact on quality and reliability of SCCP results (*see* **Note 2**). Moreover, denaturing steps should be performed just before loading the gel (e.g., during prerun)

1. Mix PCR products with denaturing solution to prepare 25 μl of DNA fractions at 0.75 ng/ml
2. Incubate 5 min at 95°C under fume hood
3. Place immediately heated fractions in wet ice for at least 10 min
4. Store the sample in wet ice until used

18.3.3.2. Gel Polymerization

To prepare homogeneous polymerized acrylamide/bis-acrylamide gel, it is important to wash glass materials, spacers, and combs with deionized water. Then cleaned materials must be rinsed consecutively with 70 and 100% ethanol. Dried material is assembled to set a gel cast according to the specific characteristics of the used device. Original gel-cast system is composed of a set of two plates maintained by thin spacers. Nowadays, more suitable and convenient spacer-containing glass plates are provided. The latter prevent leakage during the polymerization step.

1. Adjust the short plate on the spacer-containing plate. Use a plan working table to prevent gap at the edge of glasses
2. Fix position with two clamps on vertical side
3. Place the system on a stable support to keep it horizontally
4. Add 15 μl of TEMED and 205 μl of 10% ammonium persulfate solution in 20 ml of acrylamide mix
5. Pour gently the acrylamide mixture between the plates
6. Ensure there is no troubleshooting on the gel (bubbles or impurities)
7. Insert a comb (size and number of wells according to the used gel-cast system) at the top of the gel and fill empty spaces with acrylamide mix (*see* **Note 3**)
8. Wait for 2 h at room temperature for complete polymerization

18.3.3.3. Electrophoresis

Running temperature is important in SSCP procedure as it could change the conformational structure of the analyzed single-strand DNA fragments. To prevent such phenomenon, buffer and materials need to be cold (4°C). Moreover, to maintain low temperatures during electrophoresis, run is performed at 4°C (*see* **Note 4**).

1. Remove clamps and comb from the gel-cast system
2. Place the polymerized gel on the migration system
3. Fill the appropriate tanks of the used device with fresh cold 1× TBE buffer (*see* **Note 5**)
4. Wash the wells with cold 1× TBE buffer using a syringe with a needle
5. Run unloaded gel under 10 V/cm for 30–60 min (use this prerun time to prepare samples).
6. Load 8 μl of denatured samples in well Run overnight (18 h) at 10 V/cm at 4°C

18.3.4. In-Gel Silver Nitrate Staining of ssDNA Molecules

Several methods have been reported for the visualization of DNA molecules in polyacrylamide gel. Silver nitrate staining, a conventional method for nucleic acids observation, is based on the precipitation of Ag^{+} in presence of DNA, sodium carbonate, and formaldehyde. Procedure is performed at room temperature under fume hood (*see* **Note 6**).

1. Turn off gel migration
2. Remove gels from the migration system
3. Remove the short glass plate and place the gel carefully in a staining tray
4. All the following steps should be performed on a swaying machine and under a fume hood
5. Incubate gels with 10% ethanol during 30 min
6. Remove ethanol and incubate gels with 1% nitric acid for 3 min
7. Rinse twice in deionized water for 20 s
8. Incubate 20 min in an extemporaneously prepared 1 g/l silver nitrate solution. Prevent light falling on it during staining
9. Rinse two times in deionized water
10. Add 325 ml of 36% formaldehyde in 600 ml of cold sodium carbonate solution
11. Incubate gels with 300 ml of sodium carbonate-formaldehyde solution
12. Discard the solution as soon as brown stain appears
13. Repeat incubation step in sodium carbonate-formaldehyde solution
14. Discard the solution when bands present appropriate stain intensity
15. Stop the staining reaction by incubating the gel in 10% acetic acid for 5 min
16. Discard the solution and wash the gel in deionized water for 10 min

18.3.5. Gel Conservation

After the staining step, the produced patterns should be digitized to file and to analyze data. However, crude data corresponding to stained bands in the gel should improve the analysis of some faint bands. The long-term conservation of the gel is possible according to the following simple procedure

1. Dehydrate gel in 10% ethanol solution during at least 1 h
2. Place carefully the gel on a wet cellophane sheet
3. Cover the gel with a second wet cellophane sheet
4. Clamp the wrapped gel in a frame in order to tightly maintain the cellophane sheets

5. Dry the gel at room temperature overnight
6. Remove the clamps, pull off the frame, and cut out the cellophane to get the dried gel

18.3.5.1. Gel Analysis

The complexity of the produced patterns (number of observed bands) depends on the nature of the analyzed sample. In molecular population ecology *(9)*, phylogeny *(5)*, or in clinical science *(22)*, samples correspond to individuals (or clones). The genetic polymorphism of the latter reflects the complexity (haploid, diploid, or polyploid) or the redundancy (of the targeted sequence) of their genome. In consequences, the resulting SSCP patterns are generally simple as only two to four bands are observed *(23)*. In these cases, polymorphism between samples (individuals) is described by the shift, the deletion, or the addition of band(s).

SSCP patterns associated to viral analyses could present more bands. Indeed, the viral replication process in the infected host creates a genetic diversity at the single plant level. Then, all viral SSCP analyses are carried out on populations rather than on individuals. The complexity of the SSCP pattern indicates the diversity of molecular entities present in the tested sample. The proportion of each molecular entity could be monitored using the intensity (recorded using the image J software) of the corresponding band. The use of standardized samples makes possible the qualitative (shift, addition, and deletion of band) and semiquantitative (differential intensity) analysis of viral population polymorphism(s).

When applied to part of the polymerase gene of *BYDV*-PAV isolates collected from different infected hosts, the described SSCP procedure produces patterns with multiple bands (**Fig. 18.2**). Such patterns illustrate the heterogeneity of the BYDV-PAV populations and differences in population of two isolates of the same viral species.

18.4. Notes

1. In some cases, glycerol could be added in the acrylamide mix to improve migration and band separation.
2. Optimal band migration is altered with too large amount of DNA per lane. Distortion effects occur in overloaded gels.
3. Acrylamide/bis-acrylamide polymerization in the presence of ammonium persulfate and TEMED is exothermic during the process, but endothermic at the initiation of the reaction. In consequences, polymerized gel is obtained faster in warm areas. Standard room temperature (18–22°C) is convenient: lower

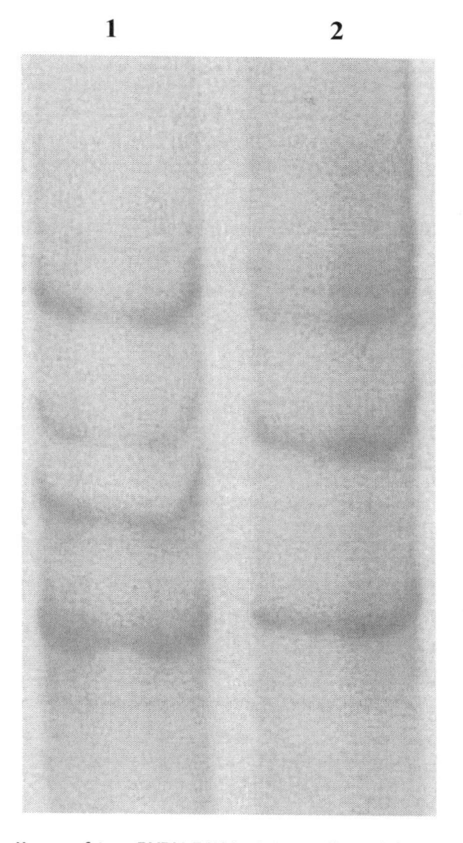

Fig. 18.2. SSCP patterns of two BYDV-PAV isolates collected from barley (*lane 1*) and wheat (*lane 2*) infected plants. Profiles illustrate the viral population polymorphism of the nt 1481–2022 region.

temperature increases polymerization delay, whereas higher temperatures could induce heterogeneous polymerization of the gel and then create band distortions during DNA migration. Gel polymerization is inhibited by alkaline molecules which deaminate acrylamide and by free radicals (e.g., oxygen).

4. SSCP running parameters influencing the migration and the resolution of bands are numerous. Some of them are already suggested in the main text. During the electrophoresis and especially for extensive running time, maximal separation is obtained with cold temperature to prevent excessive warming to keep stable single-strand conformational structures. Therefore protocol is performed at 4°C with cold buffer to regulate heating generated during electrophoresis. Moreover, it is important to note that thickness of gel must be taken into account. Indeed, the thinner the gel is, the more the system warms up.
5. The selection of an appropriate TBE concentration is very important. Due to ionic strength parameters, few variations

in the buffer composition within and outside the gel impact temperature of the whole system. Reported data suggested that 1× TBE is more suitable for higher voltage or prolonged migration *(24)*.

6. Washing steps during the silver nitrate staining allow elimination of excess of products. However, an extensive wash leads to abnormal gel staining. Silver nitrate and sodium carbonate solutions give better resolution when the solutions are very cold (0–4°C). Stop staining reaction with acetic acid could alter quality of band coloration. Do not keep the staining gel in the acetic acid solution more than 5 min.

References

1. Sunnucks, P. (2000) Efficient genetic markers for population biology. *Trends Ecol Evol* 15, 199–203.
2. Vigne, E., Bergdoll, M., Guyader, S., and Fuchs, M. (2004) Population structure and genetic variability within isolates of Grapevine fanleaf virus from a naturally infected vineyard in France: evidence for mixed infection and recombination. *J Gen Virol* 85, 2435–45.
3. Schnell, R. J., Olano, C. T., Kuhn, D. N. (2001) Detection of avocado sunblotch viroid variants using fluorescent single-strand conformation polymorphism analysis. *Electrophoresis* 22, 427–32.
4. Morrison, D. A., and Ellis, J. T. (1997) Effects of nucleotide sequence alignment on phylogeny estimation: a case study of 18S rDNAs of apicomplexa. *Mol Biol Evol* 14, 428–41.
5. Gasser, R. B., and Chilton, N. B. (2001) Applications of single-strand conformation polymorphism (SSCP) to taxonomy, diagnosis, population genetics and molecular evolution of parasitic nematodes. *Vet Parasitol* 101, 201–13.
6. Jridi, C., Martin, J.-F., Marie-Jeanne, V., Labonne, G., and Blanc, S. (2006) Distinct viral populations differentiate and evolve independently in a single perennial host plant. *J Virol* 80, 2349–57.
7. Callon, C., Delbes, C., Duthoit, F., and Montel, M.-C. (2006) Application of SSCP-PCR fingerprinting to profile the yeast community in raw milk Salers cheeses. *Syst Appl Microbiol* 29, 172–180.
8. Hayashi, K., and Yandell, D. W. (1993) How sensitive is PCR-SSCP? *Hum Mutat* 2, 338–46.
9. Sunnucks, P., Wilson, A. C. C., Beheregaray, L. B., Zenger, K., French, J., and Taylor, A. C. (2000) SSCP is not so difficult: the application and utility of single-stranded conformation polymorphism in evolutionary biology and molecular ecology. *Mol Ecol* 9, 1699–710.
10. Orita, M., Suzuki, Y., Sekiya, T., and Hayashi, K. (1989) Rapid and sensitive detection of point mutations and DNA polymorphisms using the polymerase chain reaction. *Genomics* 5, 874–79.
11. Jordanova, A., Kalaydjieva, L., Savov, A., Angelicheva, D., Jankova, S., Kremensky, I.. (1997) SSCP analysis: a blind sensitivity trial. *Hum Mutat* 10, 65–70.
12. Martins-Lopes, P., Zhang, H., and Koebner, R. (2001) Detection of single nucleotide mutations in wheat using single strand conformation polymorphism gels. *Plant Mol Biol Rep* 19, 159–62.
13. Orita, M., Iwahana, H., Kanazawa, H., Hayashi, K., and Sekiya, T. (1989) Detection of polymorphisms of human DNA by gel electrophoresis as single-strand conformation polymorphisms. *Proc Natl Acad Sci U S A* 86, 2766–70.
14. Prosser, J., Inglis, J. D., Condie, A., Ma, K., Kerr, S., Thakrar, R., Taylor, K., Cameron, J. M., and Cooke, H. J. (1996) Degeneracy in human multicopy RBM (YRRM), a candidate spermatogenesis gene. *Mamm Genome* 7, 835–42.
15. Rubio, L., Ayllon, M. A., Kong, P., Fernandez, A., Polek, M., Guerri, J., Moreno, P., and Falk, B. W. (2001) Genetic variation of Citrus Tristeza Virus isolates from California and Spain: evidence for mixed infections and recombination. *J Virol* 75, 8054–62.
16. Mastari, J., Lapierre, H., and Dessens, J. T. (1998) Asymmetrical distribution of Barley Yellow Dwarf Virus PAV variants between host plant species. *Phytopathology* 88, 818–21.

17. Tan, Z. Y., Gibbs, A. J., Tomitaka, Y., Sanchez, F., Ponz, F., and Ohshima, K. (2005) Mutations in Turnip mosaic virus genomes that have adapted to Raphanus sativus. *J Gen Virol* 86, 501–10.
18. Miller, W., Waterhouse, P., and Gerlach, W. (1988) Sequence and organization of Barley Yellow Dwarf Virus genomic RNA. *Nucleic Acids Res* 16, 6097–111.
19. Sambrook, J., Fritsch, E. F. and Maniatis T. (1989) Molecular cloning – a laboratory manual. 2nd ed. Cold Spring Harbor Lab.
20. Rubio, L., Abou-Jawdah, Y., Lin, H.-X., and Falk, B. W. (2001) Geographically distant isolates of the crinivirus Cucurbit yellow stunting disorder virus show very low genetic diversity in the coat protein gene. *J Gen Virol* 82, 929–33.
21. Turturo, C., Saldarelli, P., Yafeng, D., Digiaro, M., Minafra, A., Savino, V., and Martelli, G. P. (2005) Genetic variability and population structure of Grapevine leafroll-associated virus 3 isolates. *J Gen Virol* 86, 217–24.
22. Welsh, J. A., Castren, K., and Vahakangas, K. H. (1997) Single-strand conformation polymorphism analysis to detect p53 mutations: characterization and development of controls. *Clin Chem* 43, 2251–55.
23. Lin, H.-X., Rubio, L., Smythe, A., Jiminez, M., and Falk, B. W. (2003) Genetic diversity and biological variation among California isolates of Cucumber mosaic virus. *J Gen Virol* 84, 249–58.
24. Brody, J. R., and Kern, S. E. (2004) History and principles of conductive media for standard DNA electrophoresis. *Anal Biochem* 333, 1–13.

Chapter 19

Direct Detection of Plant Viruses in Potato Tubers Using Real-Time PCR

Neil Boonham, Lynn Laurenson, Rebecca Weekes and Rick Mumford

Summary

Virus indexing of seed potatoes can be carried out by growing eye plugs to produce small plants and then testing them by ELISA, but this method is time consuming. Direct testing of the eye plugs by ELISA is not reliable, and so a method has been developed for the routine testing of seed potatoes for virus by PCR.

Key words: Virus, ELISA, RT, PCR, Potatoes.

19.1. Introduction

The virus indexing of seed potatoes is one of the most widespread testing procedures performed by virologists. In general, a common approach is used, based on taking eye cores from dormant tubers and growing these on in a greenhouse for several weeks, before testing the sprouts produced from these cores by ELISA. By testing at least 100 tubers, individually or in small batches, the test can be used to estimate the percentage of virus-infected tubers found in a particular seed stock, hence indicating its suitability for planting or the grade at which it should be classified. Over the last 30 years or more, this method has become almost universally adopted because of the advantages it offers: not only can it be used for most common potato viruses, it is also robust, simple to perform, and is well suited to high-throughput testing. As a result, testing laboratories can routinely test hundreds or thousands of seed stocks in a season.

R. Burns (ed.), *Methods in Molecular Biology, Plant Pathology, vol. 508*

DOI: 10.1007/978-1-59745-062-1_19

However, while the growing-on test has become the established method for virus indexing, it is not without problems. In addition to requiring large amounts of greenhouse space, the whole procedure is slow, taking on average at least 6 weeks to complete. Most of this time is required to break dormancy and grow sprouts sufficiently to be tested. Of course, if tubers could be tested directly, then the growing-on step could be removed and the issues of space and time would be resolved. However, this cannot be done reliably using ELISA, as virus titres in dormant tubers are often very low (especially for late primary infections). Often the levels found are below the limit of detection for ELISA, thus making direct tuber testing using this method unreliable *(1)*.

Given the limitations of the ELISA-based system, more sensitive, alternative virus detection methods have been investigated, in particular those based on the polymerase chain reaction (PCR). More recently, real-time PCR methods have been used *(2–5)*. Real-time PCR methods have many advantages over conventional PCR-based ones and hence offer a more practical solution for developing a routine diagnostic service *(4, 6)*.

19.2. Materials

1. Sample grinder (Bioreba or Kleco).
2. Grinding bags or tubes (Bioreba or Kleco).
3. RNA extraction kit (Toyobo, MagExtractor RNA kit).
4. Lysis buffer containing additional 10 mM dithiothreitol (DTT).
5. 0.5 M NaOH.
6. Hughes and Galau extraction buffer: 1.5% lithium dodecyl sulphate (LiDS), 200 mM Tris–HCl, pH 8.5, 10 mM EDTA, 300 mM NaCl, 1.0% Na deoxycholate, 1.0% Ipegal CA-630, 10 mM DTT, 5 mM Thiourea.
7. 6 M Potassium acetate (pH 6.5).
8. 4 M Lithium chloride.
9. TE buffer containing 1% SDS.
10. 4 M NaCl.
11. Iso-propanol.
12. Real-time PCR reagents (Applied Biosystems, TaqMan Amplitaq gold and Buffer A pack, 1000rxns, 4304441).

13. Reverse transcriptase (Fermentas, RevertAid™ M-MuLV RT 5 × 10,000 units, EP0442).
14. Oligonucleotide primers (**Table** 19.1 for details).
15. Dye-labelled oligonucleotide probes (**Table** 19.1 for details).
16. Infected and non-infected control material.
17. dNTPs (WEB Scientific Set of four dNTPs 100 mM, 4 × 1 mL BI-110012).
18. PCR plates with optical caps or sealing film.
19. Real-time PCR instrument (Applied Biosystems 7900HT).
20. Kingfisher magnetic particle processor (Thermo Fisher Scientific, model mL or 96).

Table 19.1
Primers and probes used in real-time (TaqMan) assays for the detection of potato viruses

Assay name	Primer/probe name	Sequence
PVY	PVY 411F	GGG CTT ATG GTT TGG TGC A
	PVY 477R	CCG TCA TAA CCC AAA CTC CG
	PVY Probe (FAM)	TGA AAA TGG AAC CTC GCC AAA TGT CA
PLRV	PLRV F	GGC AAT CGC CGC TCA A
	PLRV R	TGT AAA CAC GAA TGT CTC GCT TG
	PLRV Probe (FAM)	CCT CGT CCT CGG GGA ACT CCA GTT
PVX	PVX 1F	ACA CAG GCC ACA GGG TCA A
	PVX 1R	GGG ATG GTG AAC AGT CCT GAA G
	PVX 2F	GGA TCC ACC AAA TCA ACT ACC AC
	PVX 2R	GGT ATG GTG AAT AGC CCT GAA TTG
	PVX Probe (FAM)	ACT GCA GGC GCA ACT CCT GC
Cox	COX F	CGT CGC ATT CCA GAT TAT CCA
	COX RW	CAA CTA CGG ATA TAT AAG RRC CRR AAC TG
	COXSOL 1511T (VIC)	AGG GCA TTC CAT CCA GCG TAA GCA

19.3. Methods

The following methods outline (**Subheading** 19.3.1) the extraction of RNA from potato tubers, (**Subheading** 19.3.2 the setting up and running of real-time reactions, and (**Subheading** 19.3.3) analysis and interpretation of results.

19.3.1. RNA Extraction

The extraction of total RNA from potato tuber is described in **Subheadings** 19.3.1.1–19.3.1.3. This includes (a) the processing of tubers, and the extraction of RNA by both (b) manual, and (c) automated methods.

19.3.1.1. Processing of Potato Tubers

Tubers for testing should be relatively soil-free and should be washed prior to sampling, if necessary. For a standard virus test, 100 tubers are tested in 25 replicates, where each replicate contains material from 4 tubers (*see* **Note 1**). Small cores or slices are taken from tubers and ground in the appropriate volume of extraction buffer. The cores/slice can be taken from around and including either the main eye ('rose-end') or the stolon-attachment point ('heel-end'); sampling and combining both ends from a single tuber might improve the reliability of detection. During sampling and grinding, steps should be taken to prevent sample-to-sample cross-contamination (*see* **Note 2**). To monitor this a sample of known healthy leaf material (e.g. healthy tomato or potato) should be extracted with every batch of samples.

The grinding can be done manually, in small, heavy-gauge, mesh-grinding bags (e.g. Bioreba AG, Switzerland) using a small hand roller or a semi-automated tissue homogeniser (e.g. Homex, Bioreba). Pre-freezing of samples in liquid nitrogen is generally unnecessary. For high-throughput testing ball mills are generally more convenient (e.g. Kleco, Kinetic Laboratory Equipment Company, California, USA). The samples are placed into a tube or canister with charge (e.g. steel ball bearings) and agitated vigorously to achieve homogenisation. If ball mills with non-disposable canisters are used they should be decontaminated between each sample by rinsing in 0.5 M NaOH for 15–30 min; the canisters should then be washed in distilled water prior to reuse (*see* **Note 3**).

19.3.1.2. Manual RNA Extraction Method

This is based on the method of Hughes and Galau *(7)*, using some modifications from the method of Spiegel and Martin *8)*. Care should be taken when working with RNA extractions, and some general guidelines should be followed (*see* **Note 4**).

1. Homogenise sample in 2.5 volumes (i.e. 0.5 mL for 200 mg of tissue) of extraction buffer (*see* **Note 5**). Decant 500 µL of homogenate into a 2-mL microfuge tube.

2. Add an equal volume of 6 M potassium acetate (pH 6.5) and incubate on ice for 15 min.
3. Centrifuge at 13,000 × *g* in a microcentrifuge for 10 min (at room temperature).
4. Carefully remove the supernatant and transfer to a fresh 2-mL microfuge tube.
5. Add an equal volume of 4 M lithium chloride, mix well, and incubate at 4°C overnight.
6. Centrifuge for 20–30 min at 13,000 × *g* (at 4°C) to pellet the RNA.
7. Decant and dispose of the supernatant and resuspend the pellet in 200 μL of TE buffer-containing 1% SDS. Add 100 μL of 5 M NaCl and 300 μL of ice-cold iso-propanol. Mix well and incubate at –20°C for 20–30 min.
8. Centrifuge for 10 min at 13,000 × *g* (4°C) to pellet nucleic acid. Decant off salt/ethanol and wash pellet by adding 500 μL 70% ethanol and spinning for 3–4 min at 13,000 × *g* (4°C).
9. Decant off the ethanol and dry the pellet to remove residual ethanol (*see* **Note 6**).
10. Resuspend pellet in 100 μL of RNase-free water (*see* **Note 7**).

19.3.1.3. Automated RNA Extraction Method

The automated extraction process uses the MagExtractor RNA extraction kit (Toyobo) and a KingFisher magnetic particle processor (Thermo Fisher Scientific). The extraction method is based around the use of silica-coated magnetic particles; nucleic acid is bound to the silica surface of the magnetic beads, which are then washed to remove contaminating plant material. Purified total nucleic acid (TNA) is eluted into molecular-biology-grade water.

1. Homogenise sample in 4 volumes (i.e. 5 mL for 1.2 g of tissue) of lysis buffer (*see* **Note 8**).
2. Decant the sample into a microfuge tube and centrifuge at around 6,000 × *g* for 1 min, in order to pellet debris.
3. Use 500 μL of cleared lysate for further processing using the Kingfisher.
4. The following setup can be used with both the Kingfisher mL (five tube strips) and the Kingfisher 96 (96-well deep well plates). The 96-well plates A–E or the strip tubes A–E (mL) are loaded as follows:
5. Tube/plate A: 500 μL of clear lysate and 50 μL of magnetic silica beads

   Tube/plate B: 1 mL of lysis buffer

Tube/plate C: 1 mL of 70% ethanol

Tube/plate D: 1 mL of 70% ethanol

Tube/plate E: 200 μL of molecular-grade water (*see* **Note 7**).

6. Samples are run on the standard Total_RNA_mL_1 (KF mL) and Total_RNA_1 (KF 96) programs which should be loaded into the instrument memory on the Kingfisher instrument. These steps should include a 5-min heating step at 65°C for eluting the RNA. The extraction stage should take around 30 min. RNA extracts are then stored at –20°C until required for testing (*see* **Note 9**).

19.3.2. Real-Time PCRs

The next step in this process involves the setting up and running of the real-time PCRs. Details of assays used are given in **Table 19.1**; care should be taken to prevent contamination in real-time PCR procedures (*see* **Note 10**) to avoid false positive amplifications, as the assay's logarithmic amplification can potentially amplify a single target molecule.

19.3.2.1. Setting-Up of Real-Time PCRs

A mastermix for each assay should be made and transferred into the wells of the reaction plate. A standard reaction mastermix is outlined in **Table 19.2** (*see* **Notes 11** and **12**).

Table 19.2
Standard reaction conditions for real-time PCR (TaqMan) assays used for the detection of potato viruses

Reaction component	Stock concentration	Volume (μL) in mastermix (final 25 μL)	Final concentration	Company
ddH_2O		11.25		
Buffer A (+ROX)	10×	2.5	1×	Applied Biosystems
F primer	7.5 μM	1.0	300 nM	MWG
R primer	7.5 μM	1.0	300 nM	MWG
TaqMan probe	5 μM	0.5	100 nM	Applied Biosystems
dNTPs	6.25 mM	2.0	500 nM	WEB Scientific
$MgCl_2$	25 mM	5.5	5.5 mM	Applied Biosystems
AmpliTaq Gold	5 U/μL	0.125	0.025 U/μL	Applied Biosystems
M-MLV Rtase[a]	4 U/μL	0.125	0.02 U/μL	Fermentas
Target RNA		1.0		

[a] *See* **Note 12**

19.3.2.2. Reaction Conditions

Real-time PCR plates are run (*see* **Note 13**) using standard cycling conditions on the 7900HT or 7500 real-time PCR instruments (Applied Biosystems). The cycling conditions are 48°C for 30 min, 95°C for 10 min followed by 40 cycles of 95°C for 15 s, then 60°C for 1 min.

19.3.3. Analysis of Results

19.3.3.1. Data Collection

Real-time reactions (*see* **Note 14**) are characterised by the point in time during cycling when fluorescence, generated as a result of amplification of a PCR product, is first detected above a defined threshold (the cycle threshold or C_T value). **Figure** 19.1 shows an amplification plot of the data output from a real-time PCR. When collecting results it is important to ensure the baseline and threshold values have been set correctly and that the three phases of the amplification curve are visible. The data from a real-time PCR test are always reported as C_T values. For example, a C_T value of 40 is given when the reaction runs for 40 cycles and no amplification is seen; this is reported as negative. In contrast, a C_T value less than 40 indicates amplification of the target and should generally be counted as positive.

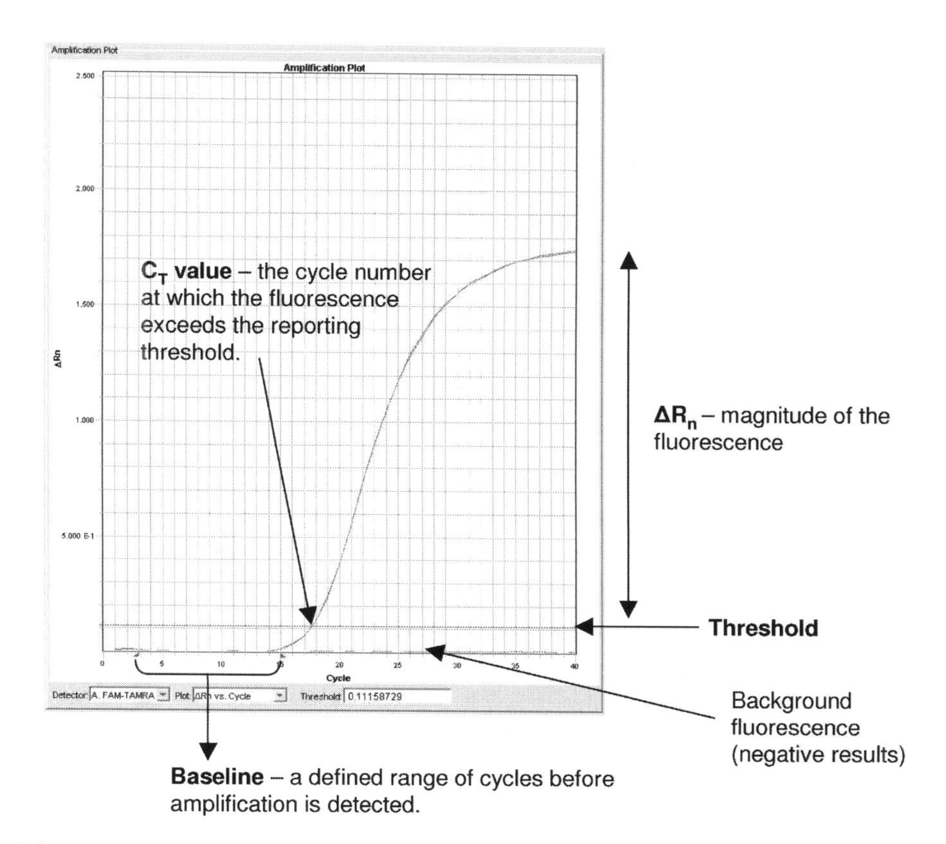

Fig. 19.1. A representative amplification plot, annotated with some terms of reference.

19.3.3.2. Interpretation of Results

A number of controls should be included in the test to aid the interpretation of results and for QC purposes. These include (1) negative controls (*see* **Note 15**), (2) positive controls to ensure each assay has worked, and (3) an endogenous control, used to monitor the success of each sample extraction, e.g. cytochrome oxidase (Cox). In addition samples should be run in duplicate (*see* **Note 16**) for each PCR test.

After checking that the negative and the positive controls have given the appropriate results, the results for the endogenous control should be interrogated. A successful extraction will give a good C_T value for the Cox assay (typically between 18 and 25). If samples do not give amplification using the Cox assay (or the C_T values are very high) then the extractions should be repeated. If all the controls have produced the appropriate results, then the results can be interrogated to determine if any of the samples are positive.

19.4. Notes

1. The practice of testing 100 tubers by combining 4 tubers per extract has been routinely used, as it provides continuity with the standard procedure used in growing-on (i.e. 100 tubers tested as 25 sub-samples, with leaves from 4 sprouts combined and tested). This allows results to be interpreted in the same way. Given the increased sensitivity of real-time PCR, bulked samples of more than four tubers could potentially be used, although this would need to be fully validated before it was used routinely.
2. To minimise sample-to-sample contamination gloves should be changed between samples (but this is not necessary between replicates). In addition, any sampling device used should be decontaminated between samples. This can be done by washing the sampling device in distilled water (dH_2O), followed by 0.5 M NaOH, followed by two further washes in dH_2O. After a sample is complete the device should be rinsed in water, soaked in 0.5 M NaOH for 15–30 min, and then rinsed in dH_2O.
3. Do not add bleach to liquid or solid waste containing guanadine lysis buffers. When bleach is mixed with buffers containing guanidine, toxic fumes of HCl gas are produced.
4. When working with RNA, always use RNase-free solutions and plasticware, etc. Water should be at least distilled and autoclaved or produced to a similar quality. It is not necessary to autoclave plastics certified as RNase-free; however, they should be handled with gloves to prevent contamination.

5. For extraction buffer in the method of Hughes and Galau, DTT and Thiourea should be added fresh to an aliquot of stock buffer (containing the first six reagents) immediately prior to extraction. The resulting buffer can be kept at room temperature and used for up to 2 weeks. Stock buffer can be autoclaved and stored at room temperature. Ipegal CA-630 is a direct replacement for the detergent NP-40.
6. RNA precipitates can be dried on the laboratory bench, in a fume, or laminar flow hood, or in a vacuum drier (e.g. Thermo Fisher Scientific Speedvac). If using the latter method, it is important not to dry the precipitate for excessive lengths of time; precipitates that are completely desiccated are difficult to resuspend.
7. Use only 'PCR-grade' or 'Molecular-biology-grade' water, which should be purified (deionised or distilled), sterile (autoclaved or 0.45 μm filtered), and nuclease-free.
8. Dithiothreitol (DTT) at a final concentration of 10 mM is used as a less hazardous (and pungent) alternative reducing agent to β-mercaptoethanol, as recommended in the kit.
9. Frozen RNA samples should be thoroughly mixed once defrosted and care taken to resuspend any precipitate that may have formed.
10. In order to minimise the risks of cross-contamination, extraction of RNA, set-up of master mix, spiking of RNA, and running of PCRs should be performed in a spatially separated fashion. The commonest forms of contamination in real-time PCR are as follows: (1) sample-to-sample contamination occurring in the extraction procedure, (2) contamination of clean reagents with DNA, and (3) well-to-well contamination in RNA spiking. Using pipettes designated for each procedure and aerosol resistant (filter) tips both minimise the risk of contamination. Although post-PCR contamination is less of an issue with real-time PCR, if the amplification products are further characterised by cloning and sequencing (for example during assay validation) extreme care should be taken in opening plates due to the sensitivity of the real-time technique.
11. For the PVX assay two sets of primers are added, each of the four primers should have a final concentration of 300 nM.
12. The reverse transcriptase is diluted to a concentration of 4 U/μL (1:50 v/v) from a concentration of 200 U/μL in water just prior to addition to mastermix.
13. The real-time PCR plates can be stored once sealed at –20°C for up to 1 month. Upon thawing the plates should be centrifuged to remove air bubbles prior to running.

14. Real-time PCR (TaqMan®) uses a probe labelled with a reported and quencher dye. During the extension phase of PCR, the probe (which is designed to anneal downstream from one of the primer sites) is cleaved by the 5′ nuclease activity of the Taq polymerase. This cleavage of the probe separates the reporter dye from the quencher dye leading to an increase in the reporter dye signal. Additional reporter dye molecules are cleaved from their respective probes with each cycle, effecting an increase in fluorescence intensity proportional to the amount of amplicon produced.
15. Negative control types: (1) no template control to monitor contamination of reagents, (2) a negative extraction control, e.g. uninfected leaf material to monitor contamination during the extraction process.
16. Duplicate samples should have very little variation, and this enables the user to monitor if amplification observed is real or spurious (e.g. well-to-well contamination). It also gives security over instrument artefacts that may affect individual wells or failed caps that may result in evaporation from the wells.

Acknowledgements

The authors would like to thank the British Potato Council and Plant Health Division of Defra for initial funding of the development work.

References

1. Hill, S. A., and Jackson, E. A. (1984). An investigation of the reliability of ELISA as a practical test for detecting potato leaf roll virus and potato virus Y in tubers. *Plant Pathology* 33, 21–26.
2. Schoen, C. D., Knorr, D., and Leone, G. (1996). Detection of potato leafroll virus in dormant potato tubers by immunocapture and a fluorogenic 5 nuclease RT-PCR assay. *Phytopathology* 86, 993–999.
3. Boonham, N., Walsh, K., Mumford, R. A., and Barker, I. (2000). The use of multiplex real-time PCR (TaqMan®) for the detection of potato viruses. *EPPO Bulletin* 30, 427–430.
4. Mumford, R. A., Walsh, K., Barker, I., and Boonham, N. (2000). The detection of potato mop-top and tobacco rattle viruses using a multiplex real-time fluorescent RT-PCR assay (TaqMan®). *Phytopathology* 90, 488–493.
5. Agindotan, B. O., Shiel, P. J., and Berger, P. H. (2007). Simultaneous detection of potato viruses, PLRV, PVA, PVX and PVY from dormant potato tubers by TaqMan® real-time RT-PCR. *Journal of Virological Methods* 142, 1–9.
6. Mumford, R. A., TomLinson, J., Barker, I., and Boonham, N. (2006). Advances in molecular phytodiagnostics – new solutions for old problems. *European Journal of Plant Pathology* 116, 1–19.
7. Hughes, D. W., and Galau, G. (1988). Preparation of RNA from cotton leaves and pollen. *Plant Molecular Biology Reporter* 6, 253–257.
8. Spiegel, S., and Martin, R. R. (1993). Improved detection of potato leafroll virus in dormant potato tubers and microplants by the polymerase cahin reaction and ELISA. *Annals of Applied Biology* 122, 493–500.

Chapter 20

An Accelerated Soil Bait Assay for the Detection of Potato Mop Top Virus in Agricultural Soil

Triona Davey

Summary

An accelerated soil bait test can be used to determine whether a field harbours virus-carrying *Spongospora subterranea*. *S. subterranea* is the causal agent of powdery scab and also the only vector of potato mop top virus (PMTV). Real-time RT-PCR can detect PMTV RNA in the roots of bait plants after 2 weeks of growth in viruliferous soil. This test may be used to assess the risk of planting potato crops in a particular field.

Key words: Real-time PCR, Reverse transcription, Potato, *Solanum tuberosum*, Potato mop top virus, *Spongospora subterranea* f. sp. *subterranea*, Detection, Powdery scab.

20.1. Introduction

Potato (*Solanum tuberosum*) is the World's fourth largest food crop, exceeded only by the cereals maize, rice, and wheat *(1)*.

Potato, an annual herbaceous dicotyledonous plant, is primarily vegetatively propagated from seed tubers. The main advantage of vegetative propagation is that a good potato clone can be maintained through a number of multiplications with a high degree of varietal purity. A major disadvantage is that many viruses and seed-borne pathogens, which affect yield and tuber quality, gradually accumulate in the tubers and are carried over repeated multiplications resulting in the gradual deterioration of the clone.

Potato is a host to many pathogenic organisms, including bacteria, fungi, viruses, viroids, nematodes, and phytoplasmas. Viruses in particular are extremely important pathogens of potato

R. Burns (ed.), *Methods in Molecular Biology, Plant Pathology, vol. 508*

DOI: 10.1007/978-1-59745-062-1_20

because they can be readily transmitted in symptomless seed tubers from generation to generation.

Strategies for disease control rely on preventive measures such as the use of resistant varieties, the control of vectors, and the detection and elimination of contaminated plants and seed.

Successful potato cultivation and production depends on the availability of disease-free, high-quality seed tubers.

The disease potato mop top is caused by the virus, potato mop top virus (PMTV). PMTV is the type member of the genus *Pomovirus* and is a *Furovirus* (*fu*ngus-transmitted, *ro*d-shaped virus). PMTV has tubular rod-shaped particles, and the genome consists of three single-stranded, positive sense RNA molecules *(2)*. Initial transmission of PMTV to potato occurs in the process of infection by the powdery scab pathogen *Spongospora subterranea* f. sp. *subterranea*; a plasmodiophorid which is found worldwide *(3)*.

PMTV is contained in *S. subterranea* resting spores capable of surviving in soil for more than 15 years *(4–6)*, which makes it possible for PMTV to remain infective in field soils susceptible to powdery scab for a long period of time *(7)*. Under wet conditions, germinating resting spores release zoospores which can infect potato roots, stolons, young shoots, and tubers *(8)*. The zoospores can acquire virus particles when *S. subterranea* develops in virus-infected host cells.

PMTV-infected tubers of sensitive potato cultivars develop necrotic arcs and circles (spraing symptoms) *(3, 9–10)* making the tubers unsuitable for the fresh potato market as well as crisp and french fry production. Spraing symptoms indicate primary infection, i.e. infection through PMTV-carrying *S. subterranea* in soil. The degree of transmission of PMTV from seed to daughter tubers is restricted, and in the absence of *S. subterranea*, may be 'self-curing' after several generations *(11)*; this indicates that soil inoculum is more important than seed inoculum.

Plants grown from PMTV-infected tubers produce two types of foliar symptoms: yellow blotches and chevrons on the leaves and/or a shortening of the internodes resulting in a bunching effect. These chevrons have been confused with those induced by potato aucuba mosaic, alfalfa mosaic, tomato black ring, or tobacco rattle viruses *(12)*. In many cases, only a proportion of the stems are affected. Symptoms of PMTV infection only occur in a growing plant if the seed is infected prior to planting.

Powdery scab usually only causes cosmetic damage to tubers, making them appear unsightly, so they are difficult to market *(13)*. Susceptibility to powdery scab and PMTV varies with cultivar. No varieties are immune to powdery scab or PMTV; however, certain varieties exhibit few or no spraing symptoms, but there are still no known sources of resistance incorporated into modern cultivars.

Some control of PMTV may be achieved by roguing plants expressing symptoms, avoiding potato cultivation in soils infected with *S. subterranea* or avoiding planting infected or contaminated seed. In addition, cultivation of infected material in virus/vector-free soil may lead to the elimination of the virus from stocks of certain cultivars within a few years.

Serological (ELISA) and molecular (RT-PCR) methods are already widely used to detect PMTV in potato leaves and tubers. However, as PMTV infection is partially systemic *(14)*, testing the leaves of a growing plant may provide false negatives. This could permit the further multiplication of infected daughter tubers, thereby facilitating the spread of the viruliferous vector and/or produce a crop with reduced marketability.

Direct tuber testing is an effective method for PMTV detection; however, this is carried out postharvest. If a crop is found to have a high incidence of PMTV infection, it may be rejected for import by some countries. This would not only be a financial burden on the grower but could also have a detrimental effect on the Scottish export trade as a whole.

A real-time PCR assay has been developed for the detection and quantification of *S. subterranea* in soil *(15)*; however, no methods exist for the detection of PMTV inside these resting spores.

This chapter outlines a method for the detection of PMTV in viruliferous soil using soil bait plants.

This assay is not quantitative; however, it could be used as a management tool to assist growers in identifying fields that could potentially infect a crop with both powdery scab and PMTV.

20.2. Materials

1. Tomato seeds (cv. Moneymaker).
2. Trough (7 × 2.5 in – B.E.F Products, Essex, UK).
3. Levington F2 (with sand) seed and modular compost (SCOTTS, UK).
4. 12 × 7 well tray (size 1.5 in at top of well – Amprica, Italy).
5. Universal extraction bag 12 × 14 cm with synthetic intermediate layer (Bioreba AG, Switzerland).
6. Homex 6 Plant Tissue Homogenizer (Bioreba, AG, Switzerland).
7. Kingfisher™ system (Thermo Labsystems, Finland).
8. MagExtractor®-RNA-reagents (Toyobo, Japan).
9. PCR plate (96 wells).

10. Plate seal.
11. JumpStart™ Taq ReadyMix™ for Quantitative PCR (Sigma, USA).
12. 3 mM $MgCl_2$.
13. PMTV Forward primer at 7.5 pmol/mL.
14. PMTV Reverse primer at 7.5 pmol/mL.
15. PMTV Probe at 5.0 pmol/mL.
16. Moloney murine leukaemia virus reverse transcriptase (Promega, USA).

20.3. Methods

The following methods are described: (1) preparation of tomato seedlings and transplanting into sample soil, (2) total RNA extraction from roots of bait plants, (3) Real-time RT-PCR for detection of viral RNA. Sample soils have to be air-dried for 1 week prior to testing (*see* **Note 1**).

20.3.1. Preparation of Tomato Seedlings and Transplanting into Sample Soil

Fill the trough (B. E. F Products, Essex, UK) with the Levington F2 compost (SCOTTS, UK), leaving a gap of approximately 1 cm from the top. Spread approximately 50 tomato seeds (cv. Moneymaker) over the surface and cover all the seeds with more compost. Grow the seeds under natural light conditions in a glasshouse at 22°C and water daily. When the seedlings are 2 weeks old they are ready to be transplanted into the sample soil.

Place approximately 20 g of air-dried sample soil into a well of a 12 × 7 well tray (size 1.5 in width at top of well – Amprica, Italy) (*see* **Note 2**). Three replicate wells are used for each soil sample. Plant a single 2-week-old tomato seedling into each replicate well. Grow the seedlings under natural light conditions in a glasshouse at temperatures between 17 and 21°C and water daily (*see* **Note 3**). Allow to grow for 2 weeks in the sample soil. After 14 days the plants are ready to be harvested.

20.3.2. Harvesting the Bait Plants

Carefully remove the plants from the wells and wash the soil from the roots using water (*see* **Note 4**). Line up the roots of the bait plant and using a scalpel (*see* **Note 5**); cut a segment across all roots (approx 100 mg root tissue). Place the root tissue from each rep into individually labelled Universal extraction bags (12 × 14 cm with synthetic intermediate layer-Bioreba, AG). Add 1 mL of lysis buffer (MagExtractor®-RNA-reagents (Toyobo, Japan) to each Universal extraction bag. Place the extraction bags containing the root tissue and lysis buffer into the frame of the Homex 6 Plant Tissue Homogenizer (Bioreba, AG, Switzerland), and grind.

20.3.3. Total RNA Extraction

Total RNA extractions are carried out using the Kingfisher™ system (Thermo Labsystems, Finland) and MagExtractor®-RNA-reagents (Toyobo, Japan), according to the manufacturer's instructions; however, the addition of 2-mercaptoethanol to the lysis buffer is omitted.

Extracted RNA template can be stored at –20°C prior to testing by real-time RT-PCR.

20.3.4. Real-Time RT-PCR for Detection and Amplification of Viral RNA

The master mix for each reaction is described as follows. The volumes should be multiplied up depending on the number of reactions to be performed (*see* **Note 6**). Each sample is analysed in triplicate. The primers and labelled probe were specific for PMTV RNA 3, which encodes the coat protein *(16)*.

Dispense 24 mL of this master mix into the required wells of a 96-well plate and add 1 mL of the RNA extract. Seal the plate and perform amplification under the following conditions: an initial hold step at 48°C for 30 min, preliminary denaturation at 94°C for 2 min, followed by 40 cycles of 95°C for 15 s and 60°C for 1 min.

Amplification and detection was performed on an Applied Biosystems 7900HT PCR System (Applied Biosystems, USA). Cycling conditions were as follows: an initial hold step at 48°C for 30 min, preliminary denaturation at 94°C for 2 min, followed by 40 cycles of 95°C for 15 s and 60°C for 1 min.

Real-time RT-PCR reagent	Volume per reaction (mL)
JumpStart™ Taq ReadyMix™ for Quantitative PCR (Sigma, USA)	12.5
3 mM $MgCl_2$	4
PMTV Forward primer at 7.5 pmol/mL	1
PMTV Reverse primer at 7.5 pmol/mL	1
PMTV Probe at 5.0 pmol/mL	0.5
M-MLV Reverse Transcriptase at 200 U/mL (Promega, USA)	0.05
H_2O	4.95

20.4. Notes

1. The soil samples can be dried on a lab bench; however, the use of a drying cabinet (set at room temperature) minimises the risk of cross-contamination.

2. Use a weighing boat to weigh the soil. One weighing boat can be used for the three reps, but a new weighing boat should be used for each sample. After weighing the soil, add a little water to dampen the soil. It is easier to plant a seedling in damp soil rather than dry soil.
3. Try to elevate the tray to prevent pooling and mixing of water from different samples, thereby eliminating the risk of cross-contamination.
4. The roots can be washed using tap water.
5. In order to increase speed, a number of scalpels can be used at one. The scalpels are sterilised by washing in the following solutions for 1 min. Water, 0.2 M NaOH, water and 96% ethanol. Flame the scalpel to remove the ethanol.
6. To allow for human error in pipetting, the master mix for the real-time RT-PCR should be prepared for more than the actual number required, i.e. make up enough master mix for 100 reactions if a full 96-well plate is being used.

Acknowledgements

This research is funded by the British Potato Council (Research & Development Project R247) and the Scottish Government Rural Payments and Inspections Directorate. We gratefully acknowledge the assistance of SGRIPD's seed potato inspectors in the collection of soil samples.

References

1. Stevenson, W. R., Loria, R., Franc, G. D., and Weingartner, D. P. (Eds). (2001). Compendium of Potato Diseases. APS Press, St. Pauls, MN, p. 144.
2. McGeachy, K. D., and Barker, H. (2000). Potato mop-top virus RNA can move long distance in the absence of coat-protein: evidence from resistant, transgenic plants. *Molecular Plant-Microbe Interactions* 13(1):125–128.
3. Calvert, E. L., and Harrison, B. D. (1966). Potato mop top, a soil-borne virus. *Plant Pathology* 15:134–139.
4. Jones, R. A. C., and Harrison, B. D. (1972). Ecological studies on potato mop-top virus in Scotland. *Annals of Applied Biology* 71: 47–57.
5. Jones, R. A. C., and Harrison, B. D. (1969). The behaviour of potato mop top virus in soil and evidence for it's transmission by *Spongosporra subterranea* (Wallr.) Lagerh. *Annals of Applied Biology* 63:1–17.
6. Campbell, R. N. (1996). Fungal transmission of plant viruses. *Annual Review of Phytopathology* 34:87–108.
7. Germundsson, A., Sandgren, M., Barker, H., Savenkov, E. I., and Valkonen, J. P. T. (2002). Initial infection of roots and leaves reveals different resistance phenotypes associated with coat-protein gene-mediated resistance to potato mop-top virus. *Journal of General Virology* 83:1201–1209.
8. Hims, M. J., and Preece, T. F. (1975). *Spongospora subterranea.* CMI Descriptions of Pathogenic Fungi and Bacteria No. 477.
9. Harrison, B. D., and Jones, R. A. C. (1970). Host range and some properties of potato mop-top virus. *Annals of Applied Biology* 65:391–402.

10. Kurppa, A. H. J. (1989). Reaction of potato cultivars to primary and secondary infection by potato mop-top furovirus and strategies for virus detection. *EPPO Bulletin* 19:593–598.
11. Cooper, J. I., Jones, R. A. C., and Harrison, B. D. (1976). Field and glasshouse experiments on the control of potato mop-top virus. *Annals of Applied Biology* 83:215–230.
12. Jones, R. A. C. (1988). Epidemiology and control of potato mop-top virus. In J. I. Cooper and M. J. C. Asher (Eds). Developments in Applied Biology 2: Viruses with Fungal Vectors, Association of Applied Biologists, Wellesbourne, UK, pp. 255–270.
13. Wale, S. J. (2000). Summary of the session on national potato production and the powdery scab workshop. Aberdeen, UK, SAC 20–22 July, 3–9.
14. Torrance, L., Cowan, G. H., Scott, K. P., Pereria, L. G., Roberts, I. M., Reavy, B., and Harrison, B. D. (1992). Detection and diagnosis of potato mop-top virus. Annual Report of Scottish Crop Research Institute for 1991, pp. 8–82.
15. Van de graaf, P., Lees, A. K., Cullen, D. W., and Duncan, J. M. (2003). Detection and quantification of *Spongospora subterranea* in soil, water and plant tissue samples using real-time PCR. *European Journal of Plant Pathology* 109:589–597.
16. Mumford, R. A., Walsh, K., Barker, I., and Boonham, N. (2000). Detection of potato mop top virus and tobacco rattle virus using a multiplex real-time fluorescent reverse-transcription polymerase chain reaction assay. *Phytopathology* 90:448–453.

Chapter 21

Detection of Phytoplasmas of Temperate Fruit Trees

Margit Laimer

Summary

Phytoplasmas are associated with hundreds of plant diseases globally. Many fruit tree phytoplasmas are transmitted by insect vectors or grafting, are considered quarantine organisms and a major economic threat to orchards. Diagnosis can be difficult, but immunochemical and molecular methods have been developed.

Key words: Phytoplasma; Fruit trees, Electron microscope, Antibodies, Primers, RFLP, ELISA.

21.1. Introduction

Phytoplasmas are non-helical mollicutes associated with several hundred diseases of plants *(1)*. These disorders were thought to be caused by viruses until 1967, when Doi et al. *(2)* recognized in diseased plants wall-less, pleomorphic bodies, which due to their morphological similarity to mycoplasmas were named mycoplasma-like organisms (MLOs).

Phytoplasmas were classified into 15 different clades, previously termed as 16 Sr groups *(3)*, and so far, with improved diagnostic tools 26 *Candidatus* Phytoplasma species have been proposed *(4)*. Phytoplasmas with their small, AT-rich genomes are thought to descend from a Gram-positive clostridial ancestor *(5, 6)*. The size of *Ca.* Phytoplasma chromosomes ranges from 530 kbp to 1,350 bp *(7)*.

Organisms of the genus *Ca. Phytoplasmas* inhabit the phloem sieve elements of vascular plants and the gut, haemolymph, salivary gland, and other organs of sap-sucking insects.

Phytoplasmas infecting temperate fruit trees in Europe, America, and Australia, spreading rapidly because of transmission by insects and grafting, are considered quarantine organisms. The poor taste

R. Burns (ed.), *Methods in Molecular Biology, Plant Pathology, vol. 508*

DOI: 10.1007/978-1-59745-062-1_21

and small size of fruits and the decline of infected trees make these diseases a major economic threat to orchards.

Phytoplasmas of the *apple proliferation* group, e.g. the closely related apple proliferation (AP), pear decline (PD), and European stone fruit yellows (ESFY) (8–10), with a linear chromosome as a unique feature (Kube et al. 2008) have mainly been reported in Europe, with the exception of the Peach yellow leaf roll (PYLR) agent, which occurs in western USA (12). X-clade phytoplasmas, like Western X-disease (WX) of cherry, peach, and nut trees, are mainly found in North America *(12)*. The chromosome size of the AP group ranges from 630 kbp for ESFY, 660 kbp for PD, and 645–690 kbp for AP, while those of WX disease agents in *Prunus* and *Vaccinium* reach 670 and 1,075 kbp, respectively *(13)*.

A physical map of the chromosome of the western X-disease phytoplasma was constructed and represents the first physical map of a phytoplasma chromosome, which is circular and comprises approximately 670 kb *(14)*.

Locations of two rDNA operons, the operon including the fus and tuf genes, and three other genes were placed on a physical map of the AP phytoplasma, strain AT, and ESFY chromosome *(15, 16)*.

The 16S rRNA gene of AP-group phytoplasmas consists of 1521 nt, with interspecific sequences similarity between 99.9 and 100%. AP, PD, and ESFY phytoplasmas, alias *Candidatus* Phytoplasma mali, *Candidatus* Phytoplasma pyri, and *Candidatus* Phytoplasma prunorum are coherent, but discrete taxa, distinguishable by molecular markers, including the 16S–23S rDNA spacer region, ribosomal protein genes and randomly cloned DNA fragments, serological comparisons, and differences in vector transmission and host-range specificity *(17)*. AP-group phytoplasmas are transmitted by members of the genus *Cacopsylla* (order *Homoptera*, family *Psyllidae*), namely AP by *C. picta (18)*, ESFY by *C. pruni(19–21)*, PD by *C. pyricola* in North America and the UK *(22)* and probably by *C. pyri* in Southern Europe *(23)*, and PYLR by *C. pyricola (24)*.

21.2. Materials

1. *Catharanthus roseus* as herbaceous host.
2. *Malus domestica* cultivar Golden Delicious for immunofluorescent diagnosis.
3. Cryomicrotome.
4. Epifluorescence microscope.
5. Electron microscope.

6. DAPI (4,6 diamidino-2-phenylindole, Sigma no. D9542).
7. FITC (fluorescein-isothiocyanate)–antimouse conjugate.
8. Microplate Reader.
9. Specific antibodies (*see* **Subheading** 21.3.4).
10. AP detection Kit (Bioreba).
11. DNA extraction based on different protocols (*see* Subheadings 21.3.5.1 and 21.3.5.2).
12. DNeasy® Plant Mini Kit (QIAGEN).
13. DNA purification kit (Quiagen).
14. Universal primers (*see* **Subheading** "Universal Primers").
15. Group-specific primers (*see* **Subheading** "Group-Specific Primers").
16. Species- and strain-specific primers (*see* **Subheading** "Species- and Strain-Specific Primers").
17. Random Fragment Length polymorphism (RFLP).
18. Restriction endonucleases, e.g. AluI, RsaI, and others.

21.3. Methods

Detection and identification of phytoplasmas is necessary for accurate disease diagnosis in both host systems, i.e. in plants, where phytoplasmas are localized in phloem cells, and in insect vectors. The choice of a particular method will vary *according to the intention/goal of research*, e.g. the degree of infection of genotypes in a resistance breeding programme, the determination of the degree of infection of an area, the distribution of different strains. Of extraordinary value is therefore in vitro reference material, which provides control samples independent of the season and prolongs the availability of strains, and has been attempted by several laboratories *(25–27)*.

Methods for phytoplasma detection are presented according to their historical development, and the respective suitability is suggested for particular applications.

21.3.1. Symptomatology and Host Range

Since phytoplasmas have not been grown in axenic culture, for a long time they were described and differentiated according to the symptoms they induce, the host plants affected, the specificity of insect vectors, and the geographic distribution. *Catharanthus roseus* (periwinkle) is a valuable experimental host, to which many phytoplasmas have been experimentally transmitted and in which they induce specific symptoms *(28)*.

In the glasshouse, test plants are grafted in spring (by chip-budding or grafting of stem or root cuttings), onto the indicator apple *Malus domestica* cv. 'Charden', 'Golden Delicious', or 'Red Boskoop'. After 2 weeks, when the graft union is formed, the plants are placed in a cold room at 5°C for 60–70 days of forced dormancy. Plants are pruned, leaving one or two buds above the graft, in order to concentrate *AP* in the young and vigorous shoots after regrowth. One or two months after pruning, the first symptoms may be observed (enlarged stipules, presence of witches' brooms) (*see* **Note 1**). In the field, indicators are inoculated at the end of summer or in autumn on indicator species and observed at least 2 years. Determination of biological properties is time consuming, laborious, and non-reliable *(1)*.

21.3.2. Electron Microscopy

Phytoplasmas in diseased plants are surrounded by a single-unit membrane, lack rigid cell walls, are pleomorphic in shape *(2)*, and in transmission electron microscopy (EM) they appear as rounded to filamentous, pleomorphic bodies with a mean diameter of 200–800 nm *(29)*. Petzold et al. *(30)* furnished convincing evidence by EM analyses on the mycoplasma character of AP. Visualisation in EM requires highly specialized lab equipment and technical skills, which make the method unsuited for routine testing. In case, specific antisera are available, e.g. AP and WX ImmunoSorbent Electron Microscopy (ISEM) might be helpful for particular verifications *(31)*.

21.3.3. DAPI Staining

This was largely applied for AP diagnosis *(31, 32)*, as a rapid and cheap, but not specific technique, which sometimes yields results difficult to interpret *(33)*. Pieces of young tissues (petioles of young leaves, phloem tissue of shoots, branches, and roots) are fixed overnight at 4°C in 4% paraformaldehyde in PBS. Longitudinal sections of 20-μm thickness are stained with 1 μg/ml DAPI (4,6 diamidino-2-phenylindole, Sigma no. D9542) and observed under an epifluorescence microscope *(31, 34)* (*see* **Note 2**).

21.3.4. Serology

Serology is an efficient detection system in plant pathology, particularly by offering the advantage of large-scale utilisation in cases of routine diagnosis. A few specific polyclonal (pAbs) and monoclonal antibodies (mAbs) have been raised against a number of fruit tree phytoplasmas, e.g. against the WX phytoplasma rabbit pAbs *(35)* and mouse mAbs *(36)*, while pAbs and mAbs were produced against AP *(33, 37, 38)*.

Limitations arose due to difficulties encountered in purifying the correct antigens from infected plant material, or insect vectors, since it is not possible to culture phytoplasmas axenically. The production of specific antisera proved to be difficult, since plant-derived immunogens often show cross-reactivity with plant antigens, and insect-derived immunogens were rarely available.

Most of the antibodies available were prepared against isolates maintained in high-titre herbaceous hosts, such as periwinkle.

21.3.4.1. Enzyme-Linked ImmunoSorbent Assay

Two mAbs against AP, 1F4/1E2 and 7H1/2C2, provide an easy, rapid, specific, and sensitive serological detection assays conveniently applied in roots, stems, and leaves of apple trees *(33)*. They react specifically by using ELISA with AP-infected periwinkles and apple trees from different regions, but not with related phytoplasma isolates PD and ESFY. ELISA should be performed according to the manufacturer's instructions (Bioreba). The most reliable results can be obtained when leaf midribs or stems collected from late spring to end of summer (June–September) are tested. Sampling is the most crucial factor for correct detection (*see* **Note 3**). In cooler climates and in case of latent infections, in Northern and Western Europe, ELISA may not be sensitive enough to detect the relatively low concentrations, so testing may be supported by polymerase chain reaction (PCR) *(39)*. A further limitation has been detected recently, since not all the isolates found by molecular diagnostic tools also react with these two monoclonal antibodies *(40)*.

21.3.4.2. Immunofluorescence

One-centimetre-long pieces of stems and roots are fixed in 4% paraformaldehyde in PBS, left overnight at 4°C, and then longitudinally cut by a cryomicrotome (Leitz Jung 1500) to obtain sections of 20-μm thickness. The sections are treated with the mab tissue culture supernatants and incubated for 1 h at 37°C. After washing, FITC (fluorescein-isothiocyanate)–antimouse conjugate (Sigma no. F1010) is added and incubated at 37°C for 30 min. Sections are observed under an epifluorescence microscope (**Fig. 21.1**). Immunofluorescence is useful for specific detection of AP in roots of apple trees throughout the year, and on leaves and twigs from fruit ripening until the dormant bud stage, and was found to be more sensitive than DAPI *(33)*.

21.3.4.3. Immunodominant Membrane Proteins

Both monoclonal and polyclonal antibodies made to several phytoplasmas, in Western blot analysis, mostly recognize only one or two abundant IMPs, thought to be involved in the transmission process by certain phloem-feeding insect vectors. Genes encoding IMPs have been cloned from a number of phytoplasmas, namely sweet potato witches' broom, AP, WX, and aster yellows (AY) and clover phyllody (CP) *(35,37,41)*, allowing the production of AS to recombinant antigens with even higher specificity and reliability. Homologues of the immunodominant membrane protein (IMP) gene from AP have been cloned and sequenced for three further members of the AP subclade, ESFY, PYLR, and PD, showing similar sizes of IMPs and clear similarities, both in the relatively conserved N-terminal domain and in the large hydrophilic domain *(42)*.

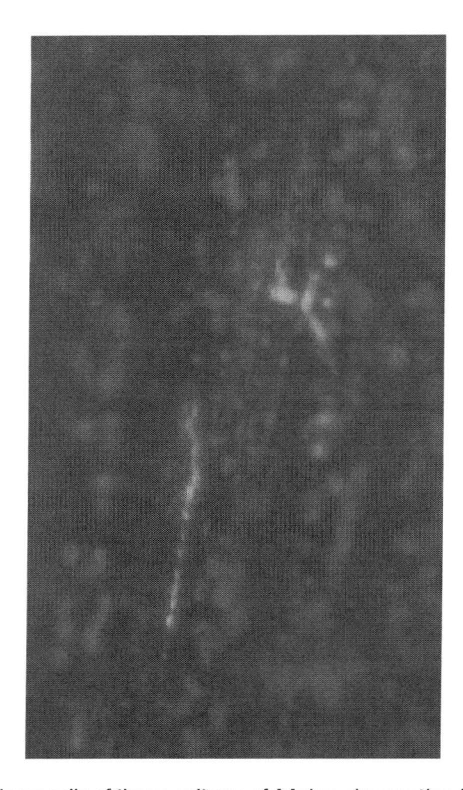

Fig. 21.1. Localization of AP in phloem cells of tissue cultures of *Malus domestica* by immunofluorescence.

An alternative approach to overcome limitations due to high strain specificity of antisera was chosen by Wei et al. *(43)*. An anti-SecA membrane protein Ab with multiple strain/phylum specificity of one phytoplasma reacts with those of phylogenetically different phytoplasmas.

21.3.5. Molecular Analyses

These require procedures to extract and enrich phytoplasmal DNA from infected plants or insects *(44, 45)* allowing the cloning and sequencing of phytoplasma DNA and methods applied for DNA amplification using PCR, followed by sequencing or RFLP analysis of ribosomal DNA. For genomic analysis requiring chromosomal phytoplasma DNA with low host DNA content, caesium chloride (CsCl) gradient centrifugation is still the best enrichment method, despite the availability of new techniques for DNA amplification, such as rolling cycle amplification (RCA), suppression subtractive hybridization (SSH), mirror orientation selection (MOS) *(46)*.

21.3.5.1. DNA Extraction from Plant Tissue

Shoots, roots, leaves, or petioles may be used. Shoots and roots are debarked, and a sample of phloem is removed with a sterile blade. For leaves, only the midrib is taken. The best results are obtained if DNA is extracted from leaf midribs or stems collected from late spring to end of summer (June–September). An efficient grinding method is pre-requisite for all described extraction procedures.

DNA of *AP* is extracted following Ahrens and Seemüller *(47)*, or Lee et al. *(45)*. The extraction buffer contains: 2% CTAB (soluble at approximately 50°C), 1.4 m NaCl, 20 mm EDTA, 100 mm Tris–HCl of pH 8. Small rigid plastic bags containing the sample and extraction buffer (5 ml for 1 g of phloem) are triturated using a ball mill. Transfer 1–2 ml extract to 2-ml microtubes. Incubate microtubes, if possible with shaking, for at least 30 min at 65°C, then centrifuge at 2,000 × *g* for 2 min in a microcentrifuge. Add to 1 ml of supernatant in a microtube 1 ml of chloroform-octanol solution (24:1). Mix the two phases to obtain an emulsion. Centrifuge at 13,000 × *g* for 5 min. Transfer the supernatant to a new microtube, and add approximately 800 µl of cold isopropanol. Mix and centrifuge at 15,000 × *g* for 10 min. Remove the supernatant. Add 500 µl of 70% ethanol and centrifuge at 15,000 × *g* for 5 min. Empty microtubes and dry residues. Add 100 µl of distilled water and shake with a vortex to help dissolution.

To increase the yield from low titre samples DNA can be isolated from 1.5 g of leaf petiole and midrib tissues following the modified phytoplasma enrichment procedure *(48)*. The extraction buffer contains (per 100 ml): K_2HPO_4 anhydrous, 1.67 g; KH_2PO_4, 0.41 g; sucrose, 10 g; bovine serum albumen, 0.15 g; PVP, 10.000 2 g; ascorbic acid, 0.53 g, adjusted to pH 7.6 with KOH. Grind 1.5 g of fresh midveins in a sterile cold mortar and pestle with 7–8 ml of freshly prepared buffer and 50 mg of sterile quartz sand (Sigma, S9887). Incubate 10–15 min on ice. Add another 5 ml buffer and homogenize thoroughly. Transfer into 15-ml tubes (Corex) and centrifuge in a precooled rotor (Beckman JA 20) at 5,000 rpm for 5 min (4°C). Transfer the supernatant to a cold clean Corex tube. Centrifuge at 19,000 rpm for 20 min. Dry the pellet and re-suspend with 2 ml of 2% CTAB buffer, prewarmed at 60°C. Incubate 10–20 min at 60°C with gentle agitation. Transfer 1 ml to a clean 2-ml microtube and proceed with the DNA extraction procedure.

DNA extraction after Kobayashi et al. *(49)*, or by a chloroform/phenol procedure *(50, 51)* is equally successful for symptomatic field samples of fruit trees *(52)* (*see* **Note 4**).

Alternatively, commercial kits (e.g. DNeasy, Qiagen, Extract-N-AmpTM Plant PCR kit (XNAP) (Sigma)) performed according to the manufacturer's instructions can be used for DNA extraction. The extracts can be stored at –80°C for 1 year.

21.3.5.2. DNA Extraction from Insect Vectors

Total DNA from potential vectors is extracted following the method of Zhang et al. *(53)* with minor modifications *(23)*: 800 µl CTAB (cetyltrimethylammonium bromide) buffer (2% CTAB,

1.4 M NaCl, 20 mM EDTA, 100 mM Tris–HCl, pH 8.0, 0.2% mercaptoethanol) at 60°C is added to the crushed tissues thoroughly mixed by vortexing in 1.5-ml microfuge tubes. Keep the mixture at 60°C for 20 min, and briefly vortex several times. After incubation, add 600 µl of chloroform: octanol (24:1), vortex vigorously, and centrifuge at maximum speed for 5 min in an Eppendorf Centrifuge. Transfer the supernatant to a clean microfuge tube, add an equal volume of ice-cold isopropanol, and place the mixture on ice for 30 min. Centrifuge at maximum speed for 8 min and discard the aqueous phase. Rinse the pellet with 80% ethanol, air-dry, and re-suspend in 50–100 µl of water. Proceed within PCR.

Alternatively nucleic acids can be extracted from leafhopper pools of about ten individuals following the protocol of Doyle and Doyle *(54)*.

21.3.5.3. Dot and Southern Blot Analysis

Molecular probes, phytoplasma-specific cloned DNA for dot and Southern hybridization, e.g. IH184, IH196 *(55)* and AT17, AT67, and AT27 *(38)*, showed an insufficient sensitivity for many phytoplasmas associate with woody species, where titres are relatively low *(1, 6,44)*. The two stone-fruit-infecting phytoplasmas, WX and ESFY, can be distinguished by Southern blot analysis using cloned chromosomal fragments of the AP and the WX agents as probes *(9, 57)*.

21.3.5.4. Polymerase Chain Reaction

This is the most versatile tool for detecting phytoplasmas in their plant and insect hosts *(58)*. PCR primers have been developed from randomly cloned fragments of the phytoplasma genome *(59, 60)*, as well as from phytoplasma-specific sequences within the 16S rRNA gene *(9, 47)*. A problem with PCR detection is the presence of putative inhibitors in phytoplasma-infected plant material *(61, 62)*. Some of the available primers exhibit some homology to chloroplast and plastid DNA, or even DNA of bacterial contaminations *(5, 52,63)*. Nested PCR greatly increase sensitivity and specificity of the detection from woody hosts, making it very suitable for routine detection, allowing detection in symptomless fruit tree and insect material. Alternatively PCR amplification and RFLP analysis of the amplified product, e.g. 16S rRNA gene, allows the rapid identification of phytoplasma isolates *(60, 64)*. To detect even cryptic or multiple phytoplasma infections, nested PCRs and RFLP analyses are required.

In some cases, however, the 16S rRNA sequences of related phytoplasmas are very similar, thus making it difficult or impossible to design PCR primers that could specifically identify a particular phytoplasma; however, the intergenic spacer region (IS) located between the 16S and 23S rRNA genes generally shows greater variation *(58)*. For example primers in the ribosomal sequences show cross-reactivity to related phytoplasmas of the AP group.

RFLP analysis of PCR-amplified rDNA was employed to comprehensively classify the phytoplasmas on a molecular basis for the first time, using frequently cutting restriction endonucleases, such as *Alu*I, *Rsa*I, and others *(60, 64)*. At present, RFLP analysis of PCR-amplified rDNA is the method of choice for routine differentiation and classification of phytoplasmas. As the information increases, differentiation can also be improved by increasing the number of restriction enzymes. However, from the data available, it appears that classification based on RFLP analysis of the 16S rRNA gene alone does not reflect the full range of phenotypic diversity. More variable are nucleic acid sequences of ribosomal protein genes, which have enabled identification of more RFLP groups than 16S rDNA data. The combined RFLP analysis of both sequences resulted in the most detailed differentiation of the AY and X-disease group phytoplasmas so far *(5)*. Increasing knowledge of molecular components and information about more isolates and sequences allowed the development of group-, species-, and even strain-specific primers. Ribosomal protein genes (*rpl22* and *rps3*) are more variable markers useful for differentiating phytoplasma strains below the genus level. However, non-ribosomal primers are too specific to detect all strains of AP *(40)*.

Alternatively, an IC-PCR method that uses the primers for AP detection and the available AP antiserum (kindly provided by Drs. N. Loi and L. Carraro) was found very sensitive and suitable for large-scale testing of apple material in vivo and in vitro *(52)*.

The choice of the primers in combination with appropriate enzymes for RFLP analysis will depend on the orientation of the research or the diagnostic purpose.

Universal Primers

For an initial overview it might be advisable to start with universal primers (**Table 21.1**). The two universal primer pairs, *R16mF2/R16mR1* and *R16F2n/R16R2 (1)*, are based on 16S rDNA gene sequences. Universal phytoplasma detection with primers *P1/Tint (58)* amplifies a product from all phytoplasma-infected plants in a portion of the tRNAIle region within the spacer region between the 16S and 23S rRNA SR. It is considerably shorter than the 16S rRNA gene (220–250 bp vs. *ca.* 1,530 bp) and less conserved than the 16S rDNA *(59, 64)*.

Sensitivity of primers might be low for fruit trees or they might amplify non-phytoplasma sequences. Further direct PCR may fail detection in 35 cycles. Combining the universal primer R16F2/R2 *(60)* designed on the basis of unique sequences of phytoplasmas of 16S rRNA sequences amplified a 1.4-kb fragment in a nested PCR with a second universal primer R16F2n/R2 allows to amplify a fragment from all known groups of phytoplasmas from woody field samples and insect vectors *(5)*. A nested polymerase chain reaction (PCR) using primer pair P1/P7 *(64,*

Table 21.1.
Most frequently used universal primer combinations for the detection of temperate fruit tree phytoplasmas

Primer	Location	Oligonucleotide sequence	Expected product (bp)	Reference
P1/P7	16S/23S	5′-AAGAGTTTGATCCTGGCTCAGGATT-3′	1,784	*(64,65)*
		5′-CGTCCTTCATCGGCTCTT-3′		
fU5/rU3	16S	5′-CGGCAATGGAGGAAACT-3′	876	*(9)*
		5′-TTCAGCTACTCTTTGTAACA-3′		
P1/Tint	16S/IS	5′-AAGAGTTTGATCCTGGCTCAGGATT-3′	1,627	*(58)*
		5′-TCAGGCGTGTGCTCTAACCAGC-3′		
R16F2n/R16R2	16S/IS	5′-GAAACGACTGCTAAGACTGG-3′	1,416	*(5)*
		5′-TGACGGGCGGTGTGTACAAACCCCG-3′		
R16mF2/R16mR1	16S/IS	5′-CATGCAAGTCGAACGA-3′	1,239	*(1)*
		5′-CTTAACCCCAATCATCGAC-3′		
P1/R16(X)F1R	16S	5′-AAGAGTTTGATCCTGGCTCAGGATT-3′	217	*(72)*
		5′-CATCTCTCAGCTACTTGCGGGTC-3′		
M1/M2	16S	5′-GTCTTTACTGACGC-3′	509	*(72)*
		5′-CTTCAGCTACCCTTTGTAAC-3′		
PA2F/PA2R	16S/IS	5′-GCC CCG GCT AAC TAT GTG C-3′	1,187	*(52)*
		5′-TTG GTG GGC CTA AAT GGA CTC-3′		
NPA2F/NPA2R	16S/IS	5′-ATG ACC TGG GCT ACA AAC GTG A-3′	485	*(52)*

		5′-GGT GGG CCT AAA TGG ACT CG-3′		
Pc399/P1694	16S/IS	5′-AACGCCGCGTGAACGATGAA-3′	1,200	*(63)*
		5′-ATCAGGCGTGTGCTCTAACC-3′		
rpF1/rpR1	S10-operon	5′-GACATAAGTTAGGTGAATTT-3′		*(68)*
		5′-ACGATATTTAGTTCTTTTTGG-3′		

65) in the first amplification followed by R16F2n/R16R2 in the second amplification allows to detect phytoplasmas in infected plants and leafhoppers *(1, 5)*.

The primers *R16F2n/R16R2 (1)* amplify a 1,239-bp fragment of 16S rDNA (Table 21.1). The 40-µl reaction mixture is composed as follows: 0.4 µm of each primer, 200 µm dNTPs, 1 unit of Taq DNA polymerase, and 5 µl DNA extract in the reaction buffer supplied by the manufacturer of the Taq DNA polymerase. PCR parameters for a DNA Thermal Cycler 480 (Applied Biosystems) are: 2 min at 94°C, 35 cycles of 1 min at 94°C, 2 min at 60°C, and 3 min at 72°C, followed by a final extension for 10 min at 72°C and cooling to 4°C. After amplification, 5–10 µl of the PCR products is subjected to electrophoresis on 1% agarose gel under stable 90 V, in TBE buffer according to standard procedures *(67)* along with a DNA size marker. PCR products are analysed by digestion with restriction enzyme *Alu*I. Each 20-µl reaction mixture is composed as follows: 2 units of restriction enzyme and 10 µl PCR product in the reaction buffer. Reactions are incubated at 37°C for at least 2 h. Digested PCR products are subjected to electrophoresis on 2% agarose gel along with a DNA size marker. The phytoplasma is identified as a member of the AP or 16SrX group if the PCR product is digested with *Alu*I to give 476-, 229-, 189-, 150-, 139-, and 56-bp fragments.

For PCR with universal primers *fU5/rU3 (9)* the 40-µl reaction mixture is composed as follows: 0.5 µm of each primer, 100 µm dNTPs, 0.2 units of Taq DNA polymerase, and 5 µl DNA extract in the reaction buffer. The PCR is performed in 0.2-ml reaction tubes in a thermocycler with the following parameters: 2 min at 94°C, 40 cycles of 20 s at 94°C, 20 s at 55°C, and 1 min at 72°C, followed by a final extension for 4 min at 72°C, and cooling to 4°C. After amplification, 10 µl of the PCR products is subjected to electrophoresis on 1% agarose gel under stable 90 V, in TBE buffer according to standard procedures *(67)* along with a DNA size marker. PCR products are analysed by digestion with restriction enzyme *Alu*I. Each 20-µl reaction mixture is composed as follows: 2 units restriction enzyme and 10 µl PCR product in the reaction buffer. Reactions are incubated at 37°C for at least 2 h. Digested PCR products are subjected to electrophoresis on 2% agarose gel along with a DNA size marker. The phytoplasma is identified as a member of the AP or 16SrX group if the PCR product is digested with *Alu*I to give 476-, 189-, 149-, and 56-bp fragments.

Primers *PA2F/2R (52)* are suitable general primers for phytoplasma detection, operating at high annealing temperatures and, thus, increasing the specificity and decreasing the risk of false positives (Table 21.1). PCR is carried out in a volume of 20 µl as follows: 0.5 µl of dNTP (10 mM each), 0.5 µl of the forward primer and 0.5 µl of the reverse primer (20 pmol/µl), 2.0

μl 10× PCR buffer, 1.4 μl $MgCl_2$ (25 mM), 0.1 μl *Taq* polymerase (5 units/μl), 12 μl sterile water, and 50 ng of DNA are used for direct PCR. PA2F/R PCR products are diluted 1:40 and 1 μl used for nested PCR. The thermal cycling programmes for PA2F/R PCR include 35 cycles of 30 s at 94°C, 1 min 15 s at 60°C and 1 min 30 s at 72°C, with a final elongation step of 10 min at 72°C. For NPA2F/R PCR, the following (shorter) PCR programme is used: 35 cycles of 30 s at 94°C, 30 s at 60°C, and 45 s at 72°C, with a final elongation of 10 min at 72°C.

Subsequent RFLP analysis with *Tai* I distinguishes ESFY from AP/PD; *Tsp509* I digestion reveals a polymorphism between AP/ESFY and PD and *Taq* I differentiates AP from ESFY and PD *(52)*.

Primers *Pc399/P1694* (**Table 21.1**) should avoid false positives due to bacterial background and are particularly useful to screen dormant budwood material for quarantine purposes *(63)*.

Another more variable region is the gene encoding the *elongation factor Tu* (*tuf* gene). However the primers derived from a conserved *tuf* gene region of culturable mollicutes only detect phytoplasmas from the X disease group infecting fruit trees *(67)*.

To increase the differentiation capacity of PCR, the primer pair rpF1/rpR1 was designed by Lim and Sears *(68)* to amplify a segment of the ribosomal protein gene operon (**Table 21.1**).

Group-Specific Primers

Fruit tree phytoplasma-specific primers were developed to detect phytoplasmas in the AP group (Table 21.2). The primer pair P1/PYLRint amplifies rDNA from all phytoplasmas in the apple proliferation group. The most specific of these primer pairs is fPD/rPDS, detecting only German pear decline, California pear decline, and peach yellow leaf roll. The primer pair generated from the European stone fruit yellows sequence (fAT/rPRUS) amplifies a product of the expected size from all phytoplasmas in the apple proliferation group except the apple proliferation phytoplasma. Conversely, the primer pair generated from the apple proliferation sequence (fAT/rAS) detects all tested apple proliferation group isolates except European stone fruit yellows. Isolates of WX disease, walnut witches' broom, and vaccinium witches' broom yield a product of the expected size only when amplified with primer pair P1/WXint (**Table 21.2**).

AP- or *16SrX-group-specific* primers *fO1* and *rO1 (9)* correspond to nucleotides 65–91 and 1135–1115 of the 16S rDNA. The 40-μl reaction mixture is composed as follows: 0.5 μm of each primer, 100 μm dNTPs, 0.2 units of Taq DNA polymerase, 5 μl DNA extract in the reaction buffer supplied by the manufacturer of the Taq DNA polymerase. The PCR is performed in 0.2-ml reaction tubes in a thermocycler with the following parameters: 2 min at 94°C, 40 cycles of 20 s at 95°C, 20 s at 55°C, and 1 min at 72°C, followed by a final extension for 4 min at 72°C

Table 21.2.
Most frequently used species-specific primer combinations for the detection of temperate fruit tree phytoplasmas

Phytoplasma	Primer	Location	Oligonucleotide sequence	Expected product (bp)	Reference
AP	fAT/rAS	16S/IS	5′-CATCATTTAGTTGGGCACTT-3′	500	*(58)*
			5′-GGCCCCGGACCATTATTTATT-3′		
	P1/PYRLint	16S/IS	5′-AAGAGTTTGATCCTGGCTCAGGATT-3′	1,550	*(58)*
			5′-CCCGGCCATTATTAATTTTTATC-3′		
	fO1/rO1	16S	5′-CGGAAACTTTAGTTTCAGT-3′	1,071	*(9)*
			5′-AAGTGCCCAACTAAATGAT-3′		
	qAP16SF/qAP16SR	16S	5′-CGAGGTGAGTAACACGTAA-3′	75	*(70)*
			5′-CCATTAGCAGTCGTTTCC-3′		
	AP5/AP4	NRL-protein gene	5′-TCTTTTAATCTTCAACCATGGC-3′	483	*(59)*
			5′-CCAATGTGTGAAATCTGTAG-3′		
	AP3/AP4	NRL-protein gene	5′-GAAACATGTCCTATTGGTGG-3′	162	*(59)*
			5′-CCAATGTGTGAAATCTGTAG-3′		
	RpAP15f/rpAP15r	S10-operon	5′-AGTGCTGAAGCTAATTTGG-3′	920	*(71, 72)*
			5′-TGCTTTTTATAGCAAAAGGTT-3′		
	RpAP15f2/rp(I)R1A	S10-operon	5′-CTCCTAAATCAGCTTCAAGT-3′	1,036	*(71)*
			5′-TTCTTTTTGGCATTAACAT-3′		
	rp(I)F1A/rp(I)R1A	S10-operon	5′-TTTTCCCCTACACGTACTTA-3′	1,200	*(71)*

			5′-GTTCTTTTTGGCATTAACAT-3′		
ESFY	fAT/rPRUS	16S/IS	5′-CATCATTTAGTTGGGCACTT-3′	500	*(58)*
			5′-GGCCCAAGCCATTATTGATT-3′		
	P1/PYRLint	16S/IS	5′-AAGAGTTTGATCCTGGCTCAGGATT-3′	1,550	*(58)*
			5′-CCCGGCCATTATTAATTTTTATC-3′		
	PA2F/PA2R	16S/IS	5′-GCC CCG GCT AAC TAT GTG C-3′	1,187	*(52)*
			5′-TTG GTGGGC CTA AAT GGA CTC-3′		
	NPA2F/NPA2R	16S/IS	5′-ATGACC TGG GCT ACA AAC GTG A-3′	485	*(52)*
			5′-GGT GGG CCT AAA TGGACT CG-3′		
	fO1/rO1	16S	5′-CGG AAACTTTTAGTTTCAGT-3′	1,071	*(9)*
			5′-AAG TGCCCAACTAAATGA T-3′		
PD	fPD/rPDS	16S/IS	5′-GACCCGTAAGGTATGCTGA-3′	1,400	*(58)*
			5′-CCCGGCCATTATTAATTTTA-3′		
	fAT/rPRUS	16S/IS	5′-CATCATTTAGTTGGGCACTT-3′	500	*(58)*
			5′-GGCCCAAGCCATTATTGATT-3′		
	fAT/rAS	16S/IS	5′-CATCATTTAGTTGGGCACTT-3′	500	*(58)*
			5′-GGCCCCGGACCATTATTTATT-3′		
	P1/PYRLint	16S/IS	5′-AAGAGTTTGATCCTGGCTCAGGATT-3′	1,550	*(58)*
			5′-CCCGGCCATTATTAATTTTTATC-3′		
	fO1/rO1	16S	5′-CGG AAACTTTTAGTTTCAGT-3′	1,071	*(9)*
			5′-AAG TGCCCAACTAAATGA T-3′		

(continued)

Table 21.2. (continued)

Phytoplasma	Primer	Location	Oligonucleotide sequence	Expected product (bp)	Reference
PYLR	P1/PYRLint	16S/IS	5′-AAGAGTTTGATCCTGGCTCAGGATT-3′	1,550	*(58)*
			5′-CCCGGCCATTATTAATTTTTATC-3′		
WX	P1/WXint	16S/IS	5′-AAGAGTTTGATCCTGGCTCAGGATT-3′	1,600	*(58)*
			5′-GACAGTGCTTATAACTTTTA-3′		
	fU2W/rWX	16S	5′-ATAATGGAGGTCATCAG-3′	430	*(73)*
			5′-CGAAGTTAGGTGACCGCTTTG-3′		
	fU3/Tmod	16S/IS	5′-CTGTTACAAAGRGTAGCT-3′	420–422	*(74)*
			5′-ATCAGGCGTGTGCTCTA-3′		

and cooling to 4°C. After amplification, 10 μl of the PCR products is subjected to electrophoresis on a 1% agarose gel under stable 90 V, in TBE buffer according to standard procedures *(68)*. Restriction enzymes *Ssp*I and *Sfe*I are used in separate reactions for RFLP analysis (*see* **Note 5**). The phytoplasma is identified as *AP* if the PCR product is digested with *Ssp*I at position 419, and with *Sfe*I at position 998.

Species- and Strain-Specific Primers

AP-specific PCR primers are *AP5/AP4 (58)* amplifying a 483-bp fragment of a nitroreductase-like protein gene. The 40-μl reaction mixture is composed as follows: 0.5 μm of each primer, 125 μm dNTPs, 0.5 units Taq DNA polymerase (e.g. Replitherm polymerase, Epicentre), 5 μl DNA in the reaction buffer supplied by the manufacturer of the Taq DNA polymerase. The PCR is performed in thin-walled 0.2-ml reaction tubes in a thermocycler (e.g. GeneAmp PCR system 9600 (Applied Biosystems) with the following parameters: 1 min at 95°C, 40 cycles of 10 s at 95°C, 15 s at 58°C, and 45 s at 72°C, followed by a final extension for 4 min at 72°C and cooling to 4°C. After amplification, 10 μl of the PCR products is subjected to electrophoresis on 1% agarose gel under stable 90 V, in TBE buffer according to standard procedures *(67)* along with a DNA ladder (e.g. 100-bp ladder, Fermentas). PCR products are viewed and photographed under UV light. If primers AP5/4 are used, then RFLP analysis is not required. However, with these specific primers, the test has a reduced sensitivity and some isolates may not be detected.

The PCR/RFLP method based on rpl22 and rps3 genes of S10 ribosomal protein (rp) operon *(4)* involves a first PCR with primer pair rpAP15f/rpAP15r (Table 21.2) in direct-PCR and nested-PCR following the amplification with primer pair rpAP15f2/rp(I)R1A *(60)*. For PCR amplification, 38 cycles were conducted in an automated thermal cycler (MJ Research DNA Thermal Cycler PTC-200) with Ampli*Taq* Gold polymerase. PCR was performed in mixtures containing 1 μl of undiluted DNA preparation, 200 μM each dNTP, and 0.4 μM each primer. The following conditions were used: denaturation at 94°C for 1 min (11 min for the first cycle), annealing for 2 min at 55°C, and primer extension for 3 min (7 min in the final cycle) at 72°C. Diluted (1:30) PCR product (1 μl) from the first amplification was used as the template in the nested PCR. The PCR products (5 μl) were electrophoresed through a 1% agarose gel, stained in ethidium bromide, and visualized with a UV trans-illuminator (*see* **Note 6**).

PCR products are digested with *Mse*I and *Tsp509*I. The restriction products then were separated by electrophoresis through a 5% (12% for rp products) polyacrylamide gel (*see* **Note 7**) for 1 h at 150 V (2 h for 12% gel), stained in ethidium bromide, and visualized with a UV trans-illuminator.

21.3.6. Real-Time PCR

RT-PCR was developed to detect and quantify the agent of AP. Depending on the primers selected it is possible to detect specifically either AP alone or the entire 16 S group (**Tables 21.1** and 21.2 *(39, 70, 75)*). RT-PCR allows the detection of AP in single vector insects, thus allowing to study the infectivity of insects and to evaluate the level of plant resistance to AP in a breeding programme *(39)*. Although it is less laborious than nested PCR/RFLP. RT-PCR is equally sensitive and allows to estimate the concentration of phytoplasmas in infected tissues or insects. With group-specific primers it is possible to detect all quarantine phytoplasmas of European fruit trees *(75)*. However, RT-PCR does not provide a means for classifying the detected phytoplasmas *(76)* and currently is not affordable for routine diagnostics and is therefore considered mainly for fundamental research purposes (*see* **Note 8**). Recently, the development of an oligonucleotide microarray-based assay was shown to have the potential both to detect and to identify a number of phytoplasmas from infected host plants, including AP *(77)*.

21.4. Notes

1. Work with infected plant material requires caution to avoid cross-contamination in the greenhouse.
2. Caution: care is needed in handling DAPI, as it is carcinogenic. Use gloves while handling your samples.
3. Leaf samples should be collected randomly all over the plant, to overcome limitations due to the uneven distribution of the pathogen in the canopy. From July until leaf fall ELISA is as sensitive as PCR, but it is more rapid and convenient than PCR (32, indications by Bioreba). During this period, the serological technique can substitute for PCR, especially for large-scale diagnosis such as health selection programmes when the number of samples for examination is high.
4. Phenol and chloroform are harmful. Wear suitable protective clothing and gloves, and work in a fume food, when manipulating phenol. Consider also appropriate waste disposal. Review a phenol Material Safety Data Sheet before handling the material.
5. Digestion should be carried out according to the manufacturers' guidelines, using appropriate restriction buffers, which are usually supplied by companies.
6. Caution: care should be taken in handling ethidium bromide, as it is carcinogenic. Wear chemical protective gloves such as

nitrile gloves, lab coats, and closed-toed shoes. UV light is harmful to skin and eyes. Wear glasses blocking UV light or better a face shield. Consider also appropriate waste disposal.

7. Caution: care should be taken against dust while preparing polyacrylamide gels.
8. A triple repetition of the assay however is recommended.

Acknowledgements

Special thanks to Dr. M. Martini for providing information at a preview stage, and to Drs. N. Loi and L. Carraro for providing the AP antiserum, Prof. A. Bertaccini, and Dr. W. Jarausch for valuable discussions.

References

1. Lee, I.-M.,Gundersen-Rindal, D.E., Davis, R.E. and Bartoszyki, I.M. (1998) Revised classification scheme of phytoplasmas based on RFLP analyses of 16S rRNA and ribosomal protein gene sequences. *Int J Syst Bacteriol* 48, 1153–1169.
2. Doi, Y.,Teranaka, M., Yora, K. and Asuyama, H. (1967) Mycoplasma or PLK grouplike microorganisms found in the phloem elements of plants infected with mulberry dwarf, potato witches' broom, aster yellows or paulownia witches' broom. *Ann Phytopathol Soc Jpn* 33, 259–266.
3. Firrao, G.,Andersen, M., Bertaccini, A., Boudon, E., Bove, J.M., Daire, X., Davis, R.E., Fletcher, J., Garnier, M., Gibb, K.S., Gundersen-Rindal, D.E., Harrison, N.A., Hiruki, C., Kirkpatrick, B.C., Jones, P., Kuske, C.R., Lee, I.M., Liefting, L., Marcone, C., Namba, S., Schneider, B., Sears, B.B., Seemüller E., Smart, C.D., Streten, C. and Wang, K. (2004) '*Candidatus* Phytoplasma', a taxon for the wall-less, non-helical prokaryotes that colonize plant phloem and insects. *Int J Syst Evol Microbiol* 54, 1243–1255.
4. Martini, M.,Lee, I.-M., Bottner, K.D., Zhao, Y., Botti, S., Bertaccini, A., Harrison, N., Carraro, L., Marcone, C. and Osler, R. (2007) Ribosomal protein gene-based phylogeny for finer differentiation and classification of phytoplasmas. *Int J Syst Evol Microbiol* 57, 2037–2051.
5. Gundersen, D.E. and Lee, I.-M. (1996) Ultrasensitive detection of phytoplasmas by nested-PCR assays using two universal primer pairs. *Phytopathol Mediterr* 35, 144–151.
6. Seemüller, E.,Marcone, C., Lauer, U., Ragozzino, A. and Göschl, M. (1998) Current status of molecular classification of the phytoplasmas. *J Plant Pathol* 80, 3–26.
7. Neimark, H. and Kirkpatrick, B.C. (1993) Isolation and characterization of full-length chromosomes from non-culturable plant-pathogenic mycoplasma-like organisms. *Mol Microbiol* 7, 21–28.
8. Jarausch, W.,Saillard, C., Broquaire, J., Garnier, M. and Dosba, F. (2000) PCR-RFLP and sequence analysis of a non-ribosomal fragment for genetic characterization of European stone fruit yellows phytoplasmas infecting various *Prunus* species. *Mol Cell Probes* 14, 171–179.
9. Lorenz, K.-H.,Schneider, B., Ahrens, U. and Seemüller, E. (1995) Detection of the apple proliferation and pear decline phytoplasmas by PCR amplification of ribosomal and nonribosomal DNA. *Phytopathology* 85, 771–776.
10. Lee, I.-M.,Bertaccini, A., Vibio, M. and Gundersen, D.E. (1995) Detection of multiple phytoplasmas in perennial fruit trees with decline symptoms in Italy. *Phytopathology* 85, 728–735.
11. Kube M., Schneider B., Kuhl H., Dandekar T, Heitmann K, Migdoll A M, Reinhardt R. and Seemüller E.2008. The linear chromosome of the plant-pathogenic mycoplasma 'Candidatus Phytoplasma mali'. BMC Genomics. 9: 306

12. Kison, H.,Kirkpatrick, B.C. and Seemüller, E. (1997) Genetic comparison of the peach yellows leaf roll agent with European fruit tree phytoplasmas of the apple proliferation group. *Plant Pathol* 46, 1–7.
13. Marcone, C.,Neimark, H., Ragozzino, A., Lauer, U. and Seemüller, E. (1999) Chromosome sizes of phytoplasmas composing major phylogenetic groups and subgroups. *Phytopathology* 89, 805–810.
14. Firrao, G.,Smart, C. D. and Kirkpatrick, B.C. (1996) Physical map of the western X-disease phytoplasma chromosome. *J Bacteriol* 178, 3985–3988.
15. Lauer, U. and Seemüller, E. (2000) Physical map of the chromosome of the apple proliferation phytoplasma. *J Bacteriol* 182, 1415–1418.
16. Marcone, C. and Seemüller, E. (2001) A chromosome map of the European stone fruit yellows phytoplasma. *Microbiology* 147, 1213–1221.
17. Seemüller, E. and Schneider, B. (2004) '*Candidatus Phytoplasma mali*', '*Candidatus Phytoplasma pyri*' and '*Candidatus Phytoplasma prunorum*', the causal agents of apple proliferation, pear decline and European stone fruit yellows, respectively. *Int J Syst Evol Microbiol* 54, 1217–1226.
18. Frisinghelli, C., Delaiti, M., Grando, S., Forti, D. and Vindimian, M.E. (2000) *Cacopsylla costalis* (Flor 1861) as a Vector of Apple Proliferation in Trentino. *J Phytopathol* 148, 314–320.
19. Carraro, L.,Loi, N., Ermacora, P. and Osler, R. (1998) High tolerance of European plum varieties to plum leptonecrosis. *Eur J Plant Pathol* 104, 141–145.
20. Carraro, L.,Osler, R., Loi, N., Ermacora, P. and Refatti, E. (2001) Fruit tree phytoplasma diseases diffused in nature by psyllids. *Acta Hortic* 550, 345–350.
21. Jarausch, W., Jarausch-Wehrheim, B., Danet, J.L., Broquaire, J.M., Dosba, F., Saillard, C. and Garnier, M. (2001) Detection and identification of European stone fruit yellows and other phytoplasmas in wild plants in the surroundings of apricot chlorotic leaf roll-affected orchards in southern France. *Eur J Plant Pathol* 107, 209–217.
22. Davies, D.L. and Clark, M.F. (1992) Production and characterization of polyclonal and monoclonal antibodies against peach yellow leaf roll MLO associated antigens. *Acta Hortic* 309, 275–283.
23. Avinent, L., Llácer, G., Almacellas, J. and Torá, R. (1997) Pear decline in Spain. *Plant Pathol* 46, 694–698.
24. Blomquist, C.L. and Kirkpatrick, B.C. (2002) Identification of phytoplasma taxa and insect vectors of peach yellow leaf roll disease in California. *Plant Dis* 86, 759–763.
25. Jarausch, W.,Saillard, C. and Dosba, F. (1996) Longterm maintenance of nonculturable apple-proliferation phytoplasmas in their micropropagated natural host plant. *Plant Pathol* 45, 778–786.
26. Bertaccini, A., Carraro, L., Davies, D., Laimer da Câmara Machado, M., Martini, M., Paltrinieri, S. and Seemüller, E. (2000) Micropropagation of a collection of phytoplasma strains in periwinkle and other host plants. In: Proceedings of the XIII Congress of IOM, July 14–19, p. 101, ACROS Fukuoka.
27. Laimer, M. (2003) Detection and elimination of viruses and phytoplasmas from pome and stone fruit trees. *Hortic Rev* 28, 187–236.
28. Marwitz, R. (1990) Diversity of yellows disease agents in plant infections. *Zentralblatt Bakteriologie* Suppl. 20, 431–434.
29. Kirkpatrick, B.C. (1992) Mycoplasma-like organisms: plant and invertebrate pathogens. In: Balows A, Truper HG, Dworkin M, Harder W and Schleifer KH(eds), The Prokaryotes, second edition, pp 4050–4067, Springer, New York.
30. Petzold, H., Marwitz, R. and Kunze, L. (1973) Elektronenmikroskopische Untersuchungen über intrazelluläre rickettsienähnliche Bakterien in triebsuchtkranken Apfelbäumen. *Phytopathol Z* 78, 170–181.
31. Musetti, R. and Favali, M.A. (2004) Microscopy techniques applied to the study of phytoplasma diseases: traditional and innovative methods. *Current Issues on Multidisciplinary Microscopy Research and Education*, pp 72–80.
32. Seemüller, E. (1976) Investigations to demonstrate mycoplasmalike organisms in diseased plants by fluorescence microscopy. *Acta Hortic* 67, 109–111.
33. Loi, N., Ermacora, P., Carraro, L., Osler, R. and Chen, T.A. (2002) Production of monoclonal antibodies against apple proliferation phytoplasma and their use in serological detection. *Eur J Plant Pathol* 108, 81–86.
34. European and Mediterranean Plant Protection Organisation PM 7/62 (1) (2006) *Candidatus Phytoplasma mali*. *EPPO Bull*36, 121–125.
35. Blomquist, C.L.,Barbara, D.J., Davies, D.L., Clark, M.F. and Kirkpatrick, B.C. (2001) An immunodominant membrane protein gene from the Western X-disease phytoplasma is distinct from those of other phytoplasmas. *Microbiology* 147, 571–580.
36. Jiang, Y.P.,Chen, T.A. and Chiykowski, L.N. (1987) Production and characterization of

monoclonal antibodies against the peach eastern X-disease agent. *Can J Plant Pathol* 11, 325–331.

37. Berg, M., Davis, D.L., Clark, M.F., Vetten, H.J., Maier, G., Marcone, C. and Seemüller, E. (1999) Isolation of the gene encoding an immunodominant membrane protein of the apple proliferation phytoplasma, and expression and characterization of the gene product. *Microbiology* 145, 1937–1943.
38. Schneider, B. and Seemüller, E. (1994) Presence of two sets of ribosomal genes in phytopathogenic mollicutes. *Appl Environ Microbiol* 141, 173–185.
39. Jarausch, W.,Peccerella, T., Schwind, N., Jarausch, B. and Krczal, G. (2004) Establishment of a quantitative real-time PCR assay for the quantification of apple proliferation phytoplasmas in plants and insects. *Acta Hortic* 657, 415–420.
40. Martini, M., Ermacora, P., Falginella, L., Loi, N. and Carraio, L. (2007) Molecular differentiation of "*Candidatus Phytoplasma mali*" and its spreading in Friuli Venezia Giulia Region (North-East Italy). *Acta Hortic* 781: 395–402.
41. Barbara, D.J.,Morton, A., Clark, M.F. and Davies, D.L. (2002) Immunodominant membrane proteins from two phytoplasmas in the aster yellows clade (chlorante aster yellows and clover phyllody) are highly divergent in the major hydrophilic region. *Microbiology* 148, 157–167.
42. Morton, A.,Davies, D.L., Blomquist, C.L. and Barbara, D.J. (2003) Characterization of homologues of the apple proliferation immunodominant membrane protein gene from three related phytoplasmas. *Mol Plant Pathol* 4, 109–114.
43. Wei, W.,Kakizawa, S., Jung, H.-Y., Suzuki, S., Tanaka, M., Nishigawa, H., Miyata, S.-I., Oshima, K., Ugaki, M., Hibi, T. and Namba, S. (2004) An anti Sec A membrane protein Ab with multiple strain/phylum specificity of one phytoplasma reacts with those of phylogenetically different phytoplasmas. *Phytopathology* 94, 683–686.
44. Kirkpatrick, B.C.,Stenger, D.C., Morris, T.J. and Purcell, A.H. (1987) Cloning and detection of DNA from a non-cultivable plant pathogenic mycoplasmalike organism. *Science* 238, 197–200.
45. Lee, I.-M.,Hammond, R.W., Davis, R.E. and Gundersen, D.E. (1993) Universal amplification and analysis of pathogen 16S rDNA for classification and identification of mycoplasmalike organisms. *Phytopathology* 83, 834–842.
46. Tran-Nguyen, L.T.T. and Gibb, K.S. (2007) Optimizing phytoplasma DNA purification for genome analysis. *J Biomol Tech* 18, 104–112.
47. Ahrens, U. and Seemüller, E. (1992) Detection of DNA of plant pathogenic mycoplasmalike organisms by a polymerase chain reaction that amplifies a sequence of the 16S RNA gene. *Phytopathology* 82, 828–832.
48. Malisano, G.,Firrao, G. and Locci, R. (1996) 16S rDNA-derived oligonucleotide probes for the differential diagnosis of plum leptonecrosis and apple proliferation phytoplasmas. *EPPO Bull* 26, 421–428.
49. Kobayashi, N.,Horikoshi, T., Katsumaya, H., Handa, T. and Takayanagi, K. (1998) A simple and efficient DNA extraction method for plants, especially woody plants. *Plant Tissue Cult Biotechnol* 4, 76–80.
50. Prince, J.P.,Davis, R.E., Wolf, T.K., Lee, I.-M., Mogen, B.D., Dally, E.L., Bertaccini, A., Credi, R. and Barba, M. (1993) Molecular detection of diverse mycoplasmalike organisms (MLOs) associated with grapevine yellows and their classification with aster yellows, X-disease, and elm yellows MLOs. *Phytopathology* 83, 1130–1137.
51. Bertheau, Y.,Frechon, D., Toth, I. and Hyman, L.J. (1998) DNA amplification by polymerase chain reaction (PCR) In: Perombelon MCM and van der Wolf JM (eds), Methods for the Detection and Quantification of *Erwinia carotovora* subsp. *atroseptica* on Potatoes, pp 39–59, Occasional Publication no. 10, Scottish Crop Research Institute, Dundee.
52. Heinrich, M.,Botti, S., Caprara, L., Arthofer, W., Strommer, S., Hanzer, V., Katinger, H., Bertaccini, A. and Laimer da Câmara Machado, M. (2001) Improved detection methods for fruit tree phytoplasmas. *Plant Mol Biol Rep* 19, 169–179.
53. Zhang, Y.-P.,Uyemoto, J.K. and Kirkpatrick, B.C. (1998) A small-scale procedure for extracting nucleic acids from woody plants infected with various phytopathogens for PCR assay *J Virol Methods* 71, 45–50.
54. Doyle, J.J. and Doyle, J.L. (1990) Isolation of plant DNA from fresh tissue. *Focus* 12, 13–15.
55. Bonnet, F.,Saillard, C., Kollar, A., Seemüller, E. and Bové, J.M. (1990) Detection and differentiation of the mycoplasmalike organism associated with apple proliferation disease using cloned DNA probes. *Mol Plant Microbe Interac* 3, 438–443.
56. Lee, I.-M.,Gundersen, D.E., Hammond, R.W. and Davis, R.E. (1994) Use of mycoplasmalike organism (MLO) group-specific oligonucleotide primers for nested-PCR assays

.ed-MLO infections in a single *Phytopathology* 84, 559–566.

U.,Lorenz, K.H. and Seemüller, 993) Genetic diversity among mycoma like organisms associated with stone uit disease. *Mol Plant Microbe Interac* 6, 686–691.

58. Smart, C.,Schneider, B., Blomquist, C., Guerra, L., Harrison, N., Ahrens, U., Lorenz, K., Seemüller, E. and Kirkpatrick, B. (1996) Phytoplasma-specific PCR primers based on sequences of the 16S–23S rRNA spacer region. *Appl Environ Microbiol* 62, 2988–2993.
59. Jarausch, W.,Saillard, C., Dosba, F. and Bové, J.M. (1994) Differentiation of mycoplasmalike organisms (MLOs) in European fruit trees by PCR using specific primers derived from the sequence of a chromosomal fragment of the apple proliferation MLO. *Appl Environ Microbiol* 60, 2916–2923.
60. Schaff, D.A., Lee, I.-M. and Davis, R.E. (1992) Sensitive detection and identification of mycoplasmalike organisms by polymerase chain reactions. *Biochem Biophys Res Commun* 186, 1503–1509.
61. Musetti, R.,Favali, M. and Pressacco, L. (2000) Histopathology and polyphenol content in plants infected by phytoplasmas. *Cytobios* 102, 133–147.
62. Lepka, P.,Stitt, M., Moll, E. and Seemüller, E. (1999) Effect of phytoplasmal infection on concentration and translocation of carbohydrates and amino acids in periwinkle and tobacco. *Physiol Mol Plant Pathol* 55, 59–68.
63. Skrzeczkowski, L.J., Howell, W.E. and Eastwell, K.C. (2001) Bacterial sequences interfering in detection of phytoplasma by PCR using primers derived from ribosomal RNA operon. *Acta Hortic* 550, 417–424.
64. Schneider, B., Seemüller, E., Smart, C.D. and Kirkpatrick, B.C. (1995) Phylogenetic classification of plant pathogenic mycoplasma-like organisms or phytoplasmas. In: Razin R and Tully JG (eds), Molecular and Diagnostic Procedures in Mycoplasmology, vol 1, pp 369–380, Academic Press, San Diego.
65. Deng, S. and Hiruki, C. (1991) Amplification of 16S rRNA genes from culturable and non-culturable mollicutes. *J Microbiol Methods* 14, 53–61.
66. Sambrook, J., Fritsch, E. F. and Maniatis, T. (1989) Molecular cloning. A laboratory manual. Cold Spring Harbor, New York.
67. Schneider, B. and Gibb, K.S. (1997) Detection of phytoplasma in decline pears in southern Australia. *Plant Dis* 81, 254–258.
68. Lim, P.-O. and Sears, B.B. (1992) Evolutionary relationships of a plant-pathogenic mycoplasmalike organism and *Acholeplasma laidlawii* deduced from two ribosomal protein gene sequences. *J Bacteriol* 174, 2606–2611.
69. Jarausch, W.,Saillard, C., Dosba, F. and Bové, J.M. (1995) Specific detection of mycoplasmalike organisms in European fruit trees by polymerase chain reaction. *EPPO Bull* 25, 219–225.
70. Baric, S. and Dalla Via, J. (2004) A new approach to apple proliferation detection: a highly sensitive real-time PCR assay. *J Microbiol Methods* 57, 135–145.
71. Lee, I.-M., Martini, M., Bottner, K.D., Dane, R.A., Black, M.C. and Troxclair, N. 2003. Ecological implications from a molecular analysis of phytoplasmas involved in an aster yellows epidemic in various crops in Texas. Phytopathology 93:1368–1377.
72. Martini, M., Lee, I.-M., Zhao, Y., Botti, S., Bertaccini, A., Carraro, L., Marcone, C. and Osler, R. 2004. Ribosomal protein gene-based phylogeny: a basis for phytoplasma classification. 15th Congress of IOM, Athens, Georgia, USA, 11–16 July: 156.
73. Ahrens U., Lorenz K.H., Kison H., Berges R., Schneider B. and Seemüller E., 1994. Universal, cluster-specific, and pathogen-specific PCR amplification of 16S rDNA for detection and identification of mycoplasma-like organisms. (Abstr.) IOM Letters 3:250.
74. Palmano S. and Firrao G. 2000. Diversity of phytoplasmas isolated from insects, determined by a DNA heteroduplex mobility assay and a length polymorphism of the 16S–23S rDNA spacer region analysis. Journal of Applied Microbiology 89: 744–750.
75. Torres, E., Bertolini, E., Cambra, M., Montón, C. and Martín, M.P. (2005) Real-time PCR for simultaneous and quantitative detection of quarantine phytoplasmas from apple proliferation (16SrX) group. *Mol Cell Probes* 19, 334–340.
76. Gibb, K.S.,Padovan, A.C. and Mogen, B.D. (1995) Studies on sweet potato little-leaf phytoplasma detected in sweet potato and other plant species in northern Australia. *Phytopathology* 85, 169–174.
77. Nicolaisen, M. and Bertaccini, A. (2007) An oligonucleotide microarray-based assay for identification of phytoplasma 16S ribosomal groups *Plant Pathol* 56, 332–336.

Chapter 22

PCR Detection of Potato Cyst Nematode

Alex Reid

Summary

Potato cyst nematode (PCN) is responsible for losses in potato production totalling millions of euros every year in the EC. It is important for growers to know which species is present in their land as this determines its subsequent use. The two species *Globodera pallida* and *Globodera rostochiensis* can be differentiated using an allele-specific PCR.

Key words: Potato cyst nematode (PCN), Internal transcribed spacer (ITS), Allele-specific amplification (ASA).

22.1. Introduction

Amongst the numerous plant pathogenic nematodes that are damaging enough to be classified as quarantine pests two *Globodera pallida* (Stone) Behrens and *Globodera rostochiensis* (Wollenweber) Behrens are collectively known as potato cyst nematodes or PCN. Damage is caused to potato plants when the second-stage juvenile nematodes (J2) enter the roots and initiate feeding sites causing a transformation in the surrounding cells and making them enlarge. Feeding from these giant cells lasts for several weeks during which the nematodes moult through third and fourth stages and then, partly depending on their nutritional status, undergo a final moult to either adult males which remain vermiform and leave the roots to find a mate or adult females which remain in the roots. The females become rounded as eggs develop inside them and eventually erupt through the surface of the root. During this early stage of maturation the cysts are either

R. Burns (ed.), *Methods in Molecular Biology, Plant Pathology, vol. 508*

DOI: 10.1007/978-1-59745-062-1_22

a pale white creamy colour, in the case of *G. pallida*; or a golden yellow, in the case of *G. rostochiensis.* Eventually the cysts harden and take on a dark brown colour and the adult female nematodes die. The cysts, which are extremely resilient, protect the eggs until conditions are suitable for hatching. In the presence of host plants (of the family Solanaceae) hatching is promoted by the presence of exudates from growing roots. Only a proportion of the eggs hatch at any given time and in this aspect the two species vary. In the case of *G. pallida* the hatching takes place over a more prolonged period and generally later than for *G. rostochiensis.* A small proportion of eggs, usually 20–30%, hatch each year so the great majority of eggs will hatch within 7–10 years, although cysts frequently remain viable for up to 25–30 years. Many commercially grown potato varieties have had the H_1 resistance gene bred into them; however, this only confers resistance to the Ro1 pathotype of *G. rostochiensis* and gives no resistance to *G. pallida*. These two factors (the extended viability of the cysts and the lack of resistance) make *G. pallida* the more serious of the two species to find in a field. As *G. pallida* also hatches later, granular nematicides applied at planting have also proved less effective in controlling this species.

The damage caused by PCN can be patchy in distribution within a field *(1)* and includes stunting of the root system and production of additional lateral roots. Severe symptoms mimic those of water stress or mineral deficiency with poor top growth with yellowed, withered leaves, although there can be many other reasons for such symptoms. Yield loss can occur at densities of 5 eggs/g of soil *(2)* It is estimated that the action of these two species to the potato growing industry in the UK alone resulted in yield losses worth £43 million, based on the mean value of the crop from 1990 to 1995 *(3)*.

The standard method for the diagnosis of PCN cysts extracted from soil samples is by morphological and morphometric measurements primarily of the J2 stylet and female perineal patterns *(4)*. However, a number of molecular tests are also available including ELISA, IEF, as well as several PCR-based methods *(5–7)*. The advantages of PCR identification methods are that they are relatively quick and enable the identification of cysts as well as individual juvenile nematodes without the knowledge and skill required for morphological identifications. PCR can also be more readily applied to establish the species composition of large numbers of PCN cysts. The primary method in use at the SASA is based on Allele-Specific Amplification (ASA) *(8)* and makes use of the sequence variation within the internal transcribed spacer region 1 (ITS1) of the ribosomal DNA repeat unit. The method uses a single universal primer (anchored within the 5.8S gene) which binds to the DNA of both species along with two species-specific primers which yield differently sized products depending on the species present (391 bp in the case of *G. pallida* and 283 bp for *G. rostochiensis*). By using this technique it is possible to reliably detect mixtures of the two species at the 1 in 10 level.

22.2. Materials

22.2.1. Cyst Extraction

1. Chelex 100 (BioRad). Make up as a 5% (w/v) solution with sterile water and stir for 30 min before use.
2. Worm Lysis Buffer *(9)*: 50 mM KCl, 10 mM Tris–HCl of pH 8.3, 2.5 mM $MgCl_2$, 0.45% NP40, 0.45% Tween 20, 0.01% gelatine, 60 µg/mL Proteinase K, store at –20°C.

22.2.2. PCR Amplification

1. AmpliTaq® DNA polymerase Stoffel Fragment (Applied Biosystems), store at –20°C. (*see* **Note 1**).
2. 10× AmpliTaq® DNA polymerase Stoffel fragment buffer (Applied Biosystems), store at –20°C.
3. GeneAmp® dNTP blend 10 mM (Applied Biosystems), store at –20°C.
4. 25 mM Magnesium chloride, store at –20°C.
5. Universal primer 5′-GCAGTTGGCTAGCGATCTTC-3′ (*see* **Note 2**), store at –20°C.
6. *G. pallida*-specific primer 5′-GGTGACTCGACGATTGCTGT-3′ (*see* **Note 2**), store at –20°C.
7. *G. rostochiensis*-specific primer 5′-TGTTGTACGTGCCGTACCTT-3′ (*see* **Note 2**), store at –20°C.

22.2.3. Gel Electrophoresis

1. Agarose.
2. 5× TBE buffer: 54 g Tris base, 27.5 g boric acid, 20 mL 0.5 M EDTA (pH 8.0) made up to 1 L with distilled water.
3. Ethidium bromide (10 mg/mL).
4. Loading buffer N.B. virtually any loading buffer can be used; however, at SASA we use the loading buffer supplied with the molecular weight marker which is the 100 bp Superladder from ABGene.
5. Molecular weight markers which have bands in the 100–500 bp range (see above).
6. Horizontal gel electrophoresis tank and power pack.
7. uv transilluminator.

22.3. Methods

22.3.1. Extraction Method (Bulked Cysts)

N.B. Aerosol resistant tips are used at each stage of extraction and PCR set up to help prevent cross-contamination.

1. Pool cysts in batches of 10 into sterile 0.5-mL microfuge tubes. Add 20 μL of ice-cold water to the cysts. If not processed immediately store tube at 4°C; otherwise, place on ice.
2. Crush the cysts with a clean, sterile 0.5-mL pellet mixer (Scotlab) for a few seconds using sufficient force to just disrupt the cysts. After the sample has been processed, add a further 480 μL of ice-cold sterile water.
3. Centrifuge at 12,000 g in a microcentrifuge for 3 min. Remove the supernatant and discard. Add 500 μL of a freshly prepared 5% Chelex 100 suspension.
4. Incubate the samples at 100°C for 8 min and place on ice to cool.
5. Centrifuge at 12,000 g in a microcentrifuge for 3 min and place on ice.
6. At this stage the samples can either be processed immediately or stored at –20°C.

22.3.2. Extraction Method (Individual J2)

1. Pipette 10 μL of worm lysis buffer on a 70% ethanol swabbed microscope slide.
2. Place a single nematode in the drop of worm lysis buffer and cut into half with a sterile needle. Transfer to a sterile 0.5-mL centrifuge tube.
3. Freeze at –20°C for 10 min. Incubate at 65°C for 1 h followed by 95°C for 10 min.
4. Cool on ice and centrifuge at 12,000 × *g* for 2 min.
5. Use 5 μL supernatant in PCR. N.B. DNA extracted by this method must be used immediately and cannot be stored.

22.3.3. PCR Amplification

1. Make up a PCR master mix sufficient for the amplification of all samples and controls. Each reaction contains sterile water (14.8 μL for the bulked cyst extraction or 10.8 μL for individual J2s), 2.5 μL 10× AmpliTaq® DNA polymerase Stoffel fragment buffer, 5 μL 25 mM $MgCl_2$, 0.4 μL 10 mM dNTP mix, 1 μL primer mix (8 pmol/μL each primer), and 0.3 μL AmpliTaq® DNA polymerase Stoffel fragment (5 units/μL).
2. Dispense the Master Mix to each PCR plate well required (24 μL for bulked cyst extract or 20 μL for individual J2s).

3. Add 1 μL of cyst extract or 5 μL individual J2 extract to each well. Always include positive controls of previously extracted *G. pallida* and *G. rostochiensis* samples.
4. Place in a thermal cycler and amplify using the following programme. 95°C for 5 min then 35 cycles of 94°C for 30 s, 55°C for 30 s, 72°C for 1 min followed by a hold step of 72°C for 10 min and an infinite hold step of 4°C.
5. After amplification (approximately 2 h) process immediately or store at –20°C.

22.3.4. Gel Electrophoresis

1. Dissolve agarose in 1× TBE buffer to give a 1% (w/v) gel.
2. Add ethidium bromide to a final concentration of 0.5 μg/mL.
3. Pour molten agarose into gel mould following the manufacturer's instructions for the electrophoresis system being used and allow to set.
4. Place the gel into the tank and fill with enough 1× TBE to just cover the gel.
5. Make up buffer/sample mix for each PCR product.
6. N.B. Using the ABGene Superladder 6× loading buffer this equates to 2.5 μL loading buffer, 10 μL sterile water, and 2.5 μL PCR product per reaction.
7. Load gel and run at 5 V/cm for approximately 1 h.
8. Place gel on uv transilluminator and examine. N.B. follow the manufacturer's safety guidelines when using uv transilluminators.
9. Samples contain *G. pallida* will yield a band of 391 bp and samples containing *G. rostochiensis* a band of 283 bp. Samples containing both species will yield both bands.

22.4. Notes

1. Stoffel fragment must be used as it is has greater thermostability than the intact polymerase, is active over a wider range of magnesium concentrations, and lacks the 5′→3′ exonuclease activity of the intact enzyme. All these factors facilitate superior performance for sequence-specific amplification.
2. Make up a primer mix containing all three primers at 8 pmol/μL. Store at –20°C.

References

1. Evans, K., Webster, R., Barker, A., Halford, P., and Russell, M. (2003) Mapping infestations of potato cyst nematodes and the potential for spatially varying applications of nematicides. *Precision Agriculture*, 4, 149–162.
2. Trudgill, D.L. (1986) Yield losses caused by potato cyst nematode: a review of the current position in Britain and prospects for improvement. *Annals of Applied Biology*, 108, 181–198.
3. Pickup, J. (2002) Potato cyst nematodes – a technical overview for Scotland. http://www.scotland.gov.uk/consultations/agriculture/PCN_Technical_Paper_Scotland_SEERAD.pdf.
4. Golden, A.M. (1986) Morphology and identification of cyst nematodes. In *Cyst Nematodes*. Eds. F. Lamberti and C.E. Taylor, Plenum Press, New York. pp 23–45.
5. Bulman, S.R., and Marshall, J.W. (1997) Differentiation of Australasian potato cyst nematode (PCN) populations using the polymerase chain reaction (PCR). *New Zealand Journal of Crop and Horticultural Science*, 25, 123–129.
6. Ibrahim, S.K., Minnis, S.T., Barker, A.D.P., Russell, M.D., Haydock, P.P.J., Evans, K., Grove, I.G., Woods, S.R., and Wilcox, A. (2001) Evaluation of PCR, IEF and ELISA techniques for the detection and identification of potato cyst nematodes from field soil samples in England and Wales. *Pest Management Science*, 57, 1068–1074.
7. Bates, J.A., Taylor, E.J.A., Gans, P.T., and Thomas, J.E. (2002) Determination of relative proportions of *Globodera* species in mixed populations of potato cyst nematodes using PCR product melting peak analysis. *Molecular Plant Pathology*, 3, 153–161.
8. Mulholland, V., Carde, L., O'Donnell, K.J., Fleming, C.C., and Powers, T.O. (1996) Use of the polymerase chain reaction to discriminate potato cyst nematode at the species level. In *BCPC Proceedings No. 65, Diagnostics in crop production*. Eds. G. Marshall. The British Crop Protection Council, Farnham. pp 247–252.
9. Joyce, S.A., Reid, A., Driver, F., and Curren, J. (1994) Application of polymerase chain reaction (PCR) methods to the identification of entomopathogenic nematodes. In *COST 812 Biotechnology: Genetics of Entomopathogenic Nematode–Bacterium Complexes, Proceedings of Symposium & Workshop, St. Patrick's College, Maynooth, Co. Kildare, Ireland*. Eds. A.M. Burnell, R.-U. Ehlers, and J.P. Masson, European Commission, DG XII, Luxembourg. pp 178–187.

Chapter 23

Potato Cultivar Genome Analysis

Alex Reid, Lysbeth Hof, Danny Esselink, and Ben Vosman

Summary

Due to the yearly increase in the numbers of new potato varieties obtaining Plant Breeders' Rights the reliable maintenance of large culture collections of reference varieties for DUS testing is becoming more and more difficult, as accidental mix ups might occur. Efficient identification methods and databases can act as an aid to overcome this problem. Identification of cultivars by morphological characteristics is a highly skilled and time-consuming task, and for these reasons a rapid and robust method for variety differentiation has become extremely desirable. By use of a set of nine microsatellite (SSR) markers we can differentiate over 1,000 cultivars, including the majority of varieties on the European Union Common Catalogue, but excluding somaclonal variants (e.g. Red King Edward and King Edward) and mutants. The whole identification process from DNA extraction to accurate identification can be carried out in a single day.

Key words: Genotyping, Microsatellite, SSR, Potato, Variety identification.

23.1. Introduction

The granting of Plant Breeders' Rights for new potato varieties is determined by Distinctness, Uniformity, and Stability (DUS) testing. The DUS test takes a minimum of 2 years, according to the UPOV guidelines *(1, 2)*. These tests are based on the comparison of the morphological and physiological characteristics of a proposed new variety with those of all other varieties in common knowledge. However, many of the characteristics used are quantitatively expressed and subjectively assessed, thus leading to potential differences amongst assessors. In addition, the expression of the characters may be influenced by environmental factors, which may lead to different results between test centres. The recent expansion of the European Union has

R. Burns (ed.), *Methods in Molecular Biology, Plant Pathology, vol. 508*

DOI: 10.1007/978-1-59745-062-1_23

resulted in over 1,000 varieties being listed on the EU Common Catalogue, a figure which is likely to rise year on year. It is clearly impractical for test centres to maintain living reference collections containing all varieties in common knowledge for comparative purposes.

Currently, varietal identification is primarily made by means of characteristics of sprouts produced under very low light intensity *(3)*. This test can take up to 3 months and can still be insufficiently discriminative for an unequivocal identification, thus necessitating growing trials and phenotypic assessments. The need for a rapid and reliable method for differentiation and identification of potato varieties is, therefore, becoming increasingly more urgent as an aid to DUS testing stations and breeders. Applications include quality control issues like tracing accidental mixing of varieties, tracing infringements on PBR, maintenance of variety collections, and tracing incorrect labelling of varieties for sale for consumption *(4–6)*.

Molecular marker methods, which analyse the DNA of biological material, are not prone to phenotypic differences resulting from changes in growing conditions or from different assessors, and, have the additional advantages that they are rapid and any part of the plant can be used for analysis. Numerous methods have been investigated for varietal differentiation in potato including RFLP *(7)*, RAPD *(8–11)*, AFLP *(12)*, ISSR *(13)*, and SSR *(14–19)*. Combinations of several of these techniques have also been trialled *(20–22)*. Many of these studies examined either small numbers of varieties, large numbers of markers or were primarily concerned with mapping of the potato genome. Of all these methods, sequence-tagged microsatellite sites (STMS), sometimes also called simple sequence repeat (SSR), markers have proved to be highly efficient and reproducible *(23, 24)*, and allow the rapid differentiation of potato varieties *(17)*. At the beginning of 2006 we embarked on a Community Plant Variety Office (CPVO)-funded project to construct an integrated microsatellite and key morphological characteristic database of potato varieties in the European Union. In addition to the CPVO and SASA the other partners on the project are Naktuinbouw (The Netherlands), Bundessortenamt (Germany), and Centralny Ośrodek Badania Odmian Roślin Uprawnych (Poland). In conjunction with Plant Research International (The Netherlands) we have developed a set of nine SSR markers used in three multiplex reactions with which we can differentiate over 1,000 varieties. The following method outlined is the one in use at SASA.

23.2. Materials

23.2.1. DNA Extraction

Unless stated store all reagents at room temperature.

1. GeneScan Lysis Buffer (Neogen Europe Ltd). Store at 4°C.
2. 20 mg/mL Proteinase K in 40% (v/v) glycerol (Sigma). Store at –20°C.
3. Chloroform/isoamyl alcohol (24:1).
4. Propan-2-ol.
5. 70% Ethanol.
6. Sterile distilled water.

23.2.2. PCR Amplification

1. AmpliTaq® Gold DNA polymerase (Applied Biosystems). Store at –20°C.
2. 10× AmpliTaq® Gold Buffer (Applied Biosystems). Store at –20°C.
3. GeneAmp® dNTP blend 10 mM (Applied Biosystems). Store at –20°C.
4. 25 mM $MgCl_2$ (Applied Biosystems). Store at –20°C.
5. Oligonucleotide primers 100 pmol/μL (**Tables 23.1** and 23.2). Forward primers 5′ labelled with either NED™, VIC®, or 6FAM™ (Applied Biosystems). Reverse primers pigtailed on the 5′ end with GTTT where necessary (*see* **Note 1**). Store at –20°C; labelled primers should be exposed to light as little as possible.

23.2.3. Fragment Electrophoresis

1. Hi-Di™ Formamide (Applied Biosystems). Store at –20°C.
2. GeneScan™ 500 LIZ® size standard (Applied Biosystems) (*see* **Note 2**). Store at 4°C.
3. POP-7™ polymer (Applied Biosystems). Store at 4°C.
4. 10× Capillary electrophoresis running buffer for automated DNA sequencing (Sigma).
5. Automated DNA sequencer (e.g. Applied Biosystems 3130*xl* Genetic Analyzer).

23.3. Methods

23.3.1. DNA Extraction

There are many DNA extraction methods and kits available for plant material. The method described here was chosen for its simplicity, robustness, yield, and purity of the DNA obtained. There

Table 23.1.
SSR primer sequences, repeat motifs, and linkage groups

Marker	Forward primer (5′–3′)	Reverse primer including pigtail (5′–3′)	Repeat	Linkage group
0019	AATAGGTGTACTGACTCTCAATG	GTTTGAAGTAAAAGTCCTAGTATGTG	$(AT)_7$ (GT)10 $(AT)_4$ $(GT)_5$ (GC)4 $(GT)_4$	VI
2005	TTTAAGTTCTCAGTTCTGCAGGG	GTTTGTCATAACCTTTACCATTGCTGGG	$(CTGTTG)_3$	XI
2028	TCTCACCAGCCGGAACAT	GTTTAAGCTGCGGAAGTGATTTTG	$(TAC)_5$.(TA)3.$(CAT)_3$	XII
3009	TCAGCTGAACGACCACTGTTC	GTTTGATTTCACCAAGCATGGAAGTC	$(TC)_{13}$	VII
3012	CAACTCAAACCAGAAGGCAAA	GTTTGAGAAATGGGCACAAAAAACA	$(CT)_4$.$(CT)_8$	IX
3023	AAGCTGTTACTTGATTGCTGCA	GTTCTGGCATTTCCATCTAGAGA	$(GA)_9$.$(GA)_8$.$(GA)_4$	IV
5136	GGGAAAAGGAAAAGCTCAA	GTTTATATGAACCACCTCAGGCAC	(AGA)5	I
5148	TCTTCTTGATGACAGCTTCG	GTTTACCTCAGATAGTTGCCATGTCA	$(GAA)_{17}$	V
SSR1	GATGAGATGAGATATGAAACAACG	GTTTCGCAATTCTCTTGACACGTGTCACT-GAAAC	(TCAC)*n*	VIII

is the added advantage that it is also extremely cheap. The method works equally well with leaf, light sprout, or tuber material.

1. Homogenize approximately 200 mg of tissue mechanically (*see* **Note 3**).
2. Add 1 mL of GeneScan lysis buffer and 10 μL Proteinase K solution, and incubate for 1 h at 65°C.
3. Centrifuge for 10 min at 16,000 g.
4. Transfer 850 μL from the supernatant into a fresh 2-mL Safe-lock tube (*see* **Note 4**). Add 650 μL chloroform/isoamyl alcohol and mix by inverting tube several times.
5. Centrifuge for 10 min at 13,000 rpm.
6. Transfer 650 μL from the upper aqueous phase into a fresh 1.5-mL reaction tube. Add 520 μL propan-2-ol and mix thoroughly. Incubate for 30 min on ice.
7. Centrifuge for 10 min at 13,000 rpm to pellet DNA and discard the supernatant.
8. Wash pellet twice with 500 μL 75% ethanol. Centrifuge for 5 min at 13,000 rpm after each wash.
9. Air dry pellet for 2–3 min and resuspend in 100 μL sterile distilled water (*see* **Note 5**).
10. Measure concentration of the DNA by uv spectrophotometry.
11. Dilute template DNA to a final concentration of 10 ng/μL with distilled water.

23.3.2. PCR Amplification

The method in use at SASA uses AmpliTaq® Gold DNA polymerase from Applied Biosystems. However, it is possible to use enzymes from other manufacturers and obtain equally good results. We routinely use the multiplex conditions with no significant loss of signal, although if problems are encountered simplex reactions should solve these. The following protocol details the multiplex conditions with simplex PCR variations where appropriate.

1. Make up sufficient PCR master mix for the amplification of all samples. Each reaction contains 3.9 μL sterile water, 1 μL 10 × AmpliTaq® Gold Buffer, 1 μL 25 mM $MgCl_2$, 1 μL 10 mM dNTP mix, 1 μL primer mix (Set 1, 2, or 3, **Table 23.2**, *see* **Note 6**), and 0.1 μL AmpliTaq® Gold DNA polymerase (5 units/μL). N.B. for simplex reactions use forward and reverse primers at a concentration of 1 pmol/μL each.
2. Dispense 8 μL PCR master mix into each well of a 96-well PCR plate and add 2 μL of the appropriate sample and seal the plate.
3. Amplify using the following program. 94°C for 10 min then 30 cycles of 94°C for 30 s, ramp at 1°C/s to 50°C for 30 s, ramp at 1°C/s to 72°C for 2 min, followed by a hold step at

Table 23.2.
SSR primer sets and multiplex dilutions (*see* Note 5)

Marker	Dye	Forward primer (μL)	Reverse primer (μL)
Set 1 (add following to 160 μL H_2O)			
0019	VIC®	8	8
3009	NED™	8	8
SSR1	6FAM™	4	4
Set 2 (add following to 164 μL H_2O)			
2005	NED™	8	8
3012	6FAM™	8	8
3023	VIC®	2	2
Set 3 (add following to 164 μL H_2O)			
2028	NED™	8	8
5136	VIC®	2	2
5148	6FAM™	8	8

72°C for 10 min and in infinite hold at 25°C. N.B. for simplex reactions the 72°C extension step can be reduced to 45 s.

23.3.3. Sample Preparation for Genotyping

It is not necessary to purify the samples before running as there should be sufficient product present in the reactions. However, if desired both ethanol precipitation and magnetic bead purification work equally well and will result in a stronger signal.

1. Prepare sufficient Hi-Di™ Formamide/GeneScan™ 500 LIZ® size standard for all the samples. Each well requires 8.7 μL Hi-Di™ Formamide and 0.3 μL GeneScan™ 500 LIZ® size standard. Aliquot 9 μL into each well of a new 96-well PCR plate.
2. Add 1 μL of the PCR product obtained in **Subheading 23.3.2** to each well and seal the plate.
3. Heat at 95°C for 5 min, chill on ice, and centrifuge at 200g for 1 min.
4. Analyse plate on an automated DNA sequencer following manufacturers' instructions.

23.3.4. Genotyping Run Parameters

The following conditions are for use with an Applied Biosystems 3130*xl* Genetic Analyzer with POP-7™ and a 36-cm array. The 3130*xl* is a five-dye system and the run conditions therefore use Dye Set G5. If a five-dye system is not available alter the conditions as appropriate and use the red channel for the size standard.

23.3.4.1. Run Module

FragmentAnalysis36_POP7_1

Oven_Temperature	60°C
Poly_Fill_Vol	6,500
Current_Stability	5.0 μAmps
PreRun_Voltage	15.0 kV
Pre_Run_Time	180 s
Injection_Voltage	1.2 kV
Injection_Time	23 s
Voltage_Number_Of_Steps	20 nk
Voltage_Step_Interval	15 s
Data_Delay_Time	60 s
Run_Voltage	15.0 kV
Run_Time	1,200 s

23.3.5. Genemapper® Analysis Parameters

At SASA we use GeneMapper® Software v3.7 with the following settings. N.B. most of these are the same as the Microsatellite default settings as our findings were that these worked well even though most commercial potato cultivars are tetraploid. Before any samples can be analysed in GeneMapper® an *Analysis Method*, *Kit*, *Panel*, *Markers*, and *Bins* must be created.

23.3.5.1. Genemapper® Analysis Method Potato SSR

The Analysis Method we use is adapted from the Microsatellite Default which is found by opening GeneMapper® Manager and clicking on the Analysis Methods tab. Open the Microsatellite Default, enter the following information, and Save As Potato SSR.

General tab

Instrument	3130*xl*	
Allele tab		
Bin Set	Potato SSR Bins	
	Trinucleotide	Tetranucleotide
Cut-off value	0.2	0.25
PlusA ratio	0.95	0.95
PlusA distance	1.6	1.6
Stutter ratio	0.95	0.15

(continued)

General tab

Stutter distance	0.0–3.5	0.0–3.5	
Peak detector tab			
Peak detection algorithm		Basic	
Minimum peak height		Automatic	
Peak quality tab			
Signal level			
Homozygous min peak height		200.0	
Heterozygous min peak height		100.0	
Heterozygote balance			
Min peak height ratio		0.25	
Peak morphology			
Max peak width (bp)		1.5	
Pull-up peak			
Pull-up ratio	0.1		
Pull-up scan	1		
Allele number			
Max expected alleles	5		
Cross-talk peak			
Cross-talk ratio	0.05		
Quality flag tab			
Quality flag settings			
Spectral pull-up	0.5	Control concordance	0.5
Broad peak	0.5	Low peak height	0.5
Single peak artefact	0.5	Off-scale	0.5
Sharp Peak	0.5	Peak height ratio	0.5
Cross-talk	0.5	One basepair allele	0.5
Out-of-bin allele	0.5	Split peak	0.5
PVQ thresholds			
	Pass range	Low-quality range	
Sizing quality	0.75–1.0	0.0–0.25	
Genotype quality	0.75–1.0	0.0–0.25	

23.3.5.2. Genemapper® Panel Manager

Open the GeneMapper® Panel Manager and create a new Kit for the *Potato SSR* data.

Open the *Potato SSR* Kit and create new panels for *Marker Set 1*, *Marker Set 2*, and *Marker Set 3*.

Open each *Marker Set* and enter the values in **Table 23.3** leaving the other fields blank.

Open each individual marker and set the allele bin sizes in **Table 23.4**.

23.3.6. Calibration of the System

The system was calibrated using reference alleles, essential as described by Bredemeijer et al. *(6)*. The use of reference alleles makes the system independent of the equipment (and laboratories) used to produce the microsatellite data. Reference alleles are defined as specific peaks produced by a marker in a specific variety. To enable calibration of the system DNA samples of a set of 29 varieties can be obtained from SASA that yield all the alleles detected to date for the 9 markers. Included in the list are 2 varieties that yield no alleles for a single marker. It should be noted that varieties that do not yield alleles appear to be extremely rare for this set of markers (indeed this has only been observed for a small number of varieties with 0019 and 3,009). In our experience the absence of alleles is often accompanied by characteristic artefacts that could be mistaken for alleles. These take the form of a peak followed by stutter peaks differing in size by a single base pair as oppose to an entire repeat unit. Upon encountering such a profile it is recommended that several further samples of the same variety should be analysed to ensure that it is truly a null allele.

Alternatively the set of 16 varieties in Table 23.5 can be used for calibration. These varieties have been chosen as they are present

Table 23.3.
GeneMapper® Marker Set information

Marker name	Dye colour	Min size	Max size	Marker repeat
Marker Set 1				
0019	Green	160	240	2
3009	Yellow	135	180	2
SSR1	Blue	195	230	4
Marker Set 2				
2005	Yellow	145	200	3
3012	Blue	160	220	2
3023	Green	170	205	2
Marker Set 3				
2028	Yellow	280	415	3
5136	Green	210	255	3
5148	Blue	400	485	2

Table 23.4.
GeneMapper® allele bin settings

Marker Set 1

0019 alleles	Size (bp)	Left offset	Right offset	3009 alleles	Size (bp)	Left offset	Right offset	SSR1 alleles	Size (bp)	Left offset	Right offset
A	167.07	0.60	0.60	A	143.00	0.70	0.70	A	201.47	0.40	0.40
B	193.42	0.75	0.75	B	146.57	0.40	0.40	B	204.71	0.40	0.40
C	196.97	0.75	0.75	C	150.48	0.86	0.68	C	206.65	0.40	0.40
D	199.27	0.60	0.60	D	153.26	0.40	0.40	D	208.77	0.40	0.40
E	204.38	0.72	0.83	E	155.35	0.40	0.40	E	212.71	0.40	0.40
F	208.12	0.60	0.60	F	159.65	0.40	0.40	F	214.77	0.40	0.40
G	234.77	0.60	0.60	G	165.48	0.40	0.40	G	216.76	0.40	0.40
H	175.30	0.95	0.95	H	168.00	0.40	0.40	H	219.00	0.40	0.40
I	169.77	1.28	1.27	I	171.00	0.70	0.70	I	220.83	0.40	0.40
J	181.50	0.60	0.60	J	175.74	0.40	0.40	J	224.83	0.40	0.40
				K	157.62	0.82	0.82	K	228.71	0.40	0.40
				L	162.00	0.40	0.40	L	202.88	0.40	0.40
				M	173.98	0.50	0.50	M	223.00	0.40	0.40
								N	210.65	0.40	0.40

Marker Set 2

2005 alleles	Size (bp)	Left offset	Right offset	3012 alleles	Size (bp)	Left offset	Right offset	3023 alleles	Size (bp)	Left offset	Right offset
A	153.24	0.40	0.40	A	165.70	0.40	0.40	A	177.00	0.40	0.40
B	159.79	0.40	0.40	B	167.77	0.40	0.40	B	178.98	0.40	0.40
C	165.98	0.40	0.40	C	195.85	0.40	0.40	C	186.80	0.40	0.40
D	172.01	0.40	0.40	D	197.94	0.40	0.40	D	196.02	0.40	0.40
E	183.26	0.40	0.40	E	199.77	0.40	0.40				
F	196.61	0.40	0.40	F	201.68	0.40	0.40				
				G	211.62	0.40	0.40				

Marker Set 3

2028 alleles	Size (bp)	Left offset	Right offset	5136 alleles	Size (bp)	Left offset	Right offset	5148 alleles	Size (bp)	Left offset	Right offset
A	286.70	0.60	0.60	A	213.68	0.40	0.40	A	402.75	0.40	0.40

(continued)

Table 23.4 (continued)

B	296.07	0.60	0.60	B	216.28	0.40	0.40	B	416.09	0.75	0.75
C	365.38	0.60	0.60	C	219.38	0.40	0.40	C	422.65	0.40	0.40
D	388.21	0.60	0.60	D	225.25	0.40	0.40	D	424.97	0.40	0.40
E	395.07	0.60	0.60	E	227.99	0.40	0.40	E	428.12	0.40	0.40
F	401.47	0.90	0.90	F	230.96	0.40	0.40	F	431.93	0.70	0.70
G	405.51	0.90	0.90	G	236.50	0.40	0.40	G	433.25	0.40	0.40
H	408.11	0.60	0.60	H	248.20	0.40	0.40	H	440.25	0.40	0.40
I	292.84	0.50	0.50	I	250.89	0.40	0.40	I	446.35	0.40	0.40
				J	244.99	0.40	0.40	J	450.70	0.40	0.40
								K	452.75	0.70	0.70
								L	457.15	0.40	0.40
								M	461.64	0.90	0.95
								N	464.88	0.50	0.50
								O	470.97	0.70	0.70
								P	474.65	0.40	0.40
								Q	477.36	0.40	0.40
								R	420.47	0.40	0.40

Table 23.5.
Alleles yielded by 16 potato varieties common to the national lists of a number of European countries in 2006

	Set 1			Set 2			Set 3		
	0019	3009	SSR1	2005	3012	3023	2028	5136	5148
Asterix	D	G	AGIJ	ABD	BC	AB	ABC	CFH	GJM
Berber	BFG	BFG	ADF	BD	B	ABD	ABC	DEF	GKMO
Désirée	BFG	FG	ADFI	ABD	BC	AD	AC	EFH	BIJ
Fresco	BG	BDG	ADI	D	BC	AB	A	EF	CIJ
Impala	CF	FG	ADFI	ABD	BCF	AD	AB	CEF	BIJO
Karlena	BGH	BG	DEF	ABF	BDF	D	ABCE	CEFH	BCP
Kennebec	BE	FG	DFI	B	BF	AD	ABC	CEFH	IJO

(continued)

Table 23.5 (continued)

	Set 1			Set 2			Set 3		
Lady Rosetta	A	G	ADIJ	BD	BD	AD	A	EF	GJ
Minerva	BG	DFG	EFJ	ABD	BDF	BD	AC	EF	JL
Nicola	BF	F	DFI	ABD	BCD	ACD	AC	CEF	GIJO
Raja	BFG	DF	DI	ABDF	BC	D	ABCE	CEF	GJNO
Remarka	FG	FG	DFI	ABD	BCD	AD	AG	EFH	JMO
Santé	ABF	DG	DIJ	ABD	BCDF	BD	ACG	EH	JL
Saturna	F	DF	BDFI	ABD	BD	ABD	ABC	CE	BCIJ
Ukama	BDE	BG	AFI	ABD	BC	BD	AC	DEFH	IJOP
Victoria	BF	G	DFI	AD	BF	ABD	ABC	EF	AJP

on the National Lists of several European countries in 2006 and therefore should be readily available. Although these varieties do not yield representatives of all the possible peaks they can be used to calibrate the sizes obtained against those in **Table 23.4**.

23.4. Notes

1. The addition of a 5′ pigtail helps to reduce the occurrence of +A artefacts.
2. Due to the properties of the POP-7™ polymer the 250-bp peak in the GeneScan™ 500 LIZ® size standard does not run accurately. This peak should therefore be excluded from the analysis of the size standard.
3. The choice of tissue disruption method is largely governed by what equipment is available. A simple method is to place the tissue sample in a 2-mL Safelock Eppendorf tube and mashing it with the handle of a metal bacterial loop spreader. This can be cleaned with ethanol and water between samples to prevent cross-contamination. Crushing in liquid Nitrogen is not necessary, although it will improve the yield.
4. The use of Safelock tubes is not strictly necessary but will reduce the possibility of chloroform leaking out of the tube when the samples are mixed.

5. The precipitated DNA should dissolve with 30 min; however, we regularly leave the samples at 4°C overnight to ensure that the DNA is completely resuspended.

6. It is possible to use alternative combinations of markers in the multiplex primer sets if so desired. At PRI in the Netherlands they regularly use Set 2 (2005, 3012, 5136) and Set 3 (2028, 3023, 5148).

References

1. Anon. (2004) Guidelines for the conduct of tests for distinctness, uniformity and stability. UPOV guideline TG/23/6.
2. Cooke, R.J. (1999) New approaches to potato variety identification. *Potato Research*, 42, 529–539.
3. Houwing, A., Suk, R. and Ros, B. (1986) Generation of lightsprouts suitable for potato variety identification by means of artificial light. *Acta Horticulturae*, 182, 359–363.
4. Anon. (2003) Survey to investigate the varietal labelling of potatoes – part 1. http://www.food.gov.uk/multimedia/pdfs/fsis4403.pdf.
5. Davey, R. (2004) Time to end this damaging practice. Eyewitness. British Potato Council. p 15.
6. Bredemeijer, G.M.M., Cooke, R.J., Ganal, M.W., Peeters, R., Isaac, P., Noordijk, Y., Rendell, S., Jackson, J., Röder, M.S., Wendehake, K., Dijcks, M., Amelaine, M., Wickaert, R., Bertrand, L. and Vosman, B. (2002) Construction and testing of a microsatellite database containing more than 500 tomato varieties. *Theoretical and Applied Genetics*, 105, 1019–1026.
7. Görg, R., Schachtschabel, U., Ritter, E., Salamini, F. and Gebhardt, T. (1992) Discrimination among 136 tetraploid potato varieties by fingerprints using highly polymorphic DNA markers. *Crop Science*, 32, 815–819.
8. Demeke, T., Kawchuk, L.M. and Lynch, D.R. (1993) Identification of potato cultivars and clonal variants by random amplified polymorphic DNA analysis. *American Potato Journal*, 70, 561–570.
9. Hosaka, K., Mori, M. and Ogawa, K. (1994) Genetic relationships of Japanese potato cultivars assessed by RAPD analysis. *American Potato Journal*, 71, 535–546.
10. Sosinski, B. and Douches, D.S. (1996) Using polymerase chain reaction-based DNA amplification to fingerprint North American potato cultivars. *Horticultural Science*, 31, 130–133.
11. Isenegger, D.A., Taylor, P.W.J., Ford, R., Franz, P., McGregor, G.R. and Hutchinson, J.F. (2001) DNA fingerprinting and genetic relationships of potato cultivars (*Solanum tuberosum* L.) commercially grown in Australia. *Australian Journal of Agricultural Research*, 52, 911–918.
12. Kim Kim, J.H., Juong, H., Kim, H.Y. and Lim, Y.P. (1998) Estimation of genetic variation and relationship in potato (*Solanum* cultivars L.) cultivars using AFLP markers. *American Journal of Potato Research*, 75, 107–112.
13. Bornet, B., Goraguer, D., Joly, G. and Branchard, M. (2002) Genetic diversity in European and Argentinian cultivated potatoes (*Solanum tuberosum* subsp. *tuberosum*) detected by inter-simple sequence repeat (ISSRs). *Genome*, 45, 481–484.
14. Kawchuk, L.M., Lynch, D.R., Thomas, J., Penner, B., Sillito, D. and Kulcsar, F. (1996) Characterization of *Solanum tuberosum* simple sequence repeats and application to potato cultivar identification. *American Potato Journal*, 73, 325–335.
15. Corbett, G., Lee, D., Donini, P. and Cooke, R.J. (2001) Identification of potato varieties by DNA profiling. *Acta Horticulturae*, 546, 387–390.
16. Norero, N., Malleville, J., Huarte, M. and Feingold, S. (2002) Cost efficient potato (*Solanum tuberosum* L.) cultivar identification by microsatellite amplification. *Potato Research*, 45, 131–138.
17. Ghislain, M., Spooner, D.M., Rodríguez, F., Villamón, F., Núñez, J., Vásquez, C., Waugh, R. and Bonierbale, M. (2004) Selection of highly informative and user-friendly microsatellites (SSRs) for genotyping of cultivated potato. *Theoretical and Applied Genetics*, 108, 881–890.

18. Coombs, J.J., Frank, L.M. and Douches, D.S. (2004) An applied fingerprinting system for cultivated potato using simple sequence repeats. *American Journal of Potato Research*, 81, 243–250.
19. Reid, A. and Kerr, E.M. (2007) A rapid SSR-based identification method for potato cultivars. *Plant Genetic Resources*, 5, 7–13.
20. Milbourne, D., Meyer, R., Bradshaw, J.E., Baird, E., Bonar, N., Provan, J., Powell, W. and Waugh, R. (1997) Comparison of PCR-based marker systems for the analysis of genetic relationships in cultivated potato. *Molecular Breeding*, 3, 127–136.
21. Milbourne, D., Meyer, R.C., Collins, A.J., Ramsey, L.D., Gebhardt, C. and Waugh, R. (1998) Isolation, characterisation and mapping of simple sequence repeat loci in potato. *Molecular and General Genetics*, 259, 233–245.
22. McGregor, C.E., Lambert, C.A., Greyling, M.M., Louw, J.H. and Warnich, L. (2000) A comparative assessment of DNA fingerprinting techniques (RAPD, ISSR, AFLP and SSR) in tetraploid potato (*Solanum Tuberosum* L.) germplasm. *Euphytica*, 113, 135–144.
23. Karp, A., Edwards, K.J., Bruford, M., Funk, S., Vosman, B., Morgante, M., Seberg, O., Kremer, A., Boursot, P., Arctander, P., Tautz, D. and Hewitt, G. (1997) Molecular technologies for biodiversity evaluation: opportunities and challenges. *Nature Biotechnology*, 15, 625–628.
24. Jones, C.J., Edwards, K.J., Castiglione, S., Winfield, M.O., Sala, F., Van de Wiel, C., Bredemeijer, G., Vosman, B., Matthes, M., Daly, A., Brettschneider, R., Bettini, P., Buiatti, M., Maestri, E., Malcevschi, A., Marmiroli, N., Aert, R., Volckaert, G., Rueda, J., Linacero, R., Vazquez, A. and Karp, A. (1997) Reproducibility testing of RAPD, AFLP and SSR markers in plants by a network of European laboratories. *Molecular Breeding*, 3, 381–390.

Chapter 24

Barley Variety Identification Using SSRs

Cathy Southworth

Summary

There is a current and developing need for rapid and accurate methods of barley varietal identification which go beyond traditional morphological analysis. Methods using DNA analysis have the capacity to fulfil this role with microsatellites being the current marker of choice. The majority of barley cultivars on the National List can be differentiated using 6 SSRs and bulk samples, using the methods described here.

Key words: Barley, Microsatellites, Variety identification, Genotyping.

24.1. Introduction

The accurate identification of barley is necessary to address issues of commercial interest. For example: assurance of seed quality, protection of intellectual property rights, and the support of plant breeders' rights and patents *(1)*. Traditionally, variety identification has been determined using a combination of morphological and physiological markers.

Identification methods based on plant phenotype require time-consuming assessments of adult plant morphology and microscopic examination of seed *(2)*. In addition, morphological characteristics can be influenced by environmental conditions. The present identification of grain uses morphological characters detailed by UPOV *(3)* and can take between 30 min and a full day depending on the degree of similarity between the morphological characteristics of the two varieties. Even then some varieties are particularly difficult to separate, and with the morphology of varieties becoming more similar an alternative method of identification is required.

R. Burns (ed.), *Methods in Molecular Biology, Plant Pathology, vol. 508*

DOI: 10.1007/978-1-59745-062-1_24

Physiological markers have been developed for barley differentiation which includes the use of isoenzymes *(4)* and gel electrophoresis of hordein proteins *(5–7)*. The use of gel electrophoresis for variety identification is recognised by The International Seed Testing Association (ISTA) which provides a standard reference method for the separation of barley hordein proteins using polyacrylamide gel electrophoresis (PAGE) *(8)*.

However, physiological markers are not ideal. Profiles can be difficult to interpret, poor resolution of bands can occur *(1)*, and there exists no one method that can differentiate a large set of barley varieties *(2)*. Reports of the use of the method at SASA have highlighted inconsistencies with control samples and a low level of reliability in producing results. These problems have led to the need for more accurate means of identifying barley varieties.

An array of DNA-based techniques have been applied to variety identification. Initially these were non-polymerase chain reaction (PCR) based and included the use of restriction fragment length polymorphisms (RFLP) *(9)* and DNA-hybridisation techniques such as minisatellites *(10)*. More recent techniques have been based on the use of the (PCR) such as random amplified polymorphic DNA (RAPD) *(11)* and amplified fragment length polymorphisms (AFLP) *(12)*.

Further methods utilise primers of a specific sequence design such as (SSR) Powell et al. *(13)*, single nucleotide polymorphisms (SNPs) Coryell et al. *(14)* , and retrotransposon-based methods (inter-retrotransposon amplified polymorphism (IRAP) and retrotransposon-microsatellite amplified polymorphism (REMAP)) *(15)*.

SSRs are at present the marker of choice and as recognised by Macaulay et al. *(16)* are the marker system, along with SNPs, that provide a possible means for transfer of data between labs and construction of databases, accessible to the community as a whole.

The development of a barley SSR marker system and database was part of a PhD research project carried out at SASA *(17)*. The findings of this work and the end methodology are described here.

24.2. Materials

24.2.1. DNA Extraction

1. Sterile pestle and mortar and liquid nitrogen or coffee mill.
2. GeneScan Lysis Buffer (Neogen, Europe Ltd) kept at 4°C.
3. 20 mg/mL Proteinase K in 40% (v/v) glycerol (sigma).
4. 30 mg/mL RNAse A (Applied Biosystems).
5. Chloroform/isoamyl alcohol (24:1) at room temperature (Sigma).

6. Isopropanol kept at 4°C (Sigma).
7. 75% Ethanol (aq. v/v) (Sigma).
8. Sterile double-distilled water.

24.2.2. PCR Amplification

1. AmpliTaq Gold® DNA polymerase (5 units/µl) (Applied Biosystems).
2. AmpliTaq Gold® Buffer (Applied Biosystems).
3. 10 mM dNTP mix (Applied Biosystems).
4. 25 mM $MgCl_2$ (Applied Biosystems).
5. 2 pmol/µl Forward primer labelled with one of FAM ™, PET™, VIC®, NED™ (Applied systems).
6. 100 pmol/µl Reverse primer (MWG).
7. Sterile double-distilled water.
8. PCR machine.

24.2.3. PCR Fragment Analysis with an ABI Prism® 3100Avant

1. GeneScan™-500 LIZ® Size standard (Applied Biosystems).
2. Hi-Di Formamide 1 (Applied Biosystems).
3. Pop-4™ Polymer (Applied Biosystems).
4. Capillary electrophoresis system and associated software (ABI Prism® 3100*Avant*).

24.3. Methods

24.3.1. Barley Samples

Due to the occurrence of non-uniformity in barley varieties *(17)* it is necessary to use bulk seed samples in order to determine the SSR allele most frequently found in each variety (termed the major allele). This is also the case when analysing seed batches to assess variety type. Samples of 100 seed were found to be sufficient and manageable to identify the major allele for that variety.

24.3.2. DNA Extraction

Several commercial kits were trialled on barley root, leaf, and seed. Comparable quantity and quality was obtained from each of these and also from using the GeneScan extraction system (Agden, Scotland). Due to its ease of use, reliability, and cost effectiveness the GeneScan method is recommended and is described in detail by Reid et al., in this volume.

24.3.2.1. Homogenation

Barley seed can be homogenised using two methods, depending on whether individual or bulk seeds are being analysed. To homogenise 1–3 seeds or if fresh leaf is being used, homogenisation using liquid nitrogen was found to produce the best results. To homogenise bulk seed samples a coffee mill was used.

Homogenation Using Liquid Nitrogen

1. Grind seeds in liquid nitrogen using a sterile pestle and mortar until what remains resembles fine flour (*see* **Notes 1** and **2**).
2. Place the barley flour into an Eppendorf tube and continue with the protocol as described by Reid et al., in this volume.

Homogenation Using a Coffee Mill

1. Grind 100 barley seeds to flour using a coffee mill (*see* **Note 3**).
2. Place the barley flour into a 50-mL cone-based centrifuge tube (Fisherband®, Abgene, Epsom). Use approximately 15 mL of the flour to continue with the DNA extraction protocol.

24.3.3. PCR Amplification

24.3.3.1. Microsatellite Primers

Of 30 microsatellite primers that were screened 6 primer sets were found to provide the highest degree of differentiation between barley cultivars on the UK National List of barleys, 2003 (*see* Table 24.1). These 6 primers are polymorphic, amplify well, and give consistently clear profiles, differentiating 96 of the 134 cultivars on the National list. If large sets of barley need to be differentiated then these 6 primers would be an ideal starting point. The full set of over 600 SSR primer sequence sets that are available for barley in the public domain are given by Ramsay et al. *(18)*.

24.3.3.2. Amplification Reaction

PCR amplification using the primers described here (*see* **Table 24.1**) is carried out in a total volume of 10 μl per reaction using the following components:

Component	Volume per reaction (μl)	Final concentration
10× buffer	1	1×
dNTP mix (10 mM)	1	1 mM
$MgCl_2$ (25 mM)	1	2.5 mM
Forward primer (2 pmol/μl)[a]	1	0.2 pmol/μl
Reverse primer (100 pmol/μl)[b]	0.5	5 pmol/μl
AmpliTaq Gold® 5 units/μl	0.1	0.05 units/μl
DDH_2O	4.4	–
DNA (10 ng/μl)	1	1 ng/μl

[a]Forward primer labelled at the 5 with one of FAM ™, PET™, VIC®, NED™ for detection in capillary electrophoresisbReverse primer 'pig-tailed' at the 5′ (*see* **Note 4**)

1. Make a master mix of all the PCR components apart from the DNA. This should be enough for the number of DNA samples you are to analyse (*see* **Note 5**).
2. Pipette the master mix into the sample wells of a 96-well PCR plate, or for smaller numbers use 0.2-mL PCR tubes.

Table 24.1.
Table showing the six robust primer sets

Primer	Motif	PCR programme	Allele no (rare alleles)	Forward primer	Reverse primer	Allele sizes on ABI 3100	PIC
Bmag0120	$(AG)_{15}$	TD_TRIAL	8(4)	ATTTCATCCCAAA-GGAGAC	GTCACATAGACAG-TTGTCTTCC	229/231/233/235/237/241/261/263	0.7
Bmac0093	$(AC)_{24}$	TD_64-55	5(1)	CGTTTGGGACGT-ATCAAT	GGGAGTCTTGAG-CCTACTG	157/160/162/164/166	0.7
Bmag0211	$(CT)_{16}$	TD_TRIAL	4(1)	ATTCATCGATCTT-GTATTAGTCC	ACATCATGTCGA-TCAAAGC	185/187/189/191	0.7
Bmag0009	$(AG)_{13}$	30@60	4(0)	AAGTGAAGCAAG-CAAACAAACA	ATCCTTCCATATTT-TGATTAGGCA	174/176/178/180	0.6
HvM36	$(GA)_{13}$	30@55	5(2)	TCCAGCCGACAA-TTTCTTG	AGTACTCCGACAC-CACGTCC	111/113/117/123/141	0.7
HvM54	$(GA)_{14}$	TD 1@55	6(3)	AACCCAGTAACA-CCGTCCTG	AGTTCCCTGACCC-GATGTC	153/157/159/161/165/169	0.6
		Mean	5.3			Mean	0.7

3. Add 1 µl of the DNA samples into the appropriate tube or plate well and seal.
4. Place the PCR samples into the PCR machine and use the PCR cycling conditions appropriate for the primers as shown in **Table 24.1**.

TD_TRIAL: 95°C for 10 min; 7 cycles of 94°C for 30 s, 65°C for 30 s decreasing to 58°C by using a ramp which decreases the annealing temperature 1°C each cycle, then 72°C for 30 s then 30 cycles of 94°C for 30 s and 58°C for 30 s; hold for 5 min at 72°C.

TD64-55: 95°C for 10 min; 10 cycles of 94°C for 1 min, 1 min at 64–55°C reducing 1°C each cycle; 72°C at 1 min then 30 cycles of 94°C for 1 min, 55°C for 1 min and 72°C for 1 min then hold at 72°C for 5 min.

30at55: 95°C for 10 min; 30 cycles of 94°C for 30 s, 55°C for 30 s then 72°C for 5 min.

The cycling conditions for the programmes 30at60 are the same as for 30at55, but replace the 55°C annealing temperature with 60°C (*see* **Note 6**).

24.3.4. PCR Fragment Analysis

24.3.4.1. Sample Preparation for Analysis

Once PCR has been carried out the samples can be prepared for analysis on a genetic analyser, so that the PCR DNA fragments can be sized using capillary electrophoresis. This method describes preparation for analysis on an ABI Prism® 3100*Avant* which has 16 capillaries and can process two 96-well plates per run. Please refer to the manufacturers' instructions with regard to sample preparation if a different system is used.

1. Make a master mix of GeneScan™-500 LIZ® size standard (0.25 µl per sample) and Hi-Di Formamide 1 (8.75 µl per sample).
2. Pipette 9 µl of the master mix into the appropriate number of wells on a 96-well plate.
3. Pipette 1 µl of each PCR sample into individual wells on the plate and seal.
4. Denature the samples by incubating at 95°C for 5 min and then cooling on ice for 5 min.

24.3.4.2. Capillary Electrophoresis

Samples are then loaded onto the analyser and run using the default parameters. The 36-cm capillary array and POP-4™ Performance Optimised Polymer were utilised on the ABI Prism™.

24.3.4.3. Manual Interpretation of Allele Scores

The GeneScan® Analysis Software version 3.7 software produces profiles for each sample with corresponding peak heights. From these data the allele is determined for each sample. An example of the profiles for two different cultivars Chariot and Carot is shown in **Fig. 24.1**.

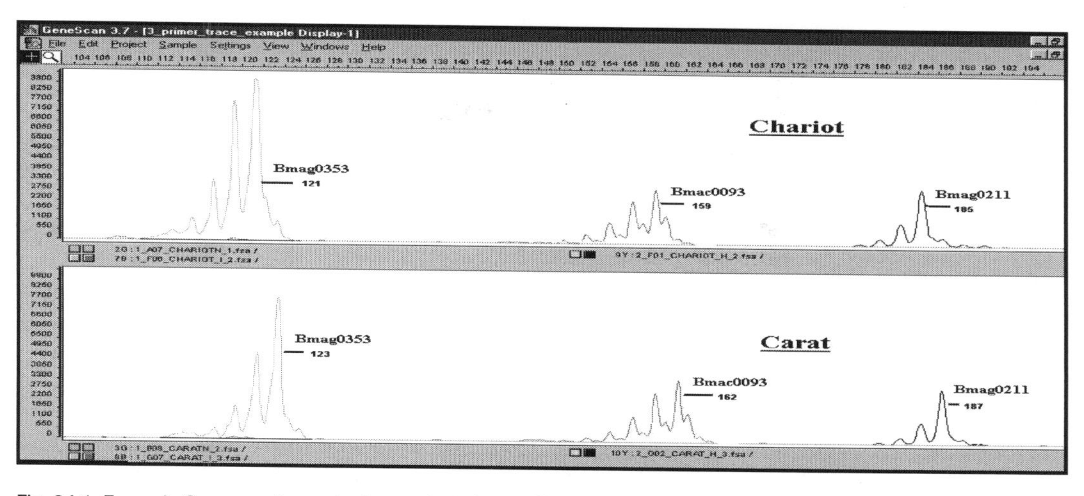

Fig. 24.1. Example Genescan traces to demonstrate how differentiation of cultivars can be achieved using microsatellite markers. The traces show allele sizes at three SSR loci: Bmag0353, Bmac0093, and Bmag0211 in two cultivars: (i) cv. Chariot and (ii) cv. Carat.

Manual interpretation of scores can present problems due to: *(1)* Stutter *(19)*, which can be seen by peaks decreasing in size from the true allele peak; *(2)* A addition *(20)*; and; (3) Inconsistencies in peak profile from one run to the next. This is how these problems can be tackled:

1. Size the main allele (the tallest peak) by rounding the peak size up or down to the nearest base pair.
2. Disregard any primers on screening that produce profiles that are difficult to interpret. This should not happen with the primer sets given here as they were screened on barley varieties and were chosen for consistency and for profiles that were easy to interpret.
3. Run a set of different cultivars using the same primers to gain a picture of the profile a particular primer gives.
4. Use 'pig-tailed' reverse primers to reduce the problems caused by 'A' addition which is represented by a small peak one base pair larger than the true allele (this can be seen in the profiles given by Bmac0093 in **Fig. 24.1**).
5. If a cultivar gives a main allele 1 bp smaller or larger than a previous cultivar using the same primer then it is possible that the same allele is being amplified. Amplify the two cultivars with the same primers in the same reaction to see if they give the same profile. If this is the case then that allele can be assigned a bin size that is 1 bp smaller and larger to enable accurate allele sizing to be carried out in future.

24.4. Notes

1. Grinding the seed is easiest when a small amount of liquid nitrogen is left in the pestle. Wearing goggles is essential as the barley seed can be very brittle and fly out of the pestle.
2. A fresh pestle and mortar is needed to homogenise each cultivar sample.
3. A Moulinex Super Junior's Coffee Mill (Birmingham) has been successfully used. Careful cleaning of the instrument needs to be carried out between uses with different samples to avoid PCR contamination.
4. To counter the problems that can occur in reading traces a 'pig-tail' sequence of GTTTCTT *(20)* can be added to the 5′ of the reverse primer.
5. To prevent the master mix running out due to pipetting errors one extra reaction amount is added to the master mix volume for samples. As a rule of thumb, it is helpful to add one extra reaction amount for every 25 reactions.
6. The PCR programme 30at55 was the standard programme used following the determination of the primer annealing temperature using a temperature gradient. Once the optimum annealing temperature had been determined it was then used as the temperature in the annealing step of this simple programme. Only when primers were more difficult to amplify or were amplifying inconsistently well were touchdown cycles used.

References

1. Lee, L.S., and R.J. Henry (2001) Commercial applications of plant genotyping. In Henry, R.J. (Ed.) *Plant Genotyping: The DNA Fingerprinting of Plants* Cabi Publishing, Oxon, England.
2. William, M., I. Dorocicz and K.J. Kasha (1997) Use of microsatellite DNA to distinguish malting and nonmalting barley cultivars. *Journal of the American Society of Brewing Chemists* 55(3): 107–111.
3. UPOV (International Union for the Protection of New Varieties of Plants) (1994) Guidelines for the conduct of tests for Distinctness, uniformity and Stability: Barley: TG/19/10: Date1994-11-04 UPOV, Geneva.
4. Anderson, H.J. (1982) Isoenzyme characters of 47 barley cultivars and their application in cultivar identification. *Seed Science Technology* 10: 405–413.
5. Gebre, H., K. Khan and A.E. Foster (1986) Barley cultivar identification by polyacrylamide gel electrophoresis of hordein proteins. *Crop Science* 26: 454–460.
6. Heisel, S.E., D.M. Peterson and B.L. Jones (1986) Identification of United States barley cultivars by sodium dodecyl sulphate polyacrylamide gel electrophoresis of hordeins. *Cereal Chemistry* 63: 500–505.
7. Neilsen, G. and H.B. Johansen (1986) Proposal for the identification of barley varieties based on the genotypes for two hordein and 39 isozyme loci for 47 reference varieties. *Euphytica* 35: 717–728.
8. ISTA (1992) *Handbook of Variety Testing: Electrophoresis Testing.* ISTA, Zurich Switzerland.
9. Botstein, D., White, R.L., Skolnick, M. and Davis, R.W. (1980) Construction of a genetic linkage map in using restriction fragment

length polymorphisms. *American Journal of Human Genetics* 32: 314–331.

10. Jeffreys, A., V. Wilson and S. Thein (1985) Hypervariable minisatellite regions in human DNA. *Nature* 314: 67–73.
11. Williams, J.G.K., A.R. Kubelik, K.J. Livak, J.A. Rafalski and S.V. Tinge (1991) DNA polymorphisms amplified by arbitrary primers are useful as genetic markers. *Nucleic Acids Research* 18: 6531–6541.
12. Vos, P., Hogers, R., Bleeker, M., Reijans, M., van de Lee, T., Hornes, M., Frijters, A., Pot, J., Peleman, J., Kuiper, M. and Zabeau, M. (1995) AFLP: a new technique for DNA fingerprinting. *Nucleic Acids Research* 23: 4407–4414.
13. Powell, W., G.C. Machray and J. Provan (1996) Polymorphism revealed by simple sequence repeats. *Trends in Plant Science* 1(7): 215–222.
14. Coryell, V.H., H. Jessen, J.M. Schupp, D. Webb and P. Keim (1999) Allele specific hybridisation markers for soybean. *Theoretical and Applied Genetics* 98: 690–696.
15. Kalendar, R., T. Grob, M. Regina, A. Suoniemi and A. Schulman (1999) IRAP and REMAP: two new retrotransposan-based DNA fingerprinting techniques. *Theoretical and Applied Genetics* 98: 704–711.
16. Macaulay, M., L. Ramsay, W. Powell and R. Waugh (2001) A representative, highly informative 'genotyping set' of barley SSRs. *Theoretical and Applied Genetics* 102: 801–809.
17. Southworth, C.L. (2007) *The use of microsatellite markers to differentiate UK barley (Hordeum vulgare) varieties and in the population genetic analysis of bere barley from the Scottish islands.* Ph.D. Thesis, Heriot Watt University, Riccarton, Edinburgh.
18. Ramsay, L., Macaulay M., degli Ivanissevich S., MacLean K., Cardle L., Fuller J., Edwards K.J., Tuvesson S., Morgante M., Massari A., Maestri E., Marmiroli N., Sjakste T., Ganal M., Powell W. and R. Waugh (2000) A simple sequence repeat-based linkage map of barley. *Genetics* 156: 1997–2005.
19. Goldstein, D. and C. Schlotterer (Eds.) (1999) *Microsatellites –Evolution and Applications.* Oxford University Press, Oxford.
20. Brownstein, M.J., Carpten, J.D. and Smith, J.R. (1996) Modulation of non-templated nucleotide addition by Taq DNA Polymerase: Primer modifications that facilitate genotyping. *BioTechniques* 20: 1004–1010.
21. Reid, A, L. Hoff, D. Esselink (2008) Potato cultivar genome analysis in Burns, R (Ed) (2008) *Plant Pathology*. Springer Publishing.

INDEX

G

H

I

L

M

N

P